T0323539

Sustainable Materials, Structures and IoT

[SMSI-2024]

First edition published 2024
by CRC Press
4 Park Square, Milton Park, Abingdon, Oxon, OX14 4RN

and by CRC Press
2385 NW Executive Center Drive, Suite 320, Boca Raton FL 33431

CRC Press is an imprint of Informa UK Limited

British Library Cataloguing-in-Publication Data
A catalogue record for this book is available from the British Library

ISBN: 9781032980409 (hbk)
ISBN: 9781032980423 (pbk)
ISBN: 9781003596776 (ebk)

DOI: 10.1201/9781003596776

Font in Sabon LT Std
Typeset by Ozone Publishing Services

Sustainable Materials, Structures and IoT

[SMSI-2024]

[First Edition]

Edited by:

Sujit Kumar Pradhan
Srinivas Sethi
Mufti Mahmud

CRC Press
Taylor & Francis Group
Boca Raton London New York

CRC Press is an imprint of the
Taylor & Francis Group, an **informa** business

Contents

Lists of Figures

List of Tables

Preface

This Taylor & Francis, CRC Press volume contains the papers presented at International Conference on Sustainable Materials, structures and IoT (SMSI-2024) being organized by the Department of Civil Engineering, Indira Gandhi Institute of Technology, Sarang (An Autonomous institute of Govt of Odisha), India, during 6th and 8th July 2024. The theme of the conference included various field of Civil, Electrical, Mechanical, Metallurgy, Chemical, Computer science, and Electronics.

The conference draws some excellent technical keynote talks and papers. Three keynote talks by Prof. Mufti Mahmud, Nottingham Trent University, UK; Prof. Bibhudatta Sahoo, NIT Rourkela, Prof. T Jothi Saravanan, IIT Bhubaneswar has been presented.

For the Conference, received of about 137 full paper and accepted only 60 papers. The contributing authors are from different parts of the globe. All the papers are reviewed by at least two independent reviewers and in some cases by as many as three reviewers. After review, papers were scrutinized by Editorial Board Member. All the papers are also checked for plagiarism and similarity index less than 10%. We had to do this unpleasant task, keeping the Taylor and Francis guidelines and approval conditions in view. We take this opportunity to thank all the authors for their excellent work and contributions and also the reviewers, who have done an excellent job.

On behalf of the technical committee, we are indebted to Prof. Mufti Mahmud, General Chair of the Conference, for his timely and valuable advice. We cannot imagine the conference without his active support at all the crossroads of decision-making process. The management of the host institute, particularly the Director Prof. Satyabrata Mohanta, HOD Civil Prof. T. K. Nath, and Convenor Dr. S. K. Pradhan have extended all possible support for the smooth conduct of the Conference. Our sincere thanks to all of them.

We would also like to place on record our thanks to all the keynote speakers, session chairs, authors, technical program committee members, website chair, publication chair, finance chair, publicity chair, registration chair, logistic chair, members of organising committee, print & digital media and above all to the volunteers. We are also thankful to Taylor and Francis publication house for agreeing to publish the accepted and presented papers.

Best Wishes,
Dr Sujit Kumar Pradhan
Prof. Srinivas Sethi

Editors Biography

Sujit Kumar Pradhan

Sujit Kumar Pradhan is currently serving as an Assistant Professor in the Department of Civil Engineering, Indira Gandhi Institute of Technology Sarang, India. He holds PhD in Transportation Engineering from Indian Institute of Technology Bhubaneswar. His area of research includes pavement material characterization, pavement analysis & design, performance evaluation of asphalt binder and recycling, and geotextile and porous concrete. He has about 14 years of academic and research experience. He has published reputed international journal and conference papers and successfully completed TEQIP-III funded research and development projects. Now, he is continuing a DST Odisha funded project. Also, he has reviewed many technical papers for various international and national journals.

Srinivas Sethi

Srinivas Sethi is a Professor in Computer Science Engineering & Application, Indira Gandhi Institute of Technology, Sarang (IGIT, Sarang), India, and has been actively involved in teaching and research since 1997. He did his Ph.D., in the area of routing algorithms in mobile ad hoc network and is continuing his research work in the wireless sensor network, cognitive radio network, and cloud computing, Big-Data, BCI, Cognitive Science. He is a Member of Editorial Board for different journal and Program Committee Member for different international conferences/workshop. He is a Book Editor of four international conference proceedings published in Springer and Taylor & Francis. He has published more than 80 research papers in international journals and conference proceedings. He completed eight research and consultancy projects funded by different funding agencies such as DRDO, DST, AICTE, NPIU, and local Govt. office. He has delivered more than 125 talks as keynote speaker, resource person in different international and national conferences/workshops, conducted by different important national reputation institutions such as SAG group of DRDO, New Delhi, JNTU GV College of Engineering Vizianagaram, Pt. Rabishankar Sukla University, Raipur, CSIT, Raipur, Utkal University, Bhubaneswar, etc.

Mufti Mahmud

Mahmud's research vision is to contribute toward a secure, smart, healthy, and better world to live in. In todays digitized world, converting the ever-expanding amount of raw data to smart data, and building predictive, secure and adaptive systems aiming personalized services are essential and challenging, which require cross-disciplinary and multi-stakeholder collaborations. Toward these goals, Dr Mahmud conducts problem-driven `Brain Informatics research where he works with problem domain experts to find multidisciplinary solutions to real-world problems. Dr Mahmud's research involves Computational, Health, and Social Sciences, and uses Neuroscience, Healthcare, Applied Data Science, Computational Neuroscience, Big Data Analytics, Cyber Security, Machine Learning, Cloud Computing, and Software Engineering and plans to develop secure computational tools to advance healthcare access in low-resource settings.

List of Contributors

- Patron-in-Chief
  Prof (Dr.) Satyabrata Mohanta, Director, IGIT Sarang

- Patron(s)
  Dr. T. K. Nath, Professor & Head, Civil IGIT Sarang, Dean (A & E)

- General Chair
  Prof. Mufti Mahmud , Nottingham Trent University , U.K

- Program Chair
  Prof Srinivas Sethi, Professor, Deptt. of CSEA

- Convenor
  Dr. Sujit Kumar Pradhan,Assistant Professor Department of Civil Engineering

- Advisory Board
Dr. Pratap Kumar Pani, Dean (FARC), Professor, CE, IGIT Sarang
Dr. B.P. Panigrahi, Dean (PGS&R), Professor, EE, IGIT Sarang
Dr. P.D.Das, Professor & Head, Deptt. of Mathematics
Dr. B.D Sahoo, Dean (Students welfare), Professor & Head, Deptt. of Mechanical
Dr. Ashima Rout, Asso. Professor & Head, ECE, IGIT Sarang
Dr. B.B. Choudhury, Professor & Head, Dept. of Production
Dr. Goutam Kumar Pothal, Professor & Head, Dept. of Architecture
Dr. P.K. Mallik, Associate Professor, Head, MME, IGIT Sarang
Dr. Sunil Kumar Tripathy, Associate Professor & Head, Dept. of Physics
Dr. Binod Bihari Panda, Associate professor & Head, Dept.of Chemistry and HSSM

- Organising Chair
Dr Urmila Bhanja, Professor, Electronics & TC Engg.
Dr Chitta Ranjan Sahoo, Prof, Department of Civil Engineering
Dr Rabindra Kumar Behera, Associate Professor, Department of Electrical Engineering
Dr Umakanta Mahanta, Assistant Professor, Electrical Engineering
Mr Prashant Ranjan Dhal, Asst. Professor, Dept. of Mechanical Engineering
Dr Sanjaya Ku. Patra, Asst. Professor, Department of CSEA
Dr Maheswar Behera, Assistant Professor, Electrical Engineering
Dr Dillip Ku. Swain, Asst. Professor, Department of CSEA
Dr Niroj Ku. Pani, Asst. Professor, Department of CSEA
Dr Bhagirathi Tripathy, Assistant Prof. Department of Civil Engineering
Dr Sandeep Kumar Sahoo, Assistant Professor, Metallurgical & Materials Engineering
Dr Supriya Sahoo, Assiatant Professor, Department of Mechanical Engineering
Mr Paresh Kumar Pasayat, Assistant Professor, Department of Electronics & TC Engineering
Dr Jogendra Kumar Majhi, Assistant Professor, Metallurgical & Materials Engineering
Dr Kashinath Barik, Assistant Professor, Chemical Engineering
Mr Aditya Kumar Bhoi, Assistant Prof. Department of Civil Engineering
Mr Himanshu Kumar Dash, Dept. of Production Engineering

Mr Ashok Kumar Pradhan, Dept. of Production Engineering
Mr Subrat Kumar Nayak, Assistant Prof. Department of Civil Engineering
Mr Ritwik Patnayak, Assistant Prof. Department of Civil Engineering
Ms Anwesha Rath, Assistant Prof. Department of Civil Engineering
Mrs Swetankita Sahoo,Assistant Prof. Department of Civil Engineering
Mrs Prangya Dipti Das, Assistant Prof. Department of Civil Engineering

- Website Chairs
Dr Sanjaya Ku. Patra, Asst. Professor, Department of CSEA
Dr Biswanath Sethi, Asst. Professor, Department of CSEA
Mr Gaurab Kumar Ghosh, Dptt. of Mechanical Engineering

- Publicity Chairs
Mr Prashant Ranjan Dhal, Dept. of Mechanical Engineering
Mr Himanshu Kumar Dash, Asst. Professor, Dept. of Production Engineering

- Registration Chairs
Mr Satyapriya Senapati, Asst. Professor, Dept. of Civil Engineering
Mrs Priyadarshini Das, Asst. Professor, Dept. of Civil Engineering

- Publication Chairs
Dr Chandradhwaj Nayak, Assistant Professor, Dept. of Chemical Engg
Dr Kshetramohan Sahoo, Assistant Professor, Dept. of Chemical Engg

- Finance Chair
Dr Sujit Kumar Pradhan, Assistant Professor, Dept. Of Civil Engineering

CHAPTER 1

Advancements in Sludge Utilization for Sustainable Construction Practices

A Comprehensive Review

Aman Jangir[1], Biswajit Acharya[2]

[1]PhD Research Scholar, Civil Engineering Department,
Rajasthan Technical University, Kota, India
E-mail: aman.phd23@rtu.ac.in
[2]Associate Professor, Civil Engineering Department,
Rajasthan Technical University, Kota, India
E-mail: bacharya@rtu.ac.in

Abstract

The focus of this comprehensive review explores the recent advancements in utilizing sludge for sustainable construction practices, reflecting the industry's growing commitment to environmental responsibility. Traditionally considered a waste product from wastewater treatment, sludge has emerged as a viable resource with significant potential for construction applications. This paper synthesizes diverse studies investigating sludge's incorporation into concrete, soil stabilization, and the development of alternative building materials. Critical analysis encompasses the environmental impact, economic feasibility, and technical performance of sludge-derived materials, providing a holistic assessment of their suitability for sustainable construction. The review addresses challenges such as regulatory constraints, public perception, and quality control, offering insights into potential solutions. By presenting a comprehensive overview of the current state of knowledge, the paper guides researchers, engineers, and policymakers toward informed decision-making for sustainable construction practices. It serves as a valuable resource by highlighting opportunities and constraints associated with sludge utilization, paving the way for future research and development in this pivotal area. The synthesis of existing literature informs and inspires a strategic approach to integrating sludge into construction projects, contributing to the broader goal of achieving environmentally responsible and sustainable building practices.

Keywords: sludge utilization, sustainable construction, soil stabilization

1. Introduction

The construction industry is undergoing a transformative shift towards sustainable practices, driven by the imperative to reduce environmental impact and promote resource efficiency. In this context, the exploration of alternative materials is gaining prominence, with a particular focus on byproducts traditionally deemed as waste. One such promising avenue is the utilization of sludge, a residual product of wastewater treatment processes,

DOI: 10.1201/9781003596776-1

as a resource for sustainable construction practices. Historically, sludge has been viewed as a challenge in terms of disposal, often incurring significant environmental and economic costs. However, recent advances and new techniques have revealed sludge's potential as a useful raw material for a wide range of construction applications. This comprehensive review aims to consolidate and analyze the latest developments in the integration of sludge into sustainable construction practices. The paper will systematically explore the incorporation of sludge in key construction aspects, including its use in concrete formulations, soil stabilization techniques, and the development of alternative building materials. An in-depth examination of the environmental, economic, and technical aspects of sludge-derived materials will be conducted to assess their viability for sustainable construction. Additionally, the review will address challenges associated with sludge utilization, such as regulatory considerations, public perception, and quality control issues, providing insights into strategies to overcome these obstacles. The exploration of sludge as a valuable resource underscores the potential for turning what was once considered waste into an asset for environmentally responsible building practices.

2. Literature Review

Research conducted by Karim et al. (2020) explored the enhancement of soil engineering properties through the incorporation of sewage sludge ash (SSA). The findings indicated that the addition of SSA significantly enhanced the soil's characteristics and engineering performance. The optimal blend was identified as soil mixed with 7.5% SSA, which demonstrated superior properties after a curing period of 28 days, making it suitable for various engineering applications. Additionally, a mixture containing 5% SSA also emerged as a viable alternative under the same curing conditions. The inclusion of SSA reduced the plasticity index of the soil, leading to a marked decrease in massive swelling. This research highlights that utilizing 7.5% SSA in the soil can effectively enhance soil properties, presenting an environmentally friendly solution by diverting SSA from landfills.

Manish Kumar et al. (2020) investigated the use of lime-stabilized sewage sludge (SS) for construction, testing samples with 0–8% lime over 7, 14, and 28 days. Results showed a significant increase in unconfined compressive strength from 207 to 1102 kPa with 6% lime after 28 days due to pozzolanic reactions. Lime also reduced swell pressure and plasticity index, enhancing workability. The study suggests lime treatment as a viable method for stabilizing SS for flexible pavement subbases, aligning with IRC 37-2012 guidelines.

Jagaba et al. (2019) introduced incinerated sewage sludge ash (ISSA) as a soil stabilizer, revealing its oxide compositions through XRF, XRD, and TCLP analyses. ISSA, with notable Al_2O_3 (23.51%), SiO_2 (61.42%), and Fe_2O_3 (4.24%), at 7% content, reduced PI, met durability requirements, and notably increased California Bearing Ratio (CBR) values. The study concludes that a 7% ISSA additive significantly enhances soft soil strength, making it a promising soil-stabilizing agent for geotechnical applications.

Kacprzak et al. (2017) examined the environmental sustainability of SS management by comparing various treatment and disposal methods. It highlights the complex composition of SS and the need for proper treatment to reduce risks and utilize it as a biosolid. Despite its small volume, SS accounts for 50% of wastewater treatment costs. The authors emphasize the waste hierarchy, favoring recovery over disposal, and propose a management scenario involving anaerobic digestion, dewatering, drying, and thermal treatment to recover materials and energy. They advocate for sustainable strategies to minimize greenhouse gas emissions, conserve resources, and reduce soil pollution, using tools like life cycle assessment (LCA) for evaluating environmental impacts.

Ayininuola and Ayodeji (2016) studied sludge ash as a soil stabilizing agent, proposing an alternative approach to sludge disposal. The investigation found that the optimal moisture content (OMC) reached its minimum at 7% ash content. Additionally, the MDDs and shear strength of soil samples peaked with 7% stabilization, attributed to the pozzolanic characteristics of the sludge ash, enhancing soil strength in the presence of moisture. Therefore, adding 7% sludge ash proved effective in strengthening the soil.

Lucena et al. (2014) investigated the viability of incorporating 10% SS by weight into pavement base layers. Test mixtures were prepared with varying additive contents (ranging from 2% to 8%) in sludge-soil combinations. Modified soil samples, treated with emulsion, demonstrated a reduction in specific properties. The findings indicated that the introduction of SS to soil is an effective strategy for applications in road construction, backfilling, and improving the bearing capacity of soil in structural foundations. Mixtures incorporating additives exhibited a lower maximum dry unit weight compared to those utilizing pure soil with added sludge. Notably, there was a gain in CBR with the incorporation of lime and cement additives. These findings emphasize the possible advantages of SS in soil anchoring for pavement base layers.

Lin et al. (2013) investigated the use of inorganic additives, i.e., fly ash, calcium bentonite, and kaolinite, to enhance the stabilization/solidification (S/S) process. Calcium bentonite was shown to be the most efficient addition, increasing compressive strength while decreasing pollutant leaching more than the others. At a weight ratio of 1:0.2:0.2 (sludge:cement:calcium bentonite), treated sludge fulfilled landfill and construction material standards in 7 and 28 days, respectively. While fly ash provided longer-term strength at greater dosages, calcium bentonite consistently increased pollutant immobilization and cement hydration. XRD and SEM/EDS tests revealed that calcium bentonite improved cement hydration and bonding. Thus, calcium bentonite is advised for increasing the cement-based S/S of SS.

Chen and Lin (2009) investigated the efficacy of ISSA/cement admixture in enhancing soil properties. Results showed significant improvements in soil strength (3–7 times), reduced swelling (10–60%), and enhanced CBR values (up to 30 times), suggesting its potential for geotechnical applications. The study highlighted ISSA/cement's role in transforming soil plasticity and its dependence on curing time for strength development. Pozzolanic reactions between cement and ISSA contributed to soil stabilization. Overall, the findings offer promising insights into utilizing SSA for sustainable geotechnical engineering practices.

Lin et al. (2005) investigated the inclusion of various amounts of cremated sludge ash (0, 15, 30, and 45%) and five different glaze values (0.03, 0.06, 0.1, 0.15, and 0.2 g/cm^2) in the manufacturing of biscuit tiles. The study's findings supported the viability of cremated sludge ash for the production of glazed tiles. Notably, the study found that different levels of ash addition had no detectable influence on the acid-alkali resistance or aging resistance of the glazed tile samples. Energy Dispersive X-ray Spectroscopy (EDS) examination revealed that glaze concentrations had less influence on the composition of the tile specimens than the addition of SS. Furthermore, a decrease in water absorption was reported for glaze concentrations of 0.06, 0.1, and 0.15.

Sharif and Attom (2000) investigated the utilization of burned sludge ash as a soil-stabilizing agent. The study found that adding 7.5% burned sludge ash based on dry weight increased both unconfined compressive strength and maximum dry density. In particular, exceeding the 7.5% threshold resulted in a decreasing pattern in both MDD and UCS.

3. Conclusion

- Fly ash and kaolinite additives fell short of meeting the 7-day UCS requirements for solidified sludge in short-term sewage-sludge landfills.
- Immediate enhancement of compressive strength is vital for landfilling, as prolonged curing periods lead to increased storage space and treatment costs.

- ISSA/cement admixture significantly improved soil properties, including strength and swelling reduction, suggesting its potential for geotechnical applications.
- Incorporating SS into soil boosts pavement base construction and enhances soil bearing capacity. Lime and cement additives bolster CBR, highlighting SS potential in soil stabilization.
- The advantages of integrating SSA into the soil were underscored, demonstrating enhanced engineering performance and minimized environmental footprint, especially with a 7.5% SSA mixture.
- Curing time is a crucial factor in selecting appropriate S/S technologies.
- Calcium bentonite outperforms fly ash and kaolinite, providing immediate improvement with minimal dosage.
- Calcium bentonite significantly enhances the cement-based stabilization/solidification of SS, improving compressive strength and reducing pollutant leaching more effectively than fly ash and kaolinite. It is the preferred additive for optimizing S/S processes.
- Calcium bentonite benefits in earth-construction applications due to its low dosage and rapid cure period.
- For purposes that require stronger compressive strength regardless of curing time, a high fly ash dosage is an ideal choice.
- Kaolinite is not suggested as an SS additive due to reduced compressive strengths found after 7 and 28 days.
- Lime-stabilized SS effectively increased soil strength and reduced swell pressure and plasticity index, proposing it as a viable method for flexible pavement subbases.

References

[1] Ayininuola, G. M., & Ayodeji, I. O. (2016). Influence of sludge ash on soil shear strength. *Journal of Civil Engineering Research, 6*(3), 72–77.

[2] Chen, L., & Lin, D. F. (2009). Stabilization treatment of soft subgrade soil by sewage sludge ash and cement. *Journal of Hazardous Materials, 162*(1), 321–327.

[3] Jagaba, A. H., Shuaibu, A., Umaru, I., Musa, S., Lawal, I. M., & Abubakar, S. (2019). Stabilization of soft soil by incinerated sewage sludge ash from municipal wastewater treatment plant for engineering construction. *Sustainable Structures and Materials, 2,* 32–44.

[4] Karim, M. A., Hassan, A. S., & Hawa, A. (2020). Enhancement of soil engineering properties with sewage sludge ash. *MOJ Ecology & Environmental Sciences, 5*(5), 230–236.

[5] Kacprzak, M., Neczaj, E., Fijałkowski, K., Grobelak, A., Grosser, A., Worwag, M., ... & Singh, B. R. (2017). Sewage sludge disposal strategies for sustainable development. *Environmental Research, 156,* 39–46.

[6] Lin, C., Zhu, W., & Han, J. (2013). Strength and leachability of solidified sewage sludge with different additives. *Journal of Materials in Civil Engineering, 25*(11), 1594–1601.

[7] Lin, D. F., Luo, H. L., Luo, H. L., & Sheen, Y. N. (2005). Glazed tiles manufactured from incinerated sewage sludge ash and clay. *Journal of the Air & Waste Management Association, 55*(2), 163–172.

[8] Lucena, L. C., Juca, J. F., Soares, J. B., & Portela, M. G. (2014). Potential uses of sewage sludge in highway construction. *Journal of Materials in Civil Engineering, 26*(9), 04014051.

[9] Sharif, M. M., & Attom, M. F. (2000). The use of burned sludge as a new soil stabilizing agent. In *Environmental and Pipeline Engineering 2000* (pp. 378–388). ASCE.

[10] Taki, K., Choudhary, S., Gupta, S., & Kumar, M. (2020). Enhancement of geotechnical properties of municipal sewage sludge for sustainable utilization as engineering construction material. *Journal of Cleaner Production, 251,* 119723.

CHAPTER 2

Properties of layered concrete made of Portland slag cement and Portland pozzolana cement in a double-layered system

Priya Kumari, Prasanna Kumar Acharya[#], Mausam Kumari Yadav, Kunal Satyam Ranjan

School of Civil Engineering, KIIT DU, Bhubaneswar, India
[#] Corresponding author mail: pkacharya64@yahoo.co.in

Abstract

To reduce the production of cement, researchers are working to find alternatives to cement and adopt technologies to make concrete structures efficient by improving the efficiency of concrete. This paper presents the improved efficiency of cement concrete with mechanical solidity in the application of an innovative technology that exhibits spatial variations in composition and properties, which is formed in more than one layer. This layered concrete is herein named functionally graded concrete (FGC). Preparation of FGC is made using slag cement (PSC) and pozzolana cement (PPC). Samples were prepared in two layers considering the quality: the first layer with concrete made of PSC and the second layer made of PPC. Four types of methodology—fresh layer on a fresh layer, fresh layer on a compacted layer, fresh layer on an initial set layer, and fresh layer on a final set layer—were followed during casting. The results were compared with single-layered control samples made of PSC and PPC. The results indicated that all FGC samples, irrespective of quality and methodology, performed better than control samples. The results also revealed that the performance of the FGC is better on parallel application of the load to the graded layer. The results of the study may encourage the construction industry to make structures more efficient concerning economy, ecology, and technical parameters.

Keywords: Layered concrete, functionally graded concrete, perpendicular load application, parallel load application, graded layer

1. Introduction

Functionally graded concrete (FGC) represents an innovative and promising construction material that exhibits spatial variations in composition and properties, providing a unique blend of functionality. FGC is made of more than one layer, each layer being different in quality and methodology from its successive layer. A new layer is formed at the interface of layers whose properties are said to be more efficient than its neighboring layers. In functional-grade composites, the material structure varies longitudinally to meet different needs in each part of the structural member and can reduce cement usage (Toreli et al., 2020). The compressive and bending capability of FGC is 12-15% and 4-5% more

DOI: 10.1201/9781003596776-2

than normal concrete (Satyanarayana & Natarajan, 2015). Functionally graded materials (FGM) are non-uniform composites composed of two or more distinct materials that are designed to have a constantly changing spatial composition (Udupa et al., 2014).

Construction materials with functional grades can be created using 3D concrete printing (3DCP). Comparing 3DCP with graded porosity to casted foam concrete with equivalent densities, the former exhibits approximately twice the specific bending strength. Graded pores and fibers enable printed composites to enhance strength and decrease self-weight by 31% and 20%, respectively (Geng et al., 2022). A three-point bending (TPB) beam design with varied fiber-added concrete and plain concrete thicknesses was numerically simulated using a cohesive zone model (CZM). The fracture energy and residual load capacity revealed that the in-layer thickness of fiber-added concrete in the lower part was ideal when it was less than half the depth (Evangelista et al., 2009). It is reported that functionally graded concrete made of normal concrete and concrete with 30% GGBFS showed better results in terms of compressive and interfacial bond strength (Sahoo et al., 2021). The flexural and post-cracking behavior of FGC can be increased with a thin layer of plain cement concrete (10% of total depth) as a top layer of FGC (Chan & Galobardes, 2022). By using graded concrete, structural elements' serviceability can be raised, and working load deflection can be decreased (Pratama et al., 2018). Combining the concepts of two-layer cement concrete with high-strength cement concrete and fiber-added slurry-infiltrated cement concrete allows for the creation of ultra-high-performance FGC composite beams. These systems have great flexural properties and impact properties, less cement usage, and steel fiber utilization capacity (Li et al., 2020).

2. Materials and Methods

2.1. Materials

PPC and PSC that satisfied the specifications of IS: 1489-Part 1 (1991, Reaffirmed 2005) and IS: 455 (2019) codes were used in the work. Granite aggregates from local quarries and bed sand from local rivers confirmed the requirements of IS: 383 (1970, Reaffirmed 2016) were used. The density of coarse and fine aggregates was 1696 and 1651 kg/m^3. The fineness modulus of the same was 6.46 and 2.38. The capacity of moisture uptake of both the aggregates was 0.15% and 0.71%. The values of specific gravity for aggregates were 2.81 and 2.72. The impact and abrasion values of coarse aggregate were 15.32% and 18.58%.

3. Mix Proportion, Test Specimens, and Test Procedure

A design mix of M20 standard concrete having the fraction of mix 1:1.5:3 with a water-binding material ratio of 45:100 was adopted for concrete preparation. For the preparation of 1 cum of concrete mix, PSC/PPC of 380 kg was used. Cubes measuring 100 mm in size were cast to evaluate strength on compression, while beams with dimensions of 500 × 100 × 100 mm were prepared to assess strength on flexure. Samples were prepared in two layers considering the quality: the first layer with concrete made of PSC and the second layer made of PPC. Apart from two control samples made of PSC and PPC, other four types of samples were prepared considering methodology—fresh layer on a fresh layer (FF), fresh layer on a compacted layer (FC), fresh layer on an initial set layer (FIS), fresh layer on a final set layer (FFS). The single-layered control samples were nomenclatured as PSC and PPC. The other four two-layered samples were named (S+P)FF, (S+P)FC, (S+P)FIS, and (S+P)FFS, respectively. After casting, the specimens were covered and left for curing at room temperature around the clock. Then the specimens were removed from casting cells and immersed in a water vat for curing. The properties of compression and flexure were determined based on IS code: 516. During the testing of compressive strength and flexural strength, two types of configurations of application of load were followed, like the application of load parallel and perpendicular to the graded layer.

4. Results and Discussion

4.1. Strength on Compression

The strength on compression of double-layered FGC concerning concrete samples (S+P) FF, (S+P)FC, (S+P)FIS, and (S+P)FFS were checked at the age of the 1st and 4th week when the applied load was perpendicular to the graded layer. The graded new layer was formed at the joint place of both layers. The results were compared to single-layered PSC and PPC samples, which are herein called controlled concretes (CC). The average 7th-day result of PSC and PPC control concretes was 15.9 MPa. The same for (S+P)FF, (S+P)FC, (S+P) FIS, and (S+P)FFS were found nearly 33%, 37%, 27%, and 8% more than the CC. The 28th-day result of CC was estimated at 26.50 MPa. The same for FGC samples concerning (S+P)FF, (S+P)FC, (S+P)FIS, and (S+P)FFS were 27%, 35%, 33%, and 8% more than CC. The results are presented in Figure 1.

The results indicate the positive effect of FGC making the concrete more efficient. Out of four workmanship/technologies adopted for making FGC, it is seen that the application of (S+P)FC brings greater advantages to the concrete, while (S+P)FFS gives a lesser benefit. However, all four workmanship/technology adopted to make the concrete efficient. It is also observed that the early age strength of FGC mixes concerning (S+P)FF and (S+P)FC are eye-catching, while (S+P)FIS and (S+P)FFS are moderate.

The compressive strength of concrete mixes of double-layered FGC concerning concrete samples was checked at the end of 7 and 28 days when the application of load was parallel to the graded new layer. The results were compared to controlled concrete (CC). The 7th-day result of CC was 15.9 MPa. The same for (S+P)FF, (S+P)FC, (S+P)FIS, and (S+P) FFS were found nearly 50%, 46%, 42%, and 34% more than the CC. The 28th-day result of CC was estimated at 26.50 MPa. The same for FGC samples concerning (S+P)FF, (S+P) FC, (S+P)FIS, and (S+P)FFS were 33%, 37%, 35%, and 14% more than CC. The outcomes are presented in Figure 2.1. The outcomes indicated the better performance of FGC. In this case, also FC offered the best results next to FIS. The results of (S+P)FF and (S+P)FIS are

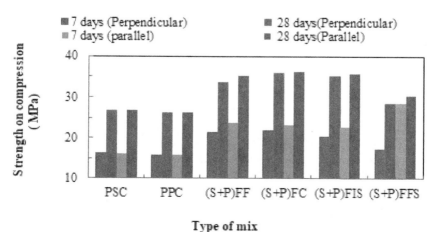

Figure 2.1: Compressive strength.

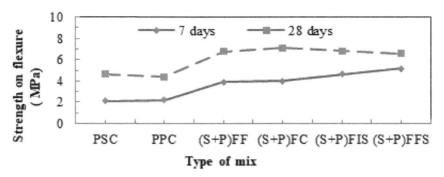

Figure 2.2: Flexural strength with load perpendicular to the graded layer.

almost found similar. In such cases of parallel application of load, the initial strength development within the 1st week is found impressive. The strength at the end of the 4th week of single-layered PSC and PPC-made concrete samples satisfies the requirements of M20-grade concrete; the same mix when provided in FGC almost satisfies the requirements of M30-grade concrete.

4.2 Strength of Flexure

The strength on flexure of single-layered control concrete (CC) samples made of blended cement (PSC and PPC) and double-layered FGC samples made through techniques like (S+P)FF, (S+P)FC, (S+P)FIS, and (S+P)FFS was tested at the end of the 1st and 4th weeks. For the strength test on flexure, the load was applied perpendicular to the graded layer only. At 7 days, the CC (average result of PSC and PPC) exhibited 2.16 MPa. But samples concerning (S+P)FF, (S+P)FC, (S+P)FIS, and (S+P)FFS offered flexural strength 1.62, 1.85, 1.85, and 1.06 times more than CC. The early-age results of FGC are very impressive as far as structural properties are concerned. The 28-day result of CC (the average result of PSC and PPC) was found to be 4.51 MPa. The same for (S+P)FF, (S+P)FC, (S+P)FIS, and (S+P)FFS was found 1.51, 1.57, 1.51, and 1.46 times more than CC. The 28-day results recognize the FGC as structural-grade concrete. The reason behind the improvement of structural properties may be attributed to the graded layer that provided more flexibility. The results are presented in Figure 2.2.

5. Conclusion

The present study led to the following outcomes:

- The strength of compression examined on completion of 7 days of curing of functionally graded concrete made in two layers on the concept of fresh on fresh, fresh on compacted, fresh on initial set, and fresh on final set was found to be nearly 8–37% higher than single-layered normal PSC or PPC concrete on the application of load perpendicular to the graded layer. The same increased by 34–50% on the application of load parallel to the graded layer. The compressive strength examined on completion of 28 days of curing of functionally graded concrete made in the above concept was calculated to be nearly 8-35% higher than single-layered normal concrete on the application of load perpendicular to the graded layer. The same increased by 14-37% on the application of load parallel to the graded layer.

- The strength on flexure at the age of 7 days of functionally graded concrete made in the above concept was calculated to be nearly 1.06–1.85 times higher than the normal concrete on the application of load perpendicular to the graded layer. The same increased by 1.46–1.57 times on the application of load parallel to the graded layer.

References

[1] Chan, R., Moy, C. K. S., & Galobardes, I. (2022). Numerical and analytical optimization of functionally graded concrete incorporating steel fibers and recycled aggregate. *Construction and Building Materials, 356*, 129249.

[2] Evangelista, F. Jr., Roesler, J., & Paulino, G. (2009). Numerical simulations of fracture resistance of functionally graded concrete materials. *Transportation Research Record: Journal of Transportation Research Board, 2113*(1), 122–131.

[3] Geng, Z., Pan, H., Zuo, W., & She, W. (2022). Functionally graded lightweight cement-based composites with outstanding mechanical performance via additive manufacturing. *Construction and Building Materials, 56*, 102911.

[4] IS: 1489 (Part 1). 1991 (Reaffirmed 2005). Portland Pozzolana Cement—Specifications. Bureau of Indian Standards, New Delhi, India.

[5] IS: 383-1970. Specifications for coarse and fine aggregates from natural sources for concrete. Bureau of Indian Standards, New Delhi, India.

[6] IS: 455: 2019. Portland Slag Cement—Specification. Bureau of Indian Standards, New Delhi, India.

[7] IS: 516: 1959 (Reaffirmed 2004). Indian standard code of practice - Methods of test for strength of concrete. Bureau of Indian Standards, New Delhi, India.

[8] Li, P. P., Sluijsmans, M. J. C., Brouwers, H. J. H., & Yu, Q. L. (2020). Functionally graded ultra-high performance cementitious composite with enhanced impact properties. *Construction and Building Materials, 183*, 107680.

[9] Pratama, M. M. A., Gan, B. S., Lie, H. A., & Putra, A. B. N. R. (2018). A numerical analysis of the modulus of elasticity. *Advances in Social Science, Education and Humanities Research, 242*, 118–120.

[10] Sahoo, S. K., Mohapatra, B. G., Patro, S. K., & Acharya, P. K. (2021). Evaluation of graded layer in ground granulated blast furnace slag-based layered concrete. *Construction and Building Materials, 276*, 122218.

[11] Satyanarayana, P., & Natarajan, C. (2015). Experimental investigation of functionally graded concrete with fly ash. *International Journal of Earth Sciences and Engineering, 8*, 143–148.

[12] Torelli, G., Fernández, M. G., & Lees, J. M. (2020). Functionally graded concrete: Design objectives, production techniques, and analysis methods for layered and continuously graded elements. *Construction and Building Materials, 242*, 118040.

[13] Udupa, G., Rao, S. S., & Gangadharan, K. V. (2014). Functionally graded composite materials. *Procedia Materials Science, 5*, 1291–1299.

CHAPTER 3

Oral Cancer Detection Using Deep Learning Architecture

Avvuru RV Naga Suneetha[1], Madarapu Indhuja[2], C. H. Purna Chandra[3], M. V. K. Sreecharan[4]

[1]Assistant Professor, Computer Science & Engineering. Vignan Institute of Technology and Science, Hyderabad, Telangana, India
suneethaavvaru@gmail.com
UG student, Computer Science & Engineering. Vignan Institute of Technology and Science, Hyderabad, Telangana, India
madarapuindhuja20891a0533@gmail.com
[3]UG student, Computer Science & Engineering. Vignan Institute of Technology and Science, Hyderabad, Telangana, India
Chandureddy7131@gmail.com
[4]UG Student, Computer Science & Engineering, Vignan Institute of Technology and Science, Hyderabad, Telangana, India
sreecharan.mathukumalli@gmail.com

Abstract

Examining deep learning methods to identify oral cancer early is the primary goal. Utilizing the unique ability of recurrent neural networks (RNNs) to handle sequential input, this research explores the application of RNNs to oral cancer diagnostics. It is critical to ascertain the best way to identify oral cancer and to remain up to date on any new advancements. It is impossible to overestimate the significance of sooner detection of oral cancer because it is essential for enhancing patient outcomes by enabling earlier treatment and higher survival rates. The results could be helpful to the field and aid in the creation of more precise and useful techniques for identifying oral cancer.

Keywords: Oral cancer, Deep Learning, Convolutional Neural Network, Recurrent Neural Network, Precancerous Lesions, Hairy Tongue, Tumor, Forward Propagation, Backward Propagation

1. Introduction

Mouth cancer, also known as OSCC, is a global health threat with millions of cases found annually. Early detection is crucial for survival, as the disease is highly treatable. Deep learning has shown potential in medical image analysis, and the main objective is to investigate optimization techniques based on deep learning for detecting oral cancer. The World Health Organization predicts oral cancer will result in 377,000 new cases and 177,000 deaths globally in 2020. Deep learning models can automatically take

DOI: 10.1201/9781003596776-3

information from and learn from medical photos, making them crucial for accurately identifying diseases, including cancer. Oral cancer is staged according to tumor size, location, and the existence of cells that have died in the mouth's lymphatic system.

The different stages of oral cancer are as follows:

- **Stage 0:** Abnormal; the outer layer of the tissue contains no damaged cells.
- **Stage 1:** A tumor is considered to be 0.2 m or smaller.
- **Stage 2:** Both the size and location of the tumor exceed 2 cm.
- **Stage 3:** The tumor is no more than 3 cm in size, yet it is larger than 4 cm. The lymph node on the same side of the neck contains more dead cells.
- **Stage 4a:** There are three possible tumor sizes: more than 3 cm (growing in the lymph node on the same side), up to 6 cm (many lymph nodes on the same side of the neck), or more than 6 cm (lymph nodes on both sides of the opposite side).
- **Stage 4b:** A lymph node that is larger than 6 cm has been created as a result of the tumor's growth.
- **Stage 4c:** The lungs, liver, and other body organs now contain a higher quantity of tumor cells (Gandhi, 2018).

2. Literature Review

Deep learning is being looked into for its potential in identifying the disease since it has demonstrated promise in enhancing early diagnosis of oral cancer, a significant global health concern.

- **Al Daief, Khaled (2023)** suggests using histopathological pictures to segment and identify oral cancer, most especially MMShift-CNN, a cutting-edge deep learning technique for treating oral squamous cell carcinoma (OSCC) studied by Dharani et al. (2023). Oral cancer and potentially malignant disorders can be automatically segmented using a deep convolution neural network algorithm.
- **Chaurasia et al. (2023)** describe the development of algorithms for the automatic segmentation of oral cancers (OCs) and potentially cancerous oral conditions using deep convolutional neural networks (. Its main focus is not on using deep learning algorithms to identify oral cancer (Ünsal et al. 2023).
- **Manikandan et al. (2023)** suggest an automated system that makes use of an enhanced Convolutional Neural Network (ICNN) for accurate oral cancer diagnosis (Santhiya et al., 2023).
- **Jun Zhang et al. (2023)** discuss how to automatically detect and visualize OCT (Optical Coherence Tomography) images of Oral cancer using deep learning algorithms (Yang et al., 2023).

3. Proposed Method

Recurrent neural networks (RNNs) are being used to identify oral cancer, a significant advancement in machine learning for improved diagnostic precision. RNNs capture sequential dependencies in data, making them useful for examining temporal elements of the disease's course. The methodology uses various datasets, including oral pictures and temporal patient data, to record changes in lesion features or clinical data over time. The RNN architecture, consisting of layers, allows the model to interpret incoming data, store past information, and generate sequential outputs, making it skilled at identifying patterns and changes over time. To optimize the RNN's parameters and predict the occurrence or course of oral cancer, temporal cross-validation techniques are employed. These metrics focus on sensitivity, specificity, recall, accuracy, and *F*1-score, with the temporal element of predictions being the main focus. Integrating RNN-based models into clinical workflows can provide significant insights and enable healthcare practitioners to make informed decisions. Continuous monitoring and updating mechanisms are implemented to adapt the model to changes in oral image distribution and patient characteristics. Ethical considerations, including bias mitigation and patient privacy, remain crucial for the responsible deployment of RNNs in oral cancer detection.

4. Methodology

In order to improve patient outcomes and early intervention, the aim is to support medical practitioners in precisely recognizing possible cases of oral cancer at various stages. Deep learning models are frequently "black boxes," but their interpretability is crucial for medical applications like cancer diagnosis. They can withstand noise and variability in images of oral cancer, ensuring they function well in various scenarios. The model is generalizable, working effectively for diverse demographics, races, and presentation differences. However, overfitting may reduce its applicability in practical situations. The model requires significant computing power, especially for intricate or high-parameter models.

5. Algorithms Used

To detect the oral condition of a person automatically, a system is trained using a dataset and models with algorithms. The project uses RNN as an algorithm, which handles sequential data. RNNs have a hidden state that contains details about the processed sequence, unlike conventional feed-forward neural networks. This allows them to display dynamic temporal behavior, making them useful for tasks like speech recognition, language modeling, and time series prediction. The concept contains algorithms that help train the data.

6. Working

RNN works on the sequential data. It always stores the previous data which is used in further computation. It also performs the same function for every member in the sequence, so it is called a recurrent neural network. RNN mainly has two steps:

i) Forward propagation

ii) Back Propagation

- **Forward Propagation:** The network embraces and feeds the data forward. After receiving the input data and processing it by the activation function, each hidden layer moves on to the next layer.

Back Propagation: It is a crucial algorithm for CNNs to learn from mistakes. It involves sending input data through the network layer by layer during the forward pass, comparing the output to genuine labels. The system architecture is shown in Figure 3.1.

System Architecture

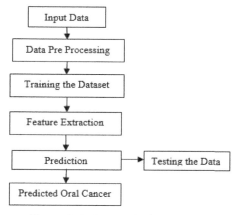

Figure 3.1: System Architecture

7. Results

Our project's primary goal is to display the patient's condition. The following methods will be used to display the output:

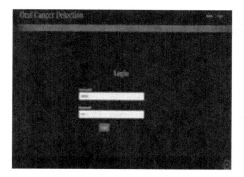

Figure 3.2: Login page.

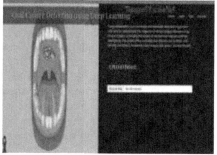

Figure 3.3: User Uploads the image.

Figure 3.4: Probability of Predicting the Output.

Results are compared to different machine learning and deep learning models, whereas our proposed method RNN gives more accuracy than among them.

Evaluation Measure	Naïvebayes (%)	KNN (%)	SVM (%)	ANN (%)	CNN (%)	RNN (%)
Precision	87.1	87.50	96.34	98.18	97.35	99
Recall	82.93	93.90	96.34	98.4	97.67	97
F1Score	85.00	90.59	96.34	98.8	97.51	98.5
Specificity	91.53	90.68	97.46	95.55	97.4	98
Accuracy	88.00	92.00	97.00	97.0	97.51	99

8. Conclusion and Future Scope

The study showcases the RNN's efficiency in early oral cancer detection. The RNN architecture, robust in forecasting temporal sequences, can identify onset signs. Integrating RNNs into clinical processes can provide real-time insights. The ethical framework emphasizes responsible use, including bias mitigation and patient privacy protection. Collaboration between data scientists, medical practitioners, and technologists is crucial.

References

[1] Dharani, R. (2023). Adaptive Coati Deep Convolutional Neural Network-based Oral Cancer Diagnosis in Histopathological Images for Clinical Applications. *Journal Name, Volume*(Issue), Page numbers.

[2] Gandhi, R. (2018). Support vector machine—introduction to machine learning algorithms. *Towards Data Science*, 7(06).

[3] Santhiya, M., Varun, N., & Krishna, B. V. (2023). An effective automated framework for oral cancer detection by enhanced convolutional neural networks. In *2023 12th International Conference on Advanced Computing (ICoAC)* (pp. 1-7). IEEE.

[4] Ünsal, G., Chaurasia, A., Akhilan, N., & Akkaya, N. (2023). Deep convolutional neural network algorithm for the automatic segmentation of oral potentially malignant disorders and oral cancers. *Proceedings of the Institution of Mechanical Engineers, Part H: Journal of Engineering in Medicine*, 237(6), 719-726.

[5] Yang, Z., Zhang, J., Shang, J., Pan, H., & Yang, Z. (2023). Deep-learning-based automated identification and visualization of oral cancer in optical coherence tomography images. *Biomedicines, 11*(3), 802.

CHAPTER 4

Effective Brain Tumor Classification Using Transfer Learning Approach with the Help of Recurrent Neural Networks

Avvaru R. V. Naga Suneetha, V. Bhavya, T. Jayanth, S. Vikas

[1,2,3,4]Computer Science & Engineering, Vignan Institute of Technology and Science
Hyderabad, Telangana State, India
Corresponding author- suneethaavvaru@gmail.com

Abstract

For premature diagnosis and treatment planning, brain tumor characterization using MRI images is vital. To use the potent attributes that a VGG-19 model pretrained on ImageNet had learned for generalized picture recognition, we used transfer learning in this study. Images from a variety of individuals with varying kinds and degrees of brain tumors compose the dataset. This study focuses on the application of convolutional neural networks in brain tumor analysis, with a specific emphasis on transfer learning using pretrained models—VGG-19, along with recurrent neural networks.

Keywords: Brain tumor, MRI, Recurrent Neural Network, Transfer Learning, Image Processing, Image Augmentation, VGG-19, Machine Learning, Deep Learning

1. Introduction

Neuronal tumors are becoming more prevalent in both industrialized and developing nations, which is a critical worldwide health problem. Brain tumors are estimated to occur in 10–18 instances per 100,000 individuals annually, which is approximately 2% of all cancer cases globally, as stated by the World Health Organization (WHO). The incidence of brain malignancies is notably high in India, where numerous instances are recorded annually. Though some variables including exposure to radiation from ionizing sources, predisposition due to genetics, and environmental factors may play a role, the root causes of brain tumors are still largely unclear.

2. Literature review

2.1. A Deep Learning Model Based on the Concatenation Approach

This study examines and compares two methods for using deep learning models to identify tumors. Initially, features from several DenseNet blocks were retrieved using a pretrained DenseNet201 model, and then the features were classified using a softmax classifier. Secondly, a pre-trained Inception v3 model's features were taken from different modules, concatenated, and categorized using softmax. A three-class brain tumor dataset was used to test both strategies, and the ensemble approach exceeded previous techniques with an accuracy of 99.51%.

DOI: 10.1201/9781003596776-4

3. A VGG Net-Based Deep Learning Framework

The need for a fast, unbiased way to evaluate large amounts of medical data has led to an increase in interest in MRI-based medical image processing for brain tumor research in recent years. Since brain tumors have a high death rate, reducing mortality requires early identification and treatment.

4. Ensemble Technique for Brain Tumor Patient Survival Prediction

The primary goal of this effort is to divide tumor regions effectively to expedite the doctors' and pathologists' diagnostic approach. Considering the efficacy and complexity of the U-Net Model, the model achieves a high accuracy of 98.28% in segmenting brain tumors.

5. Methodology and model specifications

5.1. Data Collection

The dataset used in this project was a combination of three different sources: figshare, SARTAJ, and Br35H. The figshare dataset provided a portion of MRI images, which were used for training and testing the brain tumor detection models. These images were divided into four classes: glioma, pituitary, meningioma, and no tumor. The SARTAJ dataset also contributed MRI images, categorized into the same four classes as the figshare dataset. Combining these datasets resulted in a comprehensive dataset containing 7023 MRI images of human brain scans.

6. Implementation

The preprocessed data is set for model training and testing to identify the cerebral tumor. We propose employing a transfer learning approach, utilizing pretrained VGG-19 weights refined with recurrent neural networks (RNNs).

i) Transfer learning:

Transfer learning involves reusing a model trained on one task for a similar task, leveraging its prior knowledge. By using pretrained model weights as a starting point, additional training on new data refines the model for the specific task, often yielding better performance than training from scratch. In the brain tumor identification experiment, a pretrained VGG-19 model was adapted and fine-tuned on MRI images.

ii) VGG-19:

VGG-19 is renowned for its simplicity and consistency. Using the millions of annotated photos from thousands of categories in the ImageNet dataset, the weights that were pretrained of the VGG-19 model are trained. Through the use of supervised learning, the model is taught to classify pictures into several groups by modifying its weights in response to variations in the error between its predictions and the ground truth labels.

iii) RNN:

RNNs are effective at tasks like natural language processing and time series prediction because of their capacity to store memory. RNN nuances such as Gated Recurrent Units (GRUs) and Long Short-Term Memory (LSTM) networks have been constructed to solve this problem. These network architectures can capture dependencies across prolonged spans because they include tools to selectively remember or abandon information.

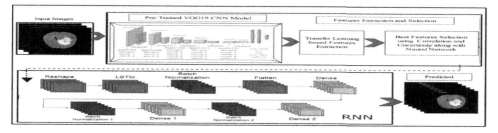

Figure 4.1: System architecture.

iv) Training procedure

In this paper, the training procedure containing the choice of the applicants of 50 training epochs has significance since it enables the model to repeatedly iterate to acquire knowledge from the dataset, hence strengthening its performance. This value finds an equilibrium between fully training the model and preventing overfitting, a phenomenon in which the model learns the training set too well but is unable to generalize to new, unobserved data. In the end, this procedure helps to improve the model's capacity to properly categorize brain tumors by arranging the data in a manner that allows it to be learned from.

v) Model construction and training:

The input layer defines the shape of the input data, which in this case is a 224×224×3 image representing an MRI scan. The weights of the VGG-19 model are loaded from a pretrained file. The input data becomes standardized within the model via batch normalization layers, which has the potential to improve training stability and speed. To determine the final classification, the fully connected (FC) layers—also referred to as dense layers—process the characteristics that the LSTM layer retrieved. Rectified Linear Unit is the activation function used in FC layers. All things considered, the combined model successfully characterizes brain cancers from MRI images by utilizing the strong feature extraction powers of the VGG-19 model along with the sequence processing skills of the LSTM layer.

7. Empirical results

Figure 4.2: Accuracy plot during training of RNN model.

The model achieved an accuracy of 98.4%.

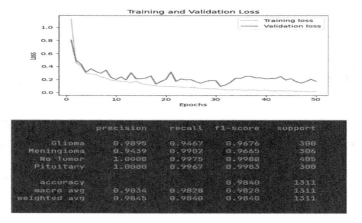

Figure 4.3: Loss Plot during Training of RNN Model

The model achieved an accuracy of 98.4%. In the confusion matrix, the model categorized 16 instances as false positives—tumors when they were not—and properly recognized 284 cases as true positives—brain tumors. Additionally, it accurately detected 1008 instances as being tumor-free (true negatives), but incorrectly classified three cases as being tumor-free when in fact they were (false negatives). The model's sensitivity, which gauges how well it can detect positive instances, is 0.9895470383275261, indicating that it is quite accurate at recognizing patients who have tumors. Overall, the confusion matrix illustrates the model's strong performance in distinguishing between cases with and without tumors, with high levels of specificity and sensitivity (Figure 4).

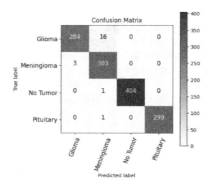

Figure 4.4: Confusion matrix

8. Conclusion

In conclusion, the study has effectively used transfer learning methods to detect brain cancer cells from MRI scans, including fine-tuning using an LSTM layer and transfer learning with a VGG-19 model. With an astounding accuracy of 98.4%, the model proved to be useful in correctly classifying various kinds of brain tumors. The efficiency and resilience of the model were also enhanced using batch normalization layers and picture augmentation approaches. The outcomes of the study have important ramifications for the field of medical imaging, especially for determining the presence of brain tumors.

References

[1] Kwon, D., Shinohara, R. T., Akbari, H., & Davatzikos, C. (2023). Combining generative models for multifocal glioma segmentation and registration. *International Conference on Medical Image Computing and Computer-Assisted Intervention*, 763–770.

[2] Li, M., Kuang, L., Xu, S., & Zhang, U. (2019). Brain tumor detection based on multimodal information fusion and convolutional neural network. *IEEE Access*.

[3] Majib, M., Rahman, M. M., Sazzad, T. M. S., Khan, N. I., & Dey, S. K. (2021). VGG-SCNet: A VGG Net-Based deep learning framework for brain tumor detection on MRI images. *IEEE Access*.

[4] Mallick, P. K., Ryu, S. H., Satapathy, S. K., Mishra, S., Nguyen, G. N., & Tiwari, P. (2019). Brain MRI image classification for cancer detection using deep wavelet autoencoder-based deep neural network. *IEEE Access*.

[5] Manogaran, G., Mohamedshakeel, P., Hassanein, A. S., Kumar, P. M., & Babu, G. C. (2018). Machine learning approach-based gamma distribution for brain tumor detection and data sample imbalance analysis. *IEEE*.

[6] Noreen, N., Muhammad Imran, S., Palaniappan, S., Qayyum, A., Ahmad, I., & Shoaib, M. (2020). A deep learning model based on a concatenation approach for the diagnosis of brain tumor. *IEEE Access*.

[7] Shah, H. A., Park, J.-H., Paul, A., Saeed, F., Yun, S., & Kang, J.-M. (2022). A robust approach for brain tumor detection in magnetic resonance images using finetuned EfficientNet. *IEEE Access*.

[8] Tran, M.-T., Yang, H.-J., Kim, S.-H., & Lee, G.-S. (2023). Prediction of survival of glioblastoma patients using local spatial relationships and global structure awareness in FLAIR MRI brain images. *IEEE Access*, 11, 37437–37449.

[9] Vinod, D. S., Heyat, M. B. B., Prakash, S. P. S., & Alsalman, H. (2024). Ensemble technique for brain tumor patient survival prediction. *IEEE Access*.

[10] Wu, W., Li, D., Du, J., Gao, X., Gu, W., Zhao, F., Feng, X., & Yan, H. (Year). An intelligent diagnosis method of brain MRI tumor segmentation using deep convolutional neural network and SVM algorithm. *Hindawi*.

CHAPTER 5

Enhancing Concrete Performance

A Study on Recron 3S Polyester Fibers and Steel-Slag Aggregate in M30 Grade Concrete

Priyanka Vishwakarma[1,*], Umesh Kumar Dhanger[2]

[1]Department of Civil Engineering, SSTC, Bhilai, India
[2]Department of Civil Engineering, BRP Govt. Polytechnic, Dhamtari, India
Emails: priyankavishwakarma209@gmail.com, uk.d11@polydmt.ac.in

Abstract

Concrete is commonly used in construction due to its cost-effectiveness, durability, and versatility. Researchers are investigating alternatives such as Recron 3S polyester fibers and Steel-Slag aggregate to enhance strength and sustainability. This study examines substituting Ordinary Portland Cement (53 grade) with Recron 3S fibers (0% and 1.25%) and natural aggregate with Steel-Slag aggregate (20%, 25%, and 30%) in M30 grade concrete. Mechanical testing revealed improvements in compressive and split tensile strength, indicating a reduction in environmental impact.

Keywords: Steel-Slag, Tensile Strength, Compressive Strength, Concrete Pavements, Recron 3S Polyester Fiber Reinforced

1. Introduction

Research in civil engineering is centered on improving strength, quality, and durability, as well as reducing costs with the use of eco-friendly materials and additives. Fiber-reinforced concrete (FRC) is utilized to enhance crack resistance and overall properties in comparison to traditional concrete. Recron3s Fiber and Steel-Slag aggregate are currently under examination for their influence on concrete performance and environmental sustainability in construction.(Vishwakarma and Suman 2024a, 2024b).There is a growing interest in the utilization of alternative materials in concrete production, such as Recron3s fiber and Steel-Slag aggregate, to partially replace cement and conventional aggregate. Fronek *et al.* (2012) identified the economic feasibility of Steel-Slag aggregate in PCC due to its nonexpansive characteristics. Subathra Devi and Gnanavel (2014) suggested replacing 40% of fine aggregate and 30% of coarse aggregate with Steel-Slag to improve workability. In RCC beams, Steel-Slag increases deflection and vertical strain while enhancing resistance to HCl and H_2SO_4, though further research is needed for better strength and acid resistance. Rondi *et al.* (2016as required by standard specifications. In addition, the leaching behaviour and the volumetric stability were investigated. Concrete mixtures were designed with aged slag as aggregate having a diameter up to 31.5 mm. The concrete mechanical properties were then compared with the properties of

DOI: 10.1201/9781003596776-5

reference mixtures having only natural aggregates. The concrete with EAF slag showed mechanical properties compatible with their use in civil constructions (compressive strength 35-45 MPa and modulus of elasticity up to 46 GPa) highlighted the benefits of slag from electric arc furnaces. Prahatheswaran & Chandrasekaran (2017) found that adding 0.5% or 1% Recron 3S fibers significantly boosts compressive and split tensile strength. Kumar & Yajdani (2017) emphasized the need for careful adjustment of fiber quantity for optimal properties. Marde *et al.* (2021) observed improved strength in M40 and M60 concrete mixes with 1% Recron 3S fibers. Nanda *et al.* (2020) examined the combined effects of various additives on concrete strength. Kumari and Sinha (2022) highlighted the benefits of Recron 3S fibers for enhancing strength and crack resistance. The literature underscores the advantages of using Steel-Slag particles and Recron 3S polyester fibers for improved strength, sustainability, and durability in concrete. Further research is needed to refine and expand these methods to enhance concrete strength. A comparative study was conducted to assess the impact on strength between traditional concrete and FRC.

Section 2 covers the experimental platform, materials used, design, and tests for FRC. Section 3 discusses rigorous tests to evaluate concrete performance. Section 4 presents research conclusions and future advancements.

2. Materials and Methodology

2.2. Materials

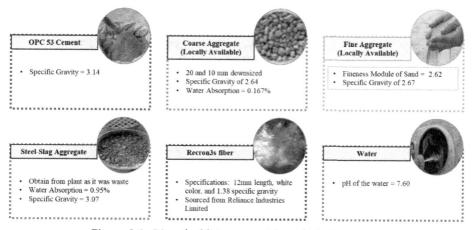

Figure 5.1: List of additive materials with their sources.

3. Sample Preparation

Four distinct concrete samples were prepared, each characterized by a specific combination of Steel-Slag aggregate and Recron 3S fibers. Sample-01 serves as plain cement concrete. The other three samples incorporate partial replacements of cement and coarse aggregate: Sample-02 contains 1.25% Recron 3S fibers and 20% Steel-Slag; Sample-03 comprises 1.25% Recron 3S fibers and 25% Steel-Slag; and Sample-04 consists of 1.25% Recron 3S fibers and 30% Steel-Slag.

Figure 5.2: Samples for experimental study.

4. Mix Design for M30-FRC

The mix design for M30-FRC involves determining the minimum strength required for the concrete mix. The development of the M30-grade concrete mix followed guidelines from IRC: 44-2017 (Indian Roads Congress 2017).

Table 5.1: Mix proportion for 1 cum for FRC concrete (SSD condition).

Material	20% SS+ 1.25% R3sF	25% SS+ 1.25% R3sF	30% SS+ 1.25% R3sF
Water	186 kg	186 kg	186 kg
Cement	418 kg	418 kg	418 kg
Recron3s fiber	5.29 kg	5.29 kg	5.29 kg
Fine aggregate	622 kg	622 kg	622 kg
Coarse aggregate	1198 kg	1198 kg	1198 kg
Graded Coarse aggregate	958 kg	898 kg	839 kg
Steel-Slag (@%Of 20mm CA)	240 kg	300 kg	359 kg

SS: Steel-Slag, R3sF: Recron3s Fiber.

5. Results and Discussion

5.1. Experimental Results and Analysis

This study replaced coarse aggregate with Steel-Slag and added Recron 3S fibers to cement, evaluating the concrete's mechanical properties, including creep, shrinkage, splitting, compressive strength, and tensile strength. Workability and strength were tested in both fresh and hardened concrete. Detailed results and discussions on these properties are provided in the subsequent sections.

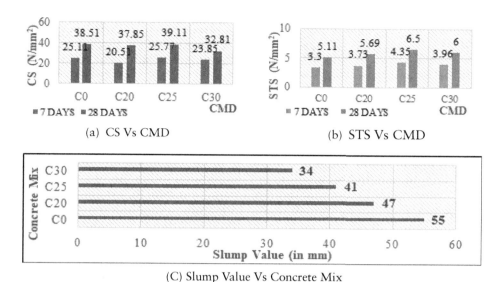

(a) CS Vs CMD

(b) STS Vs CMD

(C) Slump Value Vs Concrete Mix

Figure 5.3: Comparison of material strength and concrete mix destination.

Table 5.2: Variation in CS and STS of concrete at W/C 0.44.

CMD	CA (in %)		CS (N/mm²) After		STS (N/mm²) After	
	20 mm	Steel-Slag	7 Days	28 Days	7 Days	28 Days
C0	100	0	25.11	38.51	3.30	5.11
C20	80	20	20.51	37.85	3.73	5.69
C25	75	25	25.77	39.11	4.35	6.50
C30	70	30	23.85	32.81	3.96	6.00

CMD: Concrete Mix Destination, CA: Coarse Aggregate, CS: Compressive Strength, STS: Split Tensile Strength.

6. Results and Discussion

The strength of concrete mix specimens improved significantly after 7 and 28 days of water curing with a water-cement ratio of 0.44. The specimen with 25% Steel-Slag and 1.25% Recron 3S fiber (C25) exhibited the highest strength. Compressive strength increased with up to 25% Steel-Slag but declined with higher percentages. Split tensile strength also improved with up to 25% Steel-Slag, but decreased at higher levels. Incorporating Recron 3S fiber and Steel-Slag significantly enhanced the split tensile and compressive strength of the concrete specimens.

7. Conclusion

This study investigated the use of Recron 3S polyester fibers and Steel-Slag in M30-grade concrete to enhance strength. The optimal mix, with 25% Steel-Slag and 1.25% Recron 3S fibers, improved elasticity and load-bearing capacity. Higher Steel-Slag levels reduced slump values at a 0.44 water-cement ratio, while compressive strength increased by up to 25% with Steel-Slag, though it decreased at higher levels. The best results were obtained with 25% Steel-Slag and 1.25% Recron 3S fibers, leading to a 25% increase in split tensile strength. The use of these materials supports sustainable construction practices, with potential for further research in durability, economics, mix optimization, field applications, and environmental impact assessments for high-performance concrete.

References

[1]   Fronek, B., Bosela, P., & Delatte, N. (2012). Steel-Slag aggregate used in Portland cement concrete. *Transportation Research Record, 2267*, 37–42. https://doi.org/10.3141/2267-04

[2]   Indian Roads Congress. (2017). *Guidelines for cement concrete mix design for pavements (3rd revision)*. India Offset Press.

[3]   Kumar, K. A., Yajdani, D. S., & R. G. (2017). Study on properties of concrete using Recron 3S fiber. *International Journal of Science Technology & Engineering, 4*(3), 54–62.

[4]   Kumari, M. K., & Sinha, N. (2022). Comparative study of natural fibres coconut, polypropylene fibres Recron 3S & steel fibres on strength of concrete. *International Research Journal of Modernization in Engineering Technology and Science, 4*(5), 3170.

[5]   Marde, S., Wade, H., Jadhav, H., Thakur, A., & Mailpatil, S. C. (2021). Study on properties of concrete using Recron3S fibre. *International Journal of Innovations in Engineering and Science, 6*(6), 29–37.

[6]   Nanda, R. P., Mohapatra, A. K., & Behera, B. (2020). Influence of metakaolin and Recron 3S fiber on mechanical properties of fly ash replaced concrete. *Construction and Building Materials, 263*. https://doi.org/10.1016/j.conbuildmat.2020.120393

[7]   Prahatheswaran, V., & Chandrasekaran, P. (2017). Study on structural behaviour of fiber reinforced concrete with Recron 3S fibers. *SSRG International Journal of Civil Engineering*, (Special Issue-March 2017), 9–13.

[8]   Rondi, L., Bregoli, G., Sorlini, S., Cominoli, L., Collivignarelli, C., & Plizzari, G. (2016). Concrete with EAF steel slag as aggregate: A comprehensive technical and environmental characterisation. *Composites Part B: Engineering, 90*, 195–202.

[9]   Subathra Devi, V., & Gnanavel, B. K. (2014). Properties of concrete manufactured using steel slag. *Procedia Engineering, 97*, 95–104.

[10]   Vishwakarma, P., & Suman, D. A. (2024a). Improving M30 concrete performance with Recron 3S fiber and steel slag: A systematic review. *International Journal for Research in Applied Science and Engineering Technology, 12*(3), 2334–2343.

[11]   Vishwakarma, P., & Suman, D. A. (2024b). Experimenting with the utilization of Recron 3S fiber and steel slag in fresh concrete for analysis. *International Journal for Research in Applied Science and Engineering Technology, 12*(3), 2807–2811.

CHAPTER 6

Parametric Appraisal of Dry-Turning of EN24 using Taguchi-WASPAS and Sunflower Algorithm

Chitrasen Samantra[1], Kanchan Kumari[2], Abhishek Barua[3,4,5], Swastik Pradhan[6], Prasant Ranjan Dhal[7,*]

[1]Department of Production Engineering,
Parala Maharaja Engineering College, Brahmapur, Odisha, India
[2]Department of Mechanical Engineering,
Parala Maharaja Engineering College, Brahmapur, Odisha, India
[3]Department of Automobile Engineering,
Parala Maharaja Engineering College, Brahmapur, Odisha, India
[4]Advanced Materials Technology Department, CSIR-Institute of Minerals and Mate rials Technology, Bhubaneswar, Odisha, India
[5]Academy of Scientific and Innovative Research,
CSIR-HRD Centre Campus, Ghaziabad, Uttar Pradesh, India
[6]School of Mechanical Engineering, Lovely Professional University,
Phagwara, Punjab, India
[7]Department of Mechanical Engineering, Indira Gandhi Institute of Technology,
Dhenkanal, Odisha, India
*Corresponding E-mail: prdhal@gmail.com

Abstract

This work focuses on parametric research during the dry turning of EN24 steel, where variations in feed, spindle speed, and depth of cut were analyzed, and MRR-Material Removal rate and roughness characteristics were evaluated. The WASPAS methodology was paired with the Taguchi method to optimize surface roughness while increasing MRR. ANOVA identified the most significant factors impacting cutting force. Predictions were also generated using the sunflower algorithm. Finally, a confirmation test was conducted to ensure consistency with previous investigations.

Keywords: EN24 steel, WASPAS method, MRR, Surface roughness, Sunflower algorithm

1. Introduction

Surface finish is a crucial indicator of material machinability, impacting both item quality and manufacturing costs. Machining variable optimization can significantly improve machining economies and product quality [1, 2]. EN24 steel, a low-alloy steel, is commonly used in engineering applications, especially in manufacturing gears and crankshafts.

DOI: 10.1201/9781003596776-6

However, dry turning can exacerbate tool wear and generate more heat compared to wet machining processes [3]. This can lead to long, continuous chips during machining, causing tool breakage, surface defects, and poor surface finish. Dry turning may also result in poor surface integrity, including surface roughness, built-up edge formation, and white layer formation [4]. The absence of cutting fluids in dry turning can lead to higher residual stresses in machined components, affecting their dimensional stability, distortion, and susceptibility to stress corrosion cracking [4]. Optimization techniques can improve turning criteria, resulting in controlled tool wear and surface roughness. Many prominent optimization techniques were used to improve turning criteria, resulting in controlled results in terms of tool wear, roughness of the surface, and various other features. Despite this, surface quality remains a challenge, which necessitates improved machining procedures. The WASPAS methodology was used to optimize the results, and the sunflower optimization algorithm was used to predict better factor arrangements.

2. Experimental Methodology

In this experimental investigation, a PVD TiAlN-coated WC cutting insert (CNMG 120412) was used to machine EN24 cylindrical bars with a diameter of 25 mm and a length of 520 mm. The protective coating hardens the surface, making it suitable for coarse machining and providing wear resistance and fracture resistance. Table 6.1 presents the machine operation parameters for this research study.

Table 6.1: Input factors.

Factor	L1	L2	L3	Code
Spindle speed (rpm)	500	750	1800	A
Feed (mm/rev)	0.04	0.06	0.08	B
Depth of cut (mm)	0.3	0.4	0.5	C

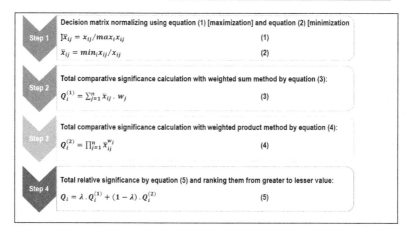

Step 1   Decision matrix normalizing using equation (1) [maximization] and equation (2) [minimization

$$\bar{x}_{ij} = x_{ij}/max_i x_{ij} \tag{1}$$

$$\bar{x}_{ij} = min_i x_{ij}/x_{ij} \tag{2}$$

Step 2   Total comparative significance calculation with weighted sum method by equation (3):

$$Q_i^{(1)} = \sum_{j=1}^{n} \bar{x}_{ij} \cdot w_j \tag{3}$$

Step 3   Total comparative significance calculation with weighted product method by equation (4):

$$Q_i^{(2)} = \prod_{j=1}^{n} \bar{x}_{ij}^{w_j} \tag{4}$$

Step 4   Total relative significance by equation (5) and ranking them from greater to lesser value:

$$Q_i = \lambda \cdot Q_i^{(1)} + (1 - \lambda) \cdot Q_i^{(2)} \tag{5}$$

Figure 6.1: WASPAS method [5].

3. WASPAS Method

The WASPAS technique, one of the most robust methods, relies on three optimality requirements as discussed (Figure 6.1) [5].

4. Sunflower Optimization

The sunflower function optimization (SFO) method takes into account the unique behavior of sunflowers in aligning with the sun. The SFO method simulates pollination by randomly generating seeds according to the shortest separation between flowers i and $i + 1$. Each genuine bloom has millions of pollination gametes. The main goal of this approach is to minimize the spacing between the vegetative tissue and the sun to maximize sunlight exposure and stabilization [6].

5. Results and Discussion

Table 6.2 displays machining parameters and their resulting responses. High MRR is achieved when cut depth and spindle speed are high. Surface roughness minimization and maximization criteria were used for normalization. Total statistical significance was calculated, with run 9 showing the highest total relative significance value (Table 6.3).

Table 6.2: Experimental design.

Run no.	A	B	C	MRR (mm³/min)	Surface roughness (µm)
1	500	0.04	0.3	22.157	3.3296
2	500	0.06	0.4	25.679	2.7123
3	500	0.08	0.5	28.179	3.3726
4	750	0.04	0.4	28.179	3.548
5	750	0.06	0.5	30.116	2.4263
6	750	0.08	0.3	25.679	3.098
7	1800	0.04	0.5	31.918	2.7826
8	1800	0.06	0.3	25.898	2.137
9	1800	0.08	0.4	29.419	2.2986

Table 6.3: WASPAS method computations.

Run no.	Q_1	Q_2	Q_i	Ranking
1	0.6684	0.6679	0.6682	9
2	0.7967	0.7967	0.7967	5
3	0.7587	0.7484	0.7535	6
4	0.7430	0.7297	0.7363	8
5	0.9127	0.9122	0.9125	2
6	0.7476	0.7455	0.7465	7
7	0.8845	0.8769	0.8807	4
8	0.9064	0.9014	0.9039	3
9	0.9263	0.9263	0.9263	1

The Taguchi approach provided precise results using the total relative value of significance. The optimal settings for machining were determined to be a spindle speed of 1800 rpm, a depth of cut of 0.5 mm, and a feed rate of 0.06 mm. The ANOVA table (Table 6.4) indicates that spindle speed (57.74%) is the most significant contributing factor for achieving good machining results with low surface roughness and high MRR, while other factors were deemed insignificant. The R-squared score of 97.80% indicates a good fit of the data and appropriate machining conditions.

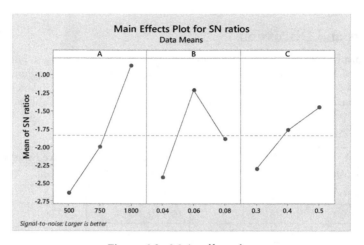

Figure 6.2: Main effect plot.

Table 6.4: ANOVA.

Factors	DF	Adj SS	% Contribution	F-value	P-value	Adj MS
A	2	4.7702	57.74	26.26	0.037	2.38508
B	2	2.1894	26.50	12.05	0.077	1.09471
C	2	1.1200	13.56	6.17	0.140	0.55999
Residual error	20	0.1816	2.20			0.09082
Total	26	8.2612				

The regression equation was derived using the alternative values as a fitness function in the sunflower optimization algorithm. The algorithm was designed to achieve ideal running parameters through 100 preset parameters and 100 iterations. Table 6.5 displays the optimal factor setup and anticipated fitness values.

$$\text{Fitness} = -0.06983 + 0.000119x_1 + 29.42x_2 - 0.9639x_3 - 0.000000x_1^2$$

$$-251.3x_2^2 + 1.467x_3^2 + 0.002245x_1x_2 + 0.000281x_1x_3 \tag{1}$$

Table 6.5: Predicted and experimental results.

	WASPAS	Taguchi-WASPAS	Sunflower optimization
Depth of cut	0.4	0.5	0.5
Feed	0.08	0.06	0.06
Spindle speed	1800	1800	1800
Surface Roughness	2.298	1.975	1.9754
Fitness	0.9263	0.9958	1.3793
MRR	29.419	31.44	31.44

6. Conclusion

This study examined the dry turning of EN24 steel by adjusting spindle speed, depth of cut, and feed rate. Factor settings of 1800 rpm spindle speed, 0.06 mm/rev feed, and 0.5 mm depth of cut were found to maintain high MRR and low surface roughness. The Sunflower algorithm and Taguchi-WASPAS method, used in combination, provided improved results. Spindle speed was identified as the primary factor for achieving high MRR and low surface roughness. Dry turning at optimal cutting parameters significantly extended tool life, improved surface smoothness, reduced machining costs, and enhanced operational efficiency.

References

[1]   Korat, M., & Agarwal, N. (2012). Optimization of different machining parameters of en24 alloy steel in CNC turning by use of taguchi method. *International Journal of Engineering Research and Applications*, 2(5): 160–164.

[2]   Varghese, L., Aravind, S., & Shunmugesh, K. (2017). Multi-objective optimization of machining parameters during dry turning of 11SMn30 free-cutting steel using grey relational analysis. *Materials Today: Proceedings*, 4(2): 4196–4203.

[3]   Thakur, A., Guleria, V., & Lal, R. (2022). Multi-response optimization in turning of EN-24 steel under MQL. *Engineering Research Express*, 4(2): 025052.

[4]   Saini, A., Dhiman, S., Sharma, R., & Setia, S. (2014). Experimental estimation and optimization of process parameters under minimum quantity lubrication and dry turning of AISI-4340 with different carbide inserts. *Journal of Mechanical Science and Technology*, 28: 2307–2318.

[5]   Barua, A., Mishra, M., Das, T., Kumari, K., Pradhan, S., Priyadarshini, M., & Saha, S. (2023). *Parametric appraisal of electrochemical machining of AISI 4140 chromoly steel using hybrid Taguchi-WASPAS-Sunflower optimization algorithm*. In E3S Web of Conferences, EDP Sciences France, 453: 01046.

[6]   Gomes, G. F., da Cunha, S. S., & Ancelotti, A. C. (2019). A sunflower optimization (SFO) algorithm applied to damage identification on laminated composite plates. *Engineering with Computers*, 35: 619–626.

CHAPTER 7

In-depth exploration of utilizing water hyacinth fibers as a sustainable method for reinforcing concrete

Tanusri Dutta

M.tech Student, Civil Engineering, Techno India University, Kolkata, West Bengal, India
E-mail:tanusriengg@gmail.com

Dr. Suman Pandey

Assistant Professor, Civil Engineering Department, Techno India University, Kolkata, West Bengal, India
E-mail: suman.p@technoindiaeducation.com

Dr. Heleena Sengupta

Associate Professor, Civil Engineering Department, Techno India University, Kolkata, West Bengal, India
E-mail: HS.square4@gmail.com

Abstract

Today, ecological integrity must be our sole aim, and we may achieve this by emphasizing the critical importance of protecting ecological integrity through sustainable construction practices, particularly regarding the use of concrete, which has a significant environmental footprint due to its high cement content. The manufacturing of cement is indeed a major source of global carbon dioxide emissions. To address this issue, utilizing admixtures and incorporating reinforcing fibers into concrete are promising strategies. Admixtures can help reduce cement consumption while maintaining or even enhancing concrete quality, and reinforcing fibers can improve concrete's tensile strength and toughness. Natural fibers, such as water hyacinth fiber (WHF), present a sustainable alternative for reinforcing concrete. WHF, being abundant and cost-effective, can partially replace fine aggregate in concrete mixtures, thereby reducing the overall environmental impact of concrete production. Research focusing on the mechanical properties of concrete reinforced with lignocellulosic fibers like WHF is crucial for assessing its effectiveness and suitability. By studying parameters like strength in compression, strength in tension, and strength in flexure, researchers can evaluate the performance of WHF-reinforced concrete and its potential for widespread application.

Keywords: Natural fiber, sustainability, geopolymer concrete, lignocellulosic fibers, phytoremediation, fiber-reinforced concrete, Water hyacinth fiber

DOI: 10.1201/9781003596776-7

1. Introduction

One significant area of focus within the movement toward sustainable construction practices is the development and utilization of "green concrete," which is produced using environmentally friendly materials and techniques. This aims to reduce its carbon footprint and enhance its sustainability compared to traditional concrete, which is a primary source of carbon dioxide emissions due to its energy-intensive manufacturing process. By incorporating alternative cementitious materials or reducing cement content altogether, green concrete can achieve CO_2 emissions reductions of up to 30% compared to traditional concrete formulations.

Concrete is brittle and has low strength against tension, so its residue has significant negative impacts on the environment, human health, or both. These materials can contribute to pollution, environmental degradation, and health hazards. Fiber reinforcement enables engineers and construction professionals to develop safer, more durable, and sustainable infrastructure across various projects and applications. Successful implementation can lead to sustainable, high-performance concrete solutions with enhanced mechanical properties and reduced environmental impact [Jamal et al., 2023].

2. Water hyacinth

The most troubling weed Water hyacinth (WH) (Family: *Pontederiaceae*) is ranked as the 8th most troublesome weed according to the list by Holm and Herberger (1969). Over time, WH has become a pervasive invasive species in tropical and subtropical regions worldwide. Its capacity for rapid expansion and formation of thick covers on the water surface has significant ecological, economic, and social impacts, requiring ongoing efforts to manage and control its spread in affected areas. Utilizing it in geopolymer concrete mixes has the potential to lower material expenses compared to conventional concrete, thereby making construction projects more economical overall [Harun, 2021].

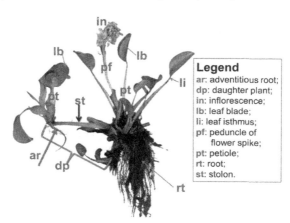

Figure 7.1: Parts of water hyacinth plant [Derese et al., 2022].

3. Water hyacinth fiber as a reinforcing material in concrete

By employing diverse extraction methods, characterization techniques, and surface treatments, micro and nano cellulose can be derived from water hyacinth fibers (WHF). Recent studies have focused on incorporating WHF reinforcements into thermoplastic, biodegradable, and thermosetting materials. These investigations showcase enhancements in mechanical attributes, thermal resilience, and resistance to water vapor and gas. This research underscores the expansive potential to employ WHF in the development of sustainable composite materials [Guna et al., 2017].

Table 7.1: Chemical components of different natural strands.

Fiber	Water hyacinth	Coconut sheath	Sisal	Jute	*Acacia arabica*
Density (Kg/m^3)	226	1375	1450	1300	1028
Cellulose (wt.%)	65.4	68	65–70		68.1
Hemicellulose (wt.%)	12.8	22	12		9.36
Lignin (wt.%)	7.2	20.6	9.9		16.86
Wax (wt.%)	0.24				0.49
Moisture (wt.%)	3.2	8.79	10		
Tensile strength (MPa)		88.63	68	393–723	
Young's modulus (GPa)		4.4	3.8	26.5	
Reference	[Ahmad et al., 2022]	[Ahmad et al., 2022]	[Ahmad et al., 2022]	[Mohan et al., 2021]	[Srivastava et al., 2023]

From the above table, it is evident that WH has the potential to fulfill the criteria for creating lightweight materials due to its low density. Moreover, due to its lower hemicellulose content, it is more durable, and its low wax content makes it more prone to bonding [Jirawattanasomku et al., 2021]. Results from tests by Prang et al. (2020) on mortar with powdered WH show better strength values than conventional mortar.

4. Fiber-reinforced concrete and WHF

The main problems affecting infrastructure include the poor durability of construction materials, insufficient workmanship, lack of user-friendly measuring equipment, absence of repair tools, and the need for affordable, high-quality materials and technology. Fiber-reinforced concrete (FRC) holds promise as a solution to these problems.

The construction industry is increasingly adopting biodegradable materials and recycling expertise. Consequently, there is a growing focus on employing WH surplus to produce FRP composites for concrete strengthening. Testing results of WHF-reinforced polymer composites indicate that they possess acceptable mechanical properties, including elastic modulus, tensile strength, and strain, for concrete strengthening applications. Additionally, they exhibit improved strength and ductility performance [Tidarut et al., 2021].

Despite its higher water absorption capacity, WH-infused concrete exhibits similar sorptivity to conventional concrete, suggesting comparable permeability characteristics. This suggests its suitability for marine structures where preventing water leakage is important. Additionally, WH extract may be used as a superplasticizer in the making of self-compacting concrete. As a water reducer and retarder, it improves the workability and setting time of the concrete mixture during construction [Ernie et al., 2023].

5. Discussion based on the previous study

Water hyacinth fiber-reinforced polymer (WHFRP) exhibits satisfactory mechanical properties, including a tensile strength of 137 MPa and an ultimate tensile strain of 1.72%, making it suitable for reinforcing concrete structures. Moreover, WHFRP confinement significantly improves the strength in compression for low-strength concrete specimens, with enhancement ratios ranging from 1.44 to 2.43. The stress–strain behavior curve of WHFRP-encased concrete exhibits an increasingly parabolic shape, indicating increased ductility. Compared to other natural fiber-reinforced polymers and carbon fiber-reinforced polymers, normal concrete confined with WHFRP demonstrates superior energy absorption, ranging from 0.60 to 0.84. WHFRP also boasts minimal environmental effects in terms of water intake compared to other fiber-reinforced polymers.

Polymer composites reinforced with WHF exhibit adequate mechanical properties, making them suitable for reinforcing concrete structures. Additionally, their utilization presents environmental benefits compared to conventional alternatives, marking them as a promising solution for sustainable construction practices. This contributes significantly to the increasing adoption of construction materials that are not harmful to the environment and can be recycled effectively, addressing environmental concerns while also considering economic factors [Tidarut et al., 2021].

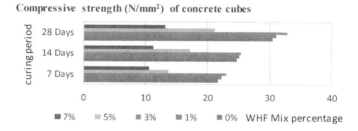

Graph 7.1: Effect of WHF% variation on compressive strength.

Akhilakumari et al. investigated the concrete mix prepared with cement containing 25% fly ash and varying proportions of WH using M25 grade concrete. Results indicated that 3% is the most effective percentage for maximum strength. Increasing WHF content from 3% to 5% enhances the concrete's compressive strength, which then declines beyond 5% [refer to Graph 7.1]. This trend is consistent from 7 to 28 days of curing.

6. Conclusion

1. WHF shows potential in the production of lightweight concrete, which is highly valued in the construction industry.
2. Concrete reinforced with WHF supports sustainable building practices, making it an excellent choice for environmentally friendly construction projects aiming to minimize their ecological footprint.
3. Examination findings demonstrate that concrete cubes with a 3-5% substitution of WHF exhibit relatively high compressive strength.
4. Concrete incorporating WH fibers has demonstrated higher compressive strength compared to regular concrete when exposed to elevated temperatures, making it

suitable as an insulating material in industrial settings and buildings at risk of explosions or high-temperature events.

5. The sorptivity test measures the durability of concrete. The study reveals that a low sorptivity value provides resistance against freeze and thaw cycles, enhancing the concrete's durability.

References

[1] Ahmad, J., et al. (2022). Concrete reinforced with sisal fibers (SSF): Overview of mechanical and physical properties. *Crystals, 12*(7), 952.

[2] Ajithram, A., Arivendan, et al. (2022). Study on characterization of water hyacinth (Eichhornia crassipes) novel natural fiber as reinforcement with epoxy polymer matrix material for lightweight applications. *Journal of Materials Science, 51*(5S), 8157S–8174S. https://doi.org/10.1177/15280837211067281

[3] Akhilakumari, P., et al. (2022). Investigation on the effect of water hyacinth in concrete. *IRJET, 09*(07). https://www.irjet.net/e-ISSN: 2395-0056/p-ISSN: 2395-0072

[4] Ernie, D. T., et al. (2023). Evaluation of water hyacinth ash, extract, and fiber in concrete: A literature review. https://doi.org/10.1007/978-981-19-8024-4_5

[5] George, S., et al. (2023). Extraction and characterization of fibers from water hyacinth stem using a custom-made decorticator. *Journal of Natural Fibers, 20*(2)

[6] Guna, V., et al. (2017). Water hyacinth: A unique source for sustainable materials and products. *ACS Sustainable Chemistry & Engineering* (IF 8.4). https://doi.org/10.1021/acssuschemeng.7b00051

[7] Harun, I., et al. (2021). Invasive water hyacinth: Ecology, impacts, and prospects for the rural economy. *Plants, 10*(8), 1613. https://doi.org/10.3390/plants10081613

[8] Hermawan, et al. (2015). Identification of source factors of carbon dioxide (CO_2) emissions in concreting of reinforced concrete. https://doi.org/10.1016/j.proeng.2015.11.107

[9] Jamal, A., et al. (2023). A comprehensive review on the use of natural fibers in cement/geopolymer concrete: A step towards sustainability.

[10] Jirawattanasomku, T., & Tidarut, et al. (2021). Use of water hyacinth waste to produce fiber-reinforced polymer composites for concrete confinement: Mechanical performance and environmental assessment. *Journal of Cleaner Production, 292*, 126041

[11] Prang, S. & Subpa-Asa, et al. (2020). The study on the effect of lightweight concrete block by water hyacinth. *IOP Conference Series: Materials Science and Engineering, 811*

[12] Srivastava, K. S., et al. (2023). A comparative study of mechanical characteristics of normal concrete with human hair fiber fabricated concrete. *IOP Conference Series: Earth and Environmental Science, 1110*, 012057. https://doi.org/10.1088/1755-1315/1110/1/012057

CHAPTER 8

Evaluating Water Hyacinth Ash as a Viable Construction Material

An In-depth Analysis of its Sustainable Potential

Arya Chakraborty[1], Heleena Sengupta[2], Suman Pandey[3]

[1]MTech Student, Civil Engineering, Techno India University, Kolkata,
West Bengal, India Chakrabortya411@gmail.com
[2]Associate Professor, Civil Engineering, Techno India University,
Kolkata West Bengal, India HS.square4@gmail.com
[3]Assistant Professor, Civil Engineering, Techno India University,
Kolkata West Bengal, India suman.p@technoindiaeducation.com

Abstract

Ordinary Portland cement is a primary binder in concrete, responsible for approximately 5%–7% of carbon dioxide (CO_2) and greenhouse gas emissions, with an annual production exceeding 4 billion tons. Reducing the carbon footprint of the concrete output while maintaining its functionality is crucial. This investigation explores using water hyacinth ash as a pozzolanic material, replacing a portion of cement in concrete mixtures. This review evaluates the strength and workability characteristics of ash-blended concrete compared to conventional concrete and provides a comprehensive outline of future research directions.

Keywords: Concrete, Greenhouse gas, Sustainable material, Water hyacinth ash

1. Introduction

Concrete consists of cement, coarse aggregates, natural sand, and water. Cement, a key ingredient in concrete, is comparatively expensive relative to other building materials (Das 2016). The production of cement utilizes a majority of natural resources and releases hazardous gases into the atmosphere, accelerating global warming (Hebert et al. 2020). The cement industry is responsible for at least 8% of global emissions caused by human activities, contributing significantly to concrete's carbon footprint (Ellis et al. 2020). Given the finite nature of Earth's natural resources, it is imperative to protect and recycle them for the benefit of current and future generations.

Current research focuses on replacing cement in concrete production with waste materials that exhibit pozzolanic behavior. Utilizing waste materials in concrete can reduce production costs and is environmentally sustainable (Farahani et al. 2017). Water hyacinth (WH) is one of the fastest-growing plants and is globally recognized as a destructive invasive species that disrupts electricity, irrigation, and navigation. Prior studies have shown that water hyacinth fibers, containing approximately 68.4% cellulose, exhibit higher strength compared to sisal and banana fibers (Wiryikfu et al. 2023).

DOI: 10.1201/9781003596776-8

This suggests potential applications of WH in the textile and construction industries. Additionally, several researchers have investigated the key physical and chemical properties of water hyacinth ash (WHA) to understand its behavior [Table 8.1].

Table 8.1: Key properties of water hyacinth ash: [Murugesh et al. (2019), Sukarni et al. (2019)].

Specific gravity	2.44
Fineness	10%
Water absorption	15%
Water content	4.9 wt%
Ash	20.1 wt%

WHA, known for its pozzolanic properties, offers a sustainable alternative to cement in concrete production. This paper investigates recent research on incorporating WHA into concrete mixtures, aiming to mitigate the environmental impacts associated with cement production (adapted from Das et al. 2016).

2. Literature review

Studies highlight the effectiveness of WH as a reinforcing agent, enhancing tensile strength and durability in concrete structures. Current research emphasizes WHA as a pozzolanic material. The research is categorized into the following groups. Material characterization and microstructural examination: This involves studying the physical, chemical, and mechanical properties of water hyacinth fibers or ash to assess their suitability as a concrete additive. Galvez et al. (2015) conducted elemental analysis on water hyacinth roots using scanning electron microscopy and energy-dispersive X-ray analysis, revealing the elemental composition at a microscopic level. Ajithram et al. (2020) further examined the chemical constituents of WH, noting its high content of hemicellulose and cellulose, which are utilized in producing biogas and biofuels. This high content of hemicellulose supports the use of WH as a reinforcing agent in concrete. Mix Design and Optimization: Research focuses on developing concrete mixtures with varying proportions of WHA to achieve desired properties such as workability, strength, and durability.

3. Observations and Discussion

Several studies have explored the effects of incorporating WHA into both fresh and hardened concrete.

Effect on the workability of concrete: The workability of freshly prepared concrete increases with the addition of WHA. This trend is consistent across various concrete grades (Graph 8.1).

Impact of WHA as a natural admixture: WH extract is a cost-effective co-superplasticizer, substituting chemical admixtures in self-compacting concrete. This extract slows down the hardening process and rate of hydration, extending the duration of concrete flow and enhancing its flowability and filling capacity (Okwadha et al. 2018).

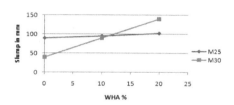
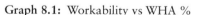

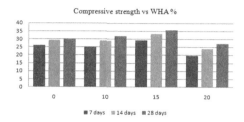

Graph 8.1: Workability vs WHA %

Graph 8.2: Effect of % inclusion of WHA on compressive. strength

[Abiramiet al. (2018), Murugesh et al. (2017)].

Impact on strength of concrete: Research indicates that the strength of concrete with WHA improves with extended curing durations, peaking after 28 days (Graph 8.2). However, although the strength generally rises to a certain threshold with an augmentation in the ash percentage, a subsequent decline has been noted for all grades beyond this threshold. Hence, additional research is needed to find the ideal ash percentage, along with exploring other enhancing materials.

4. Conclusion

The study leads to the following conclusion:

1. 1. As the proportion of water hyacinth ash in the blended cement paste increases, the specimen exhibits extended initial and final setting times, which can be advantageous for concrete placement in hot weather conditions.
2. 2. Mortar specimens with WHA show reduced sorptivity and water absorption after 28 days compared to control specimens. These characteristics are crucial indicators of mortar's durability in aggressive environments.
3. 3. The optimal replacement level for cement with WHA is 10%, although in some cases it can be extended to 15%.
4. 4. Due to its lightweight nature and pozzolanic properties, WHA can be used as an auxiliary cementitious material in lightweight concrete formulations. When used in appropriate proportions, it can contribute to the desired properties of lightweight concrete.

References

[1] Abirami, U., Kayalvizhi, K., Pavithra, K., & Govandan, A. (2018). Experimental study on the behaviour of concrete replaced with water hyacinth ash. *International Journal of Engineering Research and Technology*. ISSN: 2278-0181.

[2] Ajithrama, A., Jappesa, J. T. W., & Brintha, N. C. (2020). Investigation on utilization of water hyacinth aquatic plants towards various bio products–Survey. *Materials Today: Proceedings*. In press. https://doi.org/10.1016/j.matpr.2020.09.498.

[3] Akhilakumari, P., Hari, A., Devika, S., Shaji, S., & Chaithra, S. (2022). Investigation on the effect of water hyacinth in concrete. *International Research Journal of Engineering and Technology (IRJET)*, 9(7), 1160–1163.

[4] Antonio. (2015). Effectiveness of water hyacinth (*Eichhornia crassipes*) as partial cement replacement in load bearing masonaryblocks. An undergraduate thesis

presented to the faculty of the College of Engineering and Industrial Technology, Camarines Norte State College.

[5]   Das, N., & Singh, S. (2016). Evaluation of water hyacinth stem ash as pozzolanic material for use in blended cement. Journal of Civil engineering Science and Technology 7(1): 1-8, DOI: 10.33736/jcest.150.2016.

[6]   Ellis, L. D., Badel, A. F., Chiang, M. L., Park, R. J.-Y., & Chiang, Y.-M. (2020). Toward electrochemical synthesis of cement—An electrolyzer-based process for decarbonating $CaCO_3$ while producing useful gas streams. *Proceedings of the National Academy of Sciences of the United States of America*, 117, 12584–12591.

[7]   Ekasilp, W., & Boonthanomwong, C. (2014). Thermal insulator performance of water hyacinth and sawdust hollow concrete block, *National Research Conference*, Rangsit University, Thailand, p 179-186.

[8]   Farahani, J. N., Shafigh, P., Alsubari, B., Shahnazar, S., & Mahmud, H. B. (2017). Engineering properties of lightweight aggregate concrete containing binary and ternary blended cement. *Journal of Cleaner Production*, 149, 976–988.

[9]   Galvez, M., Guzmán, J., Gueco, S. D., Camelo, J., Castilla, R., & Vallar, E. (2015). SEM/EDX analysis of the roots of water hyacinths (*Eichhornia crassipes*) collected along Pasig River in Manila, Philippines. *Environmental Science*. ARPN Journal of Agricultural and Biological Science, Vol. 10, No. 12, 458-463.

[10]   Habert, G., Miller, S. A., John, V. M., Provis, J. L., Favier, A., Horvath, A., & Scrivener, K. L. (2020). Environmental impacts and decarbonization strategies in the cement and concrete industries. *Nature Reviews Earth & Environment*, 1 559–573.

[11]   Hodhod, H., & Anwar, M. (2008). Durability of water hyacinth ash concrete. *Engineering Research Journal*, 87, 144–156.

[12]   Khandagale, A. S., Aher, G. A., & Pharate, K. G. (2023). Experimental study of use of water hyacinth ash in concrete. *Journal of Emerging Technologies and Innovative Research (JETIR)*, 10(4), 561–564.

[13]   Murugesh, V., & Balasundaram, N. (2017). Experimental investigation on water hyacinth ash as the partial replacement of cement in concrete. *International Journal of Civil Engineering and Technology (IJCIET)*, 8(9), 1013–1018.

[14]   Murugesh, V., & Balasundaram, N. (2019). Bonding characteristics of water hyacinth ash on concrete by replacing cement: Durability studies. *International Journal of Recent Technology and Engineering (IJRTE)*, 8(2), 4977–4979.

[15]   Murugesh, V., Suriya, Boopathi, Ramkumar, & Manikandan. (2019). Experimental behaviour of water hyacinth ash as the partial replacement of cement in concrete. *International Journal of Scientific & Engineering Research*, 10(3), 207–211.

CHAPTER 9

Structural and Thermal Analysis of Gas Turbine Blade

Simulation by CFD

Krushnashree Sushree Sangita Sahoo[1], Brahmotri Sahoo[2], Chandradhwaj Nayak[2*]

[1]Assistant Professor, Department of Mechanical Engineering, Indira Gandhi Institute of Technology, Dhenkanal, Odisha, India
[2]Assistant Professor, Department of Chemical Engineering, Indira Gandhi Institute of Technology, Odisha, India
E-mail address: krishnashree2007@gmail.com, brahmotri.sahoo@igitsarang.ac.in, chandradhwaj@gmail.com

Abstract

A gas turbine is a rotating heat engine that operates by drawing air from the atmosphere and burning fuel at constant pressure to raise the temperature of the gas. The most crucial parts of a gas turbine plant responsible for extracting energy from high-temperature, high-pressure gas are the turbine blades. The material properties and inlet temperatures are the two most important parameters considered for the efficiency of gas turbine blades. Therefore, the structural stability and thermal analysis of these components need to be analyzed. This paper describes the deformation and heat flux for blades made from different materials such as Inconel 625, Inconel 718, and Rene 77, with the analysis performed using the finite element method in Ansys. The results conclude that Inconel 625 is best suited for blades in the gas turbine rotor.

Keywords: Heat flux, Total deformation, Inconel 625, Inconel 718, Rene 77

1 Introduction

In a gas turbine power plant, the prime mover is the turbine, and the blades are responsible for its operation. The turbine blades convert the energy from high-temperature and high-pressure gas into mechanical shaft work. These blades are made from stainless steel, titanium, nickel alloys, and ceramic materials. External factors affecting blade performance include static loads and temperature. Mechanical stress and thermal stress analysis are very important for blades in a gas turbine plant. Material selection for the gas turbine is crucial; materials must withstand elevated temperatures. The efficiency of the plant increases if the selected material can resist the specified elevated temperatures during service. The material should be selected to maximize the temperature-to-weight ratio for low weight. This research focuses on selecting materials

DOI: 10.1201/9781003596776-9

for turbine blades that can tolerate mechanical and thermal loads under various conditions. First, different materials are chosen based on their properties. Then, using SolidWorks, the blade is designed with proper dimensions, and the model is imported into Ansys Workbench for simulation analysis with computational fluid dynamics (CFD) to analyze structural and thermal performance.

Literature Review

N155 and Inconel 718 nickel–chromium alloys were used to determine steady-state thermal and structural performance using the finite element method (FEM) and Ansys software [Prasad et al. 2013]. The blades were analyzed with holes 5, 9, and 13 to determine the optimal number of cooling holes. Inconel 718 performed better in high-temperature regions. The blade with 13 holes was the optimal value for the analysis. FEM was also used for analyzing turbine blades in gas turbines for structural and thermal behavior, with CATIA V5 used for design and Ansys software for simulation [Krishnakanth et al. 2013]. Inconel 625 showed better thermal properties, while N155 exhibited lower deformation compared to others. Blade tip temperatures and elongations were maximum, while values at the blade root were minimum. Jabbar et al. [2014] conducted work emphasizing structural and thermal analysis using FEM with ANSYS 12 to find thermal stress and temperature distribution in rotor blades made of titanium alloy, stainless steel alloy, and Aluminum 2024 alloy. The maximum temperature decreased linearly from the blade tip to the root.

Boyaraju et al. [2015] analyzed thermal behavior for gas turbine blades, designing the blades in CATIA V5 and analyzing them with Ansys. Al2024 and T6 materials were used. Al2024 provided the best heat flux results, making it more efficient and offering a better life for the turbine. Titanium alloy was also used for turbine blade analysis using Ansys software and FEM. The fatigue stress developed in the blade due to speed changes ranging from 8500 to 19000 rpm, with six frequency modes obtained. At 19000 rpm, the frequency increased with mode, indicating the critical speed [Sa et al. 2015]. Balaji et al. [2016] studied the cooling effect of gas turbine blades, finding that increased hole orientation improved cooling effectiveness, with the highest value at 45 degrees. Cooling effectiveness increased by 14% in the range of 300–400 hole orientations. Different stress and deformation values for the blade materials were determined, with CAD modeling software used for model development and subsequent analysis done with FEA software on various materials such as Inconel 625, Inconel 718, Inconel 738, Ni–Cr, Ti–Al alloy, and MAR-M 246 [Mazarbhuiya et al. 2017].

2. Materials and Methods

2.1 Materials used

One important factor for turbine blades is material creep strength for gas turbine engines, for which superalloys are cast. The following describes the materials used in manufacturing turbine blades.

2.2 Rene 77

Rene 77 is a nickel–cobalt superalloy suitable for high-temperature applications up to 1050°C. It is commonly used for gas turbine blades and aero-engine components due to its high tensile strength and high-temperature corrosion resistance. The chemical composition of Rene 77 is listed in Table 9.1.

Table 9.1: Chemical composition of Rene 77.

Constituents	Ni	Cr	Co	Mo	Ta	Al	Ti	C	B	Zr	Fe	S	Si
Percentage	57.08	14.61	15.32	4.52	0.05	4.73	3.49	0.07	0.015	0.01	0.08	0.001	0.017

2.3 Inconel 718

Inconel 718 is a nickel-based alloy with high yield, tensile, and creep-rupture properties at 1300°F. Inconel forms a thick, stable, passivating oxide layer that shields the surface from further damage when heated. The chemical composition of Inconel 718 is presented in Table 9.2.

Table 9.2: Chemical makeup of Inconel 718.

Constituents	Ni	Cr	Co	C	Mn	Si	Ti	Al	Mb	Cd	Fe
Percentage	52.82	19.0	1.0	0.08	0.35	0.35	0.6	0.8	3.0	5.0	17.0

2.4 Inconel 625

Inconel 625, an alloy of nickel, chromium, molybdenum, and niobium, is suitable for a wide range of temperatures, from extremely hot to very cold. It can function at cryogenic temperatures up to 980°C and is used in heat exchangers, furnace hardware, jet engines, and chemical plant equipment .

Table 9.3: Chemical makeup of Inconel 625.

Constituents	Ni	Cr	Co	C	Mn	Si	Ti	Al	Mo	Nb	Fe
Percentage	58.0	20.0	1.0	0.1	0.5	0.5	0.4	0.4	8.0	3.15	5.0

The mechanical and physical characteristics of each material used for the thermal analysis of turbine blade materials are listed in Table 9.4.

Table 9.4: Mechanical and physical properties of the materials.

Materials	Density (kg/m³)	Young's Modulus (MPa)	Poisson's ratio	Thermal conductivity (W/mK)	Specific heat (J/kg K)
Rene 77	7700	200000	0.30	25	460
Inconel 718	8249	149000	0.344	11.4	435
Inconel 625	8400	150000	0.331	9.8	410

3. Methodology

To study the heat flux and overall deformation of gas turbine blades made from different materials, the modeling is initially done using SolidWorks. The model is saved as a .stp file and imported into Ansys 16 software for further analysis. The dimensions of the model are specified in Table 9.5.

Table 9.5: Model dimension.

Parameter Details	Blade height (m)	Chord width (m)	Pitch (m)	Inlet blade angle (degree)	Outlet blade angle (degree)	Mean radius (m)
Value	0.818	0.027	0.022	18.3	54.56	0.247

3.2 Model Development

The model with the specified dimensions is created in SolidWorks and then imported into Ansys Workbench for analysis, as depicted in Figure 9.1.

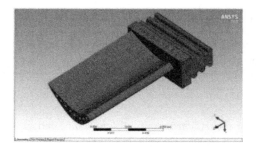

Figure 9.1: Developed model.

4. Results and Discussion

4.1 Heat Flux Analysis

Table 9.6 shows the values of heat flux at different time intervals. The analysis finds that Inconel 625 has the most efficient heat flux compared to the other two materials.

Table 9.6: Values of temperature distribution and heat flux.

Materials	Time				
	2 secs	4 secs	6 secs	8 secs	10 secs
Inconel 625	0.0133	0.0398	0.0664	0.0930	0.1196
Inconel 718	0.0043	0.0131	0.0218	0.0305	0.0392
Rene 77	0.0093	0.0280	0.0467	0.0654	0.0841

Figure 9.2 shows the temperature and heat flux distribution for Rene 77, Inconel 625, and Inconel 718, and the comparison was done. Figure 9.3 shows the heat flux comparison of different materials used. It is observed that the trend for heat flux varies linearly with time. The maximum heat flux value is observed for Inconel 625 compared to other materials. Therefore, Inconel 625 is recommended as the best material for designing blades in a gas turbine.

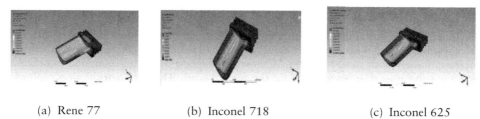

(a) Rene 77   (b) Inconel 718   (c) Inconel 625

Figure 9.2: Heat flux images (a) Rene 77, (b) Inconel 718, (c) Inconel 625.

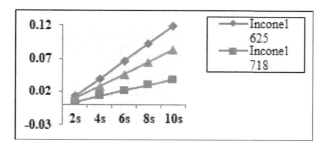

Figure 9.3: Comparative plot based on heat flux.

4.2 Total Deformation Analysis

The total deformation analysis of Inconel 625, Inconel 718, and Rene 77 is shown in Figure 9.4. The results indicate that Inconel 625 and Inconel 718 have almost the same value of deformation, while Rene 77 shows the maximum deformation due to heat and pressure, making Inconel 625 and Inconel 718 better suited as blade materials compared to Rene 77.

(a) Rene 77   (b) Inconel 718   (c) Inconel 625

Figure 9.4: Images of total deformation: (a) Rene 77, (b) Inconel 718, (c) Inconel 625.

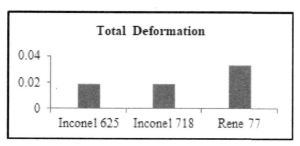

Figure 9.5: Comparative plot of based on total deformation.

5. Conclusion

The heat flux and total deformation analyses of Inconel 625, Inconel 718, and Rene 77 were conducted, and the results were compared. For all materials analyzed, heat flux values increase linearly with time. Inconel 625 shows the highest heat flux compared to others, making it the most suitable material for the current work. The total deformation analysis shows that Inconel 625 and Inconel 718 have similar deformation values compared to Rene 77, making them suitable for gas turbine blade materials.

References

[1]   Prasad, R. D. V., Narasa Raju, G., Srinivas Rao, M. S. S., & Vasudeva Rao, N. (2013). Steady state thermal & structural analysis of gas turbine blade cooling system. *International Journal of Engineering Research & Technology*, 2(1), 1–6.

[2]   Krishnakanth, P. V., Narasa Raju, G., Prasad, R. D. V., & Saisrinu, R. (2013). Structural & thermal analysis of gas turbine blade by using F.E.M. *International Journal of Scientific Research Engineering & Technology*, 2(2), 60–65.

[3]   Jabbar, A., Rai, A. K., Reedy, P. R., & Dakhil, M. H. (2014). Design and analysis of gas turbine rotor blade using finite element method. *International Journal of Mechanical and Production Engineering and Development*, 4(1), 73–94.

[4]   Boyaraju, G., Rajesekhar, S., Sridhar, A. V., & Rao, J. H. N. (2015). Thermal analysis of a gas turbine rotor blade. *International Journal of Science Engineering and Advance Technology*, 3(12).

[5]   Sa, P. G., & Byregowda, H. V. (2018). Dynamic analysis of a steam turbine with numerical approach. *Materials Today: Proceedings*, 5, 5414–5420.

[6]   Balaji, K., Azhagesan, C., Pandian, R. V., Murugan, N. M., & Aravinth, N. (2016). Review of a cooling effect on gas turbine blade. International Conference on Systems, Science, Control, Communication, Engineering, and Technology, Vol 2, page 478–482, Tamilnadu.

[7]   Mazarbhuiya, H. M. S. M., & Pandey, K. M. (2017). Steady-state structural analysis of high-pressure gas turbine blade using finite element analysis. *IOP Conference Series: Materials Science and Engineering*, 225, page 012113–012122, Elsevier.

CHAPTER 10

Flood risk assessment of Brahmani River Basin, Odisha, India

A geospatial study using multicriteria decision analysis

Tanmoy Majumder*, Sushant Kumar Biswal

Department of Civil Engineering, National Institute of Technology Agartala, Tripura, India
Corresponding author: tanmoymajumder5991@gmail.com

Abstract

Floods are highly devastating natural disasters that repeatedly inflict damage on people, property, and resources, posing substantial risks throughout the monsoon season in India. This study investigates the potential for flooding in the Brahmani River Basin in Odisha by employing multicriteria decision analysis, specifically the analytic hierarchy process and geographic information system approaches. The basin is susceptible to significant rainfall during monsoons and spans over seven state districts. The analysis employs 22 years of satellite data from Landsat-7 and Sentinel-1 and socio-economic data from the 2011 Census. A flood hazard zonation map and a flood risk map are created, categorizing the basin into several flood risk zones according to the intensity of hazards and vulnerability aspects. The findings ascertain that nearly 18% of the villages in Kendrapara, Jajapur, and Cuttack are at high risk of flooding. The study provides empirical evidence on strategies for effectively implementing land use rules in flood-prone areas to mitigate potential negative impacts.

Keywords: Brahmani River Basin, Disaster Management, Flood Risk, Multicriteria Decision Analysis, Geographic Information System

1. Introduction

India is prone to natural disasters, with floods being the most significant. Floods can cause fatalities and disrupt economic activities. Thus, flood mitigation and danger zone identification are critical. Flood risk assessment incorporates various factors like population, land use, and infrastructure. Determining the importance and weightage of each of these factors in the risk assessment is not possible using conventional methods (Wu et al., 2022). Multicriteria decision analysis (MCDA) is one such technique that can break down complex problems into more straightforward, manageable problems, enabling effective analysis (Mishra & Sinha, 2020). This study employs the MCDA approach, specifically the analytic hierarchy process (AHP) for flood risk assessment. The AHP facilitates decision-making by assigning weights to each factor depending on its relative importance.

In recent years, the amount of rainfall has increased in the Brahmani River Basin (BRB), resulting in the inundation of the majority of shallow areas in the river basin during the monsoon season. This has caused damage to infrastructure and agriculture (Majumder et al., 2022b). The current study couples geographic information systems (GIS) with

DOI: 10.1201/9781003596776-10

socio-economic and land use factors to accurately identify potential flood risk areas in the BRB. The main objectives are to create a flood hazard zonation map using satellite data and classify the basin into different flood risk zones based on threat severity.

2. Study Area and Datasets

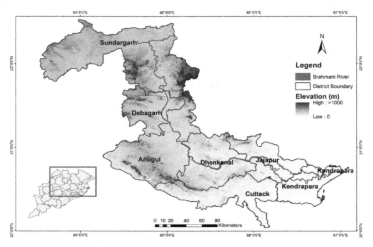

Figure 10.1: Study area.

The BRB is located in eastern India, in the state of Odisha, between latitudes 20° to 23° North and longitudes 83° to 87° East. Before reaching the Bay of Bengal, the river passes through seven districts of Odisha: Sundargarh, Debagarh, Anugul, Dhenkanal, Jajapur, Cuttack, and Kendrapara. The basin receives an average of 1148 mm of monsoonal precipitation, which is consistent with a tropical climate (Majumder et al., 2022a). This study uses satellite data from Landsat-7 and Sentinel-1, encompassing the years 2000 to 2021. Socio-economic statistics were collected from the Census of India 2011 (Census, 2011).

3. Methodology

3.1. Flood Hazard Zones

The flood hazard zones for the BRB were created using spatial and multi-temporal satellite data. The data collection consisted of Landsat 7 pictures obtained from 2000 to 2014, together with microwave synthetic aperture radar data from Sentinel-1 collected from 2015 to 2021. During this period each instance of a flood resulted in the creation of a corresponding layer of flooding, and these layers were then merged to form an annual flood inundation layer. The final flood hazard zone map for the BRB was generated by merging all the annual inundation layers.

3.2. Flood Risk Zones

Flood risk assessment involves analyzing geographical and local factors. According to Ha-Mim et al. (2022), it is crucial to include socio-economic and infrastructural data for a thorough understanding of vulnerabilities and risks. GIS and MCDA can help create flood risk maps in data-limited areas (Ray, 2024). This study in the BRB assesses flood risk by considering social (population), land use (cropland), and infrastructural (roads) vulnerabilities, and develops a vulnerability index by assigning weights to each element based on its relative importance to society. The flood risk assessment was conducted using the formula (Van Westen et al. 2009),

$$\text{Flood risk} = \text{Hazard} \times f(\text{social, infrastructure, land use}) \text{ Vulnerabilities}$$

4. Results and Discussion

4.1. Flood Hazard Zonation

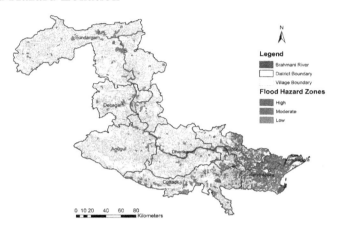

Figure 10.2: Flood hazard map of the Brahmani River Basin.

The flood hazard zonation map for the BRB was assessed by taking into account the severity of flood hazards and categorizing them into zones of low, moderate, and high hazard. There are around 1845 villages situated in regions with a low susceptibility to flooding, 78 villages in regions with a moderate susceptibility, and 48 villages in regions with a high susceptibility. Figure 10.2 shows the areas of the BRB that have been impacted by flooding; the majority of these affected villages are located in the districts of Cuttack, Jajapur, and Kendrapara.

4.2. Flood Risk Assessment

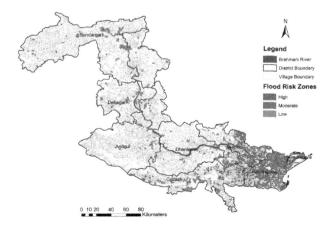

Figure 10.3: Flood risk map of the Brahmani River Basin

The study found that there are 1802 villages located in areas with a high susceptibility to infrastructure damage during flooding. In addition, there exist 80 villages within moderate to high social vulnerability zones. Furthermore, a significant 60% of communities that rely on agriculture as their primary means of sustenance are situated in areas with a high susceptibility to adverse flood conditions. Around 18% of villages in the BRB are situated in regions characterized by a significant vulnerability to floods. Based on the data presented in Figure 10.3, a substantial number of villages in Kendrapara, Jajapur, and Cuttack are vulnerable to a significant risk of flooding. These regions have a high population density with extensive agricultural land and infrastructure, which makes them extremely vulnerable to floods.

5. Conclusion

This study aimed to evaluate flood risk by integrating hazards with social, infrastructure, and land-use vulnerabilities at the village scale. Nearly 18% of all settlements are sited in highly prone flooding areas. The government should therefore enforce zoning laws related to land use planning, encourage the cultivation of flood-resistant crops, and flood-proofing of houses, and establish flood insurance schemes. However, this research is limited by factors such as the constraints on resolution in satellite imagery which could have left out some small-scale vulnerabilities. To assess potential risks in future studies, more detailed images and climate change projections could be used.

References

[1]   Census. (2011). *Population Census 2011*. https://www.census2011.co.in/

[2]   Ha-Mim, N. M., Rahman, M. A., Hossain, M. Z., Fariha, J. N., & Rahaman, K. R. (2022). Employing multi-criteria decision analysis and geospatial techniques to assess flood risks: A study of Barguna district in Bangladesh. *International Journal of Disaster Risk Reduction, 77,* 103081. https://doi.org/10.1016/j.ijdrr.2022.103081

[3]   Majumder, T., Das, B., & Chakraborty, P. (2022a). Spatio-temporal analysis of monsoon rainfall over Odisha. *Lecture Notes in Civil Engineering.* https://doi.org/10.1007/978-981-16-7509-6_40

[4]   Majumder, T., Das, B., & Padhi, J. (2022b). Trend analysis of monsoon rainfall over Odisha. *Lecture Notes in Civil Engineering.* https://doi.org/10.1007/978-981-16-7509-6_39

[5]   Mishra, K., & Sinha, R. (2020). Flood risk assessment in the Kosi megafan using multi-criteria decision analysis: A hydro-geomorphic approach. *Geomorphology, 350,* 106861. https://doi.org/10.1016/j.geomorph.2019.106861

[6]   Ray, S. K. (2024). Flood risk index mapping in data-scarce regions by considering GIS and MCDA (FRI mapping in data-scarce regions by considering GIS and MCDA). *Environment, Development and Sustainability.* Springer Netherlands. https://doi.org/10.1007/s10668-024-04641-2

[7]   Van Westen, C., Alkema, D., Damen, M. C. J., Kerle, N., & Kingma, N. C. (2009). Multi-hazard risk assessment. Distance education course guidebook. *United Nations University-ITC*, Bangkok.

[8]   Wu, J., Chen, X., & Lu, J. (2022). Assessment of long and short-term flood risk using the multi-criteria analysis model with the AHP-Entropy method in the Poyang Lake basin. *International Journal of Disaster Risk Reduction, 75,* 102968. https://doi.org/10.1016/j.ijdrr.2022.102968

CHAPTER 11

Experimental Investigation on Reactive Powder Concrete Incorporating Expanded Clay Sand

[1*]Ahmadshah Abrahimi, [2]V. Bhikshma

[1]Ph.D Scholar, University College of Engineering, Osmania University, Hyderabad, India
[2]Senior Professor, University College of Engineering, Osmania University, Hyderabad, India
*E-mail:ahmadshah.eng786@yahoo.com, v.bhikshma@osmania.ac.in

Abstract

This study examines the utilization of lightweight expanded clay sand (LECS) in reactive powder concrete (RPC) in combination with supplementary cementitious materials (microsilica and alccofine1203). The study substitutes 20%, 30%, and 40% of the RPC, as well as 60%, 80%, and 100% of the quartz sand volume for LECS. An experimental program was carried out to measure compressive strength, flow diameter, and density. The study found that replacing 80% and 100% quartz sand with LECS produced high-strength lightweight concrete with 28-day compressive strengths of 101 and 98 MPa and densities of 1940 and 1890 kg/m³, respectively. It was also found that a mix with 100% LECS and a blend of microsilica and alccofine (10% and 20%) gained 95% early compressive strength (93 MPa) in only 7 days. It shows the viability of microsilica and alccofine blends for producing early high-strength lightweight concrete.

Keywords: compressive strength, lightweight expanded clay sand, microsilica, and alccofine1203

1. Introduction

As per today's demand for high-rise structures and long-span bridges, developing high-strength lightweight concrete (HSLWC) is the major interest of researchers. However, the porous nature of lightweight aggregates raises concerns about HSLWC reliability and suitability, which reduces confidence in employing HSLWC for such applications (Real et al., 2015). To develop lightweight durable concrete, the RPC concept has been utilized for the development of lightweight reactive powder concrete (LWRPC). In past studies, LWRPC has been produced by using lightweight aggregates such as perlite (Al-jumaily, 2011), polystyrene beads (Allahverdi et al., 2018), pumice sand (Gökçe et al., 2017), pollytag, and lightweight expanded clay aggregate (Grzeszczyk & Janus, 2020), along with high-content cement and silica fume. According to existing literature (Yang et al., 2016; Ghafari et al., 2016), a significant proportion of cement and silica fume not only escalates production expenses but may also induce thermal cracking and autogenous shrinkage owing to the heat of hydration and internal water consumption in RPC. To address these issues and produce lightweight reactive

DOI: 10.1201/9781003596776-11

powder concrete (LRPC) as an eco-friendly material, this research seeks to examine various fresh and hardened properties of RPC by substituting cement with a blend of microsilica and alccofine as well as quartz sand with lightweight sand (LECS).

2. Materials

This study utilized ordinary Portland cement (OPC) of grade 53, characterized by a specific gravity of 3.11 and a surface area of 2830 cm²/g. The accompanying microsilica surface area was determined to be 19886 m²/kg, with a silica content higher than 90%. For obtaining a low w/b ratio, polycarboxylate ether superplasticizer (PCE), specifically "Tam cem11," was used. Cement substitution was carried out with alccofine1203, possessing a Blaine fineness of 12000 cm²/g, and particle size distribution parameters: D (0.1) = 1.38 µm, D (0.5) = 4.13 µm, and D (0.9) = 11 µm. Additionally, two distinct sizes of quartz sand, QB: (4–600 µm), and QC: (100–1000 µm), with the particle size distribution curves shown in Figure 1 were utilized in this study, along with lightweight expanded clay sand (LECS). The chemical composition and physical properties of RPC ingredients have been detailed in Table 11.1.

Table 11.1: Physical properties and chemical composition of RPC ingredients.

Component	Cement	Alccofine (Wt. %)	LECS	Microsilica	Quartz sand
SiO_2	21.03	34.45	61.18	98.73	98.1
Fe_2O_3	5.67	0.31	13.19	0.07	0.05
Al_2O_3	4.03	13.45	17.59	0.1	0.66
CaO	64.19	40.8	1.96	0.1	<0.1
MgO	0.88	6.1	1.53	0.2	<0.1
SO3	2.86	2.73	-	0.1	-
K_2O	0.2	0.31	1.14	0.4	<0.1
Na_2O	0.12	0.14	1.24	-	<0.1
Loss of Ignition	1.02	-	0.16	0.3	-
Specific gravity	3.11	2.86	1.33	2.2	2.6
Water absorption (%)	-	-	14.9	-	0.01

(Source: Author)

3. Research Method

This research was implemented in two stages. In the first stage, normal RPC mix optimization was done to optimize the packing density of the mixture considering past research articles (Grzeszczyk & Janus, 2020) and the "modified Andreassen model" using the Elkem Material Mix Analyzer (EMMA). Considering particle size distribution data produced by Mastersizer 2000 Ver. 5.60 plotted in Figure 11.1, the proportions were adjusted to the modified Andresen model curve. In the second stage, after final adjustment, a blend of 10% microsilica and 20% alccofine1203 mix was used as a reference mix for producing lightweight concrete by replacing the total dry volume of RPC by 20%, 30%, and 40% of LECS, as well as quartz sand by 60%, 80%, and 100%, as shown in Table 11.2. Fresh and hardened properties, including 7 and 28-day compressive strength, density, and flow diameter, were investigated by casting (100×100×100) mm cubic specimens. The four stages' mixing procedure was followed, shown in Figure 11.2.

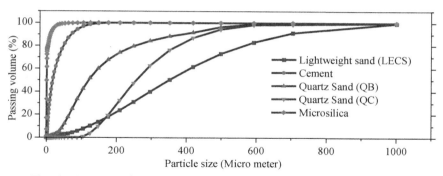

Figure 11.1: Particle size distribution of RPC ingredients. (*Source: Author*)

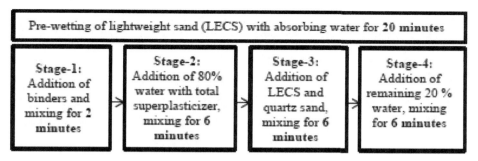

Figure 11.2: Mixing procedure. (Source: Author)

Table 11.2: Mix proportions for replacing RPC ingredients with LECS.

Ingredients	Ref, Mix	Quartz sand replacement to LECS (QL)			Dry volume replacement to LECS (VL)		
		QL-60	QL-80	QL-100	VL-20	VL-30	VL-40
		60%	80%	100%	20%	30%	40%
Cement (kg/m³)	850	850	850	850	680	595	510
Microsilica (kg/m³)	85	85	85	85	68	60	51
Alccofine (kg/m³)	170	170	170	170	136	119	102
Sand (QB)	350	140	70	0	280	245	210
Sand (QC)	690	276	138	0	552	483	414
LECS (kg/m³)	-	295	430	540	255	390	520
SP (% Binders)	0.8	0.8	0.8	0.8	0.8	0.8	0.8
W/B	0.18	0.18	0.18	0.18	0.18	0.18	0.18
Water (kg/m³)	199	199	199	199	160	140	120
Absorbing water by LECS (kg/m³)	-	39	52	65	280	245	210

(Source: Author)

4. Results and discussion

According to the graph in Figure 11.3, substituting quartz sand with LECS significantly reduced the dry density of RPC. In particular, 80% and 100% replacement mixes (QL-80 and QL-100) showed 17% and 20% reductions compared to the reference mix, respectively. Additionally, the mixes QL-80 and QL-100 achieved high-strength lightweight concrete with 7-day compressive strengths of 89 and 93 MPa, as well as 28-day compressive strengths of 101 MPa and 98 MPa, respectively. Additionally, the results show a 95% compressive strength within 7 days for both QL-100 and reference mixes; this suggests a blend of microsilica and alccofine for early strength improvement in RPC. Furthermore, utilizing 100% LECS does not impact the early strength of RPC. Mixes with insufficient binder content, such as VL-20, VL-30, and VL-40, demonstrate lower strengths and higher densities, respectively. Notably, maintaining a constant w/b ratio of 0.18, the substitution of 20%, 30%, and 40% RPC ingredients with LECS resulted in a decrease in flow diameter by 7%, 15%, and 28%, respectively, due to the reduction in alccofine content compared to the "Ref. mix". In RPC, the mixing procedure is a critical aspect, particularly when employing a low-speed mixer. In this study, the addition of lightweight sand posed another challenge due to water absorption from the mixture. Prewetting of lightweight sand for 20 min showed better performance, coupled with a 20-min mixing duration following a four-stage mixing procedure.

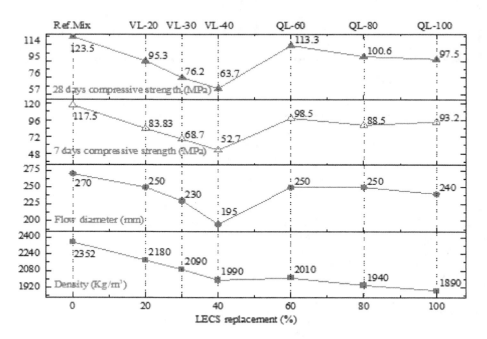

Figure 11.3: Fresh and hardened properties of RPC using LECS.

(Source: Author)

5. Conclusion

In light of the results, the following findings can be summarized:

1. The results show that utilizing 80% and 100% of lightweight sand (LECS) instead of quartz sand could produce high-strength lightweight concrete with compressive strengths (28 days) of 98 and 101 MPa and densities of 1940 and 1890 kg/m³, respectively.
2. It was also found that the blend of microsilica and alccofine1203 in RPC allowed for achieving 95% of the compressive strength within 7 days. Furthermore, the inclusion of 100% LECS did not adversely affect early strength development.
3. The results indicate that substituting quartz sand with LECS had better performance compared to total RPC volume substitution because low-binder content considerably reduced the strength and workability.

References

[1] Al-jumaily, I. a S. (2011). Some mechanical properties of reactive powder lightweight concrete. 4(1), 47–61.
[2] Allahverdi, A., Azimi, S. A., & Alibabaie, M. (2018). Development of multi-strength grade green lightweight reactive powder concrete using expanded polystyrene beads. *Construction and Building Materials*, *172*, 457–467.
[3] Ghafari, E. *et al.* (2016). Effect of supplementary cementitious materials on autogenous shrinkage of ultra-high performance concrete. *Construction and Building Materials*, *127*, 43–48.
[4] Gökçe, H. S., Sürmelioğlu, S., & Andiç-Çakir, Ö. (2017). A new approach for production of reactive powder concrete: lightweight reactive powder concrete (LRPC). *Materials and Structures/Materiaux et Constructions*, *50*(1).
[5] Grzeszczyk, S., & Janus, G. (2020). Reactive powder concrete with lightweight aggregates. *Construction and Building Materials*, *263*, 120164.
[6] Real, S., Bogas, J. A., & Pontes, J. (2015). Chloride migration in structural lightweight aggregate concrete produced with different binders. *Construction & Building Materials*, *98*, 425–436.
[7] Yang, K.H., Moon, G.D. and Jeon, Y.S. (2016) 'Implementing ternary supplementary cementing binder for reduction of the heat of hydration of concrete', *Journal of Cleaner Production*, 112, pp. 845–852.

CHAPTER 12

A study on multi-fabric weaved bi-directional laminated composites

Smruti Ranjan Sahoo[1], Abhijit Mohanty[1], Sarada Prasad Parida[2*], Rati Ranjan Dash[1]

[1]Odisha University of Technology and Research, Bhubaneswar, Odisha, India
[2]Konark Institute of Science and Technology, Bhubaneswar, Odisha, India
E-mail: sarada800@gmail.com

Abstract

This work aims to draw a breakthrough by hybridizing two fibers in a single fabric layer using E-glass and carbon fiber in C-X-G-Y, C-Y-G-X, C-G-A, C-X-C-Y, and G-X-G-Y patterns in an epoxy base by the hand layup process. Mechanical strengths are experimented with using tensile, flexural, impact, and hardness tests. Further, experimental and numerical simulations are performed to study the modal responses. It is observed that RPCs made by C-X-C-Y and G-X-G-Y reinforcement have maximum and minimum strength in mechanical analysis and modal analysis.

Keywords: Weaved composite, RPC, Modal analysis, hybrid composite

1. Introduction

Reinforced polymer composites (RPCs) are widely used in industrial and household applications. Primarily, reinforcement is the major load carrier of RPCs. Strength and cost are two vital factors in deciding the fiber to be used in RPCs. Generally, carbon fibers have high strength and cost in contrast to glass fiber. Many researchers are still working to reduce the cost of fibers using multiple fibers as reinforcement. Jagannatha and Harish (2015) reinforced carbon and glass fiber in alternative layers and reported increased mechanical strength. It is reported that using multiple fabrics in RPC increases dynamic response (Aydin et al., 2022) and mechanical strength (Hung et al., 2018). Nanoparticles can also be used to enhance the strength of the composite specimens (Sahoo et al., 2023; Parida & Jena, 2023, 2021). In this work, carbon and E-glass fibers are weaved in five patterns: C-X-G-Y, C-Y-G-X, C-G-A, C-X-C-Y, and G-X-G-Y. RPCs are fabricated by the hand layup process, tensile, flexural, impact, and hardness tests are conducted, followed by modal analysis using experimentation and finite element analysis (FEA).

2. Material preparation and characterization

A steel mold (250×250×10 mm) as shown in Figure 12.1 is used for fabrication by the open layup process following ASTM standards. Reinforcement fabric is hand-weaved to C-X-G-Y, C-Y-G-X, C-X-C-Y, and G-X-G-Y patterns and cut into the required size.

DOI: 10.1201/9781003596776-12

Four fabric layers are laid, followed by the addition of an epoxy-based matrix (prepared by mixing LY556 and MS91 in a 1:10 ratio) in a steel mold as shown in Figure 12.1. The test specimens prepared are machined to the required sizes following ASTM test standards. For tensile testing, a low feed of 1 mm/min is given to accurately follow the load vs deformation curve. For bending analysis, the feed rate is maintained at 2 mm/min. Similarly, in the hardness test by Brinell's method, the C-scale is followed with a preload setting of 60 kg. During the impact test, to follow directional failure, a notch is made, and the impact strength is measured in the form of absorbed strained energy before failure. For each test, five readings were taken, and the average test results were then analyzed. Further, an experimental modal test setup is used (Parida & Jena, 2022) to determine the modal response. The time displacement curve is then plotted, and the corresponding fundamental frequency is counted. The experimental results were found to deviate by 7–8% with FEA (Table 12.1).

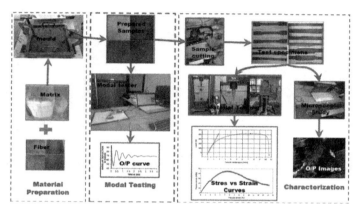

Figure 12.1: Process layout of work conducted in the study.

Table 12.1: First Natural frequencies (Hz) for composite variants.

Boundary condition	Method	RPCs with reinforced weave variants				
		G-X-G-Y	C-X-C-Y	C-X-G-Y	C-Y-G-X	C-G-A
C-F-F-F	Exp.	16.21	27.56	25.3	25.12	21.3
	FEA	18.47	31.165	28.548	28.51	24.741
C-F-C-F	Exp.	112.4	192.18	176.54	178.53	152.45
	FEA	117.54	197.04	181.23	183.23	157.16
C-C-C-F	Exp.	116.35	198.23	179.23	177.55	159.32
	FEA	120.28	200.12	183.82	182.82	162.52
C-C-C-C	Exp.	171.23	294.35	223.53	221.35	231.15
	FEA	176.69	299.18	228.58	227.99	239.49

3. Finite element modeling for vibration analysis

FEA is conducted by ANSYS 16.0 on specimens of size 250×250×2 mm. The material properties are calculated by the rule of mixture (Parida & Jena, 2021). In designing the LCP, it is divided into four layers by offsetting the work plane and meshed later with a refined global mesh size of 2 mm with 8 solid brick 185 elements by convergence analysis (Parida & Jena, 2021). Boundary conditions are then introduced, and modal analysis is performed to measure the frequencies and mode shapes, as shown in Figure 12.3.a–c) for C-C-C-C end support. The mechanical properties of the test specimen constituents, E-glass fiber, carbon fiber, and epoxy, are considered (Parida & Jena, 2023). The material properties are determined through the micromechanical approach (Sahoo et al., 2023). The effectiveness of FEA is verified by free vibration analysis of a square isotropic plate of 300×300×5 mm dimension with an elastic modulus of 80 MPa and Poisson's ratio of 0.3 in clamped-free-free-free end support. The nondimensional form of natural frequency (NDNF) is compared with other theories (Table 12.2).

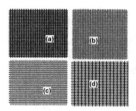 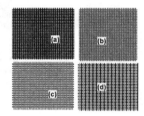

Figure 12.2: Scheme of the fiber reinforcements in the composite; (a) C-X-C-Y, (b) G-X-G-Y, (c) C-X-G-Y and C-Y-G-X, (d) C-G-A.

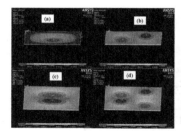 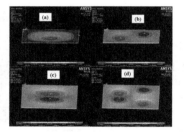

Figure 12.3: Modal displacement; (a) 1st mode, (b) 2nd mode, (c) 3rd mode.

Table 12.2: NDNF for an isotropic square plate.

Mode of vibration	CPT (Kirchoff, 1850)	FSDT (Mindlin, 1951)	HSDT (Parida and Jena, 2023)	Present study (FEA)
1	0.0955	0.0940	0.0928	0.093
2	0.2360	0.2219	0.215	0.225
3	0.4629	0.4495	0.392	0.445

4. Results and discussion

From the tensile test, it is observed that the tensile strength (strength and extensions) at the breakage of C-X-C-Y reinforced RPC and G-X-G-Y reinforced RPC are observed to be 285.81 MPa and 11.65 mm, and 183.5 MPa and 9.5 mm, respectively. C-X-G-Y, C-Y-G-X, and C-G-A reinforced RPCs have intermediate values, as shown in Figure 12.4a. Likewise, the RPCs made with C-X-C-Y fabric show maximum flexural strength and extension at breakage (269.67 MPa, 56.9 mm), while RPCs made with G-X-G-Y show minimum values (109.4 MPa, 39.2 mm). Meanwhile, C-X-G-Y, C-Y-G-X, and C-G-A reinforced RPCs have nearly equal flexural strength and extension, as shown in Figure 4b. The impact strength, impact energy, and hardness of RPCs with C-X-C-Y and G-X-G-Y are maximum and minimum. A remarkable measure in impact strength and hardness of RPC by reinforcing C-X-G-Y, C-Y-G-X, and C-G-A fabrics is achieved. However, it is observed that the weave pattern of the fabric does not affect the impact strength and hardness as all C-X-G-Y, C-Y-G-X, and C-G-A reinforced RPCs show equal impact strength (60.1–60.8 kN/mm²) and hardness (84.8–85.48 HRm) as shown in Figure 12.4c, d.

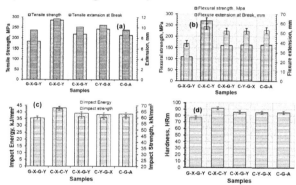

Figure 12.4: Test results; (a) Tensile, (b) Flexural, (c) Impact, (d) Hardness.

Table 12.3: Natural frequencies of the considered samples obtained by FEM in HZ.

Boundary condition	Mode of Vibration	Weave variant				
		G-X-G-Y	C-X-C-Y	C-X-G-Y	C-Y-G-X	C-G-A
C-F-F-F	1st	18.47	31.165	28.548	28.51	24.741
	2nd	29.68	39.976	38.024	38.40	34.597
	3rd	110.88	184.60	132.19	132.49	147.17
C-F-C-F	1st	117.54	197.04	181.23	183.23	157.16
	2nd	124.08	200.81	186.08	187.08	162.03
	3rd	193.18	307.43	240.13	241.35	249.47
C-C-C-F	1st	120.28	200.12	183.82	182.82	162.52
	2nd	185.4	302.28	234.09	231.09	223.56
	3rd	324.51	541.05	407.20	406.75	401.50
C-C-C-C	1st	176.69	299.18	228.58	227.99	239.49
	2nd	366.41	614.29	401.47	401.17	491.84
	3rd	512.14	614.29	530.24	532.47	491.84

The prepared RPCs are then subjected to modal testing. It is observed that natural frequencies are maximum for RPC with C-X-C-Y and minimum for G-X-G-Y reinforcement. For instance in C-F-F-F support, the fundamental frequencies (experimental and FEM) for RPC with C-X-C-Y and G-X-G-Y reinforcement are (27.56Hz, 31.165Hz) and (16.21Hz, 18.47Hz) respectively. Similar observations are made in other end supports. It is further observed that RPCs with C-G-A reinforcement show a little lower stiffness than C-X-G-Y and C-Y-G-X reinforced RPCs(25.3Hz for C-X-G-Y, 25.12Hz for C-Y-G-X, and 21.3Hz for C-G-A in C-F-F-F support obtained from experimentation). Observations for natural frequencies in other end supports are presented in Table 1 and the deviation of experimental values with FEM is only limited to 8% approx. The experimental module is only capable to examine the fundamental frequency, to determine the natural frequencies in the other three modes of vibration FEM is used (Table 12.3). The FEM observation for three modes of frequencies in all of the boundary conditions is similar to the observation made for fundamental frequencies. Hence it can be said that weaving the fabrics in C-X-C-Y and C-X-G-Y patterns gives a remarkable attainment of the natural frequencies depicting a considerable rise in the plate stiffness.

5. Conclusion

In this work, carbon and E-glass fibers were weaved into G-X-G-Y, C-X-C-Y, C-X-G-Y, C-Y-G-X, and C-G-A patterns. RPCs are fabricated, and modal tests are conducted, followed by mechanical testing and numerical modal simulation. From the study, it is observed that RPCs with C-X-G-Y and C-Y-G-X reinforcement have almost equal mechanical strength and RPCs with C-G-A reinforcement attain 60–63% of the strength of RPCs reinforced by C-X-C-Y. The modal response shows that the frequency and strength of RPCs made from carbon fiber have the peak value, followed by RPCs made from C-X-G-Y, C-Y-G-X, and C-G-A.

References

[1] Aydin, M. R., Acar, V., Cakir, F., Gündoğdu, Ö., & Akbulut, H. (2022). Comparative dynamic analysis of carbon, aramid, and glass fiber reinforced interply and intraply hybrid composites. *Composite Structures, 291*, 115595.

[2] Hung, P. Y., Lau, K. T., Cheng, L. K., Leng, J., & Hui, D. (2018). Impact response of hybrid carbon/glass fiber reinforced polymer composites designed for engineering applications. *Composites Part B: Engineering, 133*, 86–90.

[3] Jagannatha, T. D., & Harish, G. (2015). Mechanical properties of carbon/glass fiber reinforced epoxy hybrid polymer composites. *International Journal of Mechanical Engineering and Robotics Research, 4*(2), 131–137.

[4] Kirchhoff, G. (1850). Über das Gleichgewicht und die Bewegung einer elastischen Scheibe. *Journal für die reine und angewandte Mathematik, 1850*(40), 51–88.

[5] Mindlin, R. (1951). Influence of rotatory inertia and shear on flexural motions of isotropic, elastic plates.

[6] Parida, S. P., & Jena, P. C. (2021). Static analysis of GFRP composite plates with filler using higher-order shear deformation theory. *Materials Today: Proceedings, 44*, 667–673.

[7] Parida, S. P., & Jena, P. C. (2023). Multi-fillers GFRP laminated composite plates: Fabrication & properties. *Indian Journal of Engineering & Materials Sciences*, 817–825.

[8] Sahoo, S., Parida, S. P., & Jena, P. C. (2023). Dynamic response of a laminated hybrid composite cantilever beam with multiple cracks & moving mass. *Structural Engineering and Mechanics, 87*(6), 529–540.

[9] Song, J. H. (2015). Pairing effect and tensile properties of laminated high-performance hybrid composites prepared using carbon/glass and carbon/aramid fibers. *Composites Part B: Engineering, 79*, 61–66.

CHAPTER 13

Buckling analysis of glass fiber reinforced polyester composite with variable stiffness and central cut-outs

Suryakanta Panigrahy[1], Sarada Prasad Parida[2*], Rati Ranjan Dash[1]

[1]Odisha University of Technology and Research, Bhubaneswar, Odisha, India
[2*]Konark Institute of Science and Technology, Bhubaneswar, Odisha, India
E-mail: sarada800@gmail.com

Abstract

This work aims to understand the load-carrying capacity of woven glass fabric reinforced glass polyester composites (GPC) with variable weave patterns. Four kinds of GPCs (50:50 weight fraction) are made using the hand layup process. The orientation of the warp direction is fixed, and the orientation of the weft direction is rotated to produce fabrics of 0°±90°, 15°±90°, 30°±90°, and 45°±90°. Mechanical tests are conducted, followed by an analysis of buckling strength due to the presence of cutouts at the center. Three types of holes (square, circular, and triangular) with dimensions of 10, 12, 15, and 20 mm are made. Further, numerical simulation is conducted and compared with experimental findings. It is observed that fiber orientation and hole geometry play vital roles in buckling strength.

Keywords: variable stiffness, cutouts, buckling, tensile test

1. Introduction

In a fiber-reinforced composite (FRC), most of the load is carried by the fiber. During layup, the fiber may be twisted, leading to variable stiffness. This variation sometimes contributes positively or negatively. Therefore, researchers are keen to study the effect of fiber orientation on the mechanical behavior of FRC. Almeida et al. (2015) studied the performance of GPC concerning weave angle (WA) through shear tests and concluded that 0°±90° WA has double the shear strength of 90°±90°. Vummadisetti and Singh (2020) observed that GPC with (−45/+45/0/90)₂s WA has maximum strength. Further, it has been investigated that the presence of discontinuity influences the buckling strength of GPCs. Besides tensile strength, buckling and shear strength are also affected (Aslan & Daricik, 2016). The buckling strength of GPC cylinders depends on volume fraction and fabrication process (Parida & Jena, 2019). Here, four kinds of GPCs from bidirectional woven glass with variable weave patterns of the fabrics (0°±90°, 15°±90°, 30°±90°, and 45°±90°) are made, and tensile, flexural, hardness, and impact tests are conducted. Further, the buckling strength due to the presence of cutouts at the center is studied. Three types of holes (square, circular, and triangular) with dimensions of 10, 12, 15, and 20 mm are made at the center, and the buckling strength is analyzed.

DOI: 10.1201/9781003596776-13

Further, numerical simulation (ANSYS 16.0) is conducted and compared with experimental findings. The buckling strength is further analyzed for fabric orientations of 0°±90°, 15°±90°, 30°±90°, 45°±90°, 60°±90°, 75°±90° and 90°±90° by FEA.

2. Theory

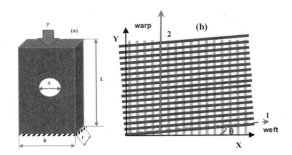

Figure 13.1: Schematic diagram; (a) Column buckling (b) fiber in GPC.

Following the buckling of the column structures, the critical load is given by (Parida and Jena, 2019);

$$P_{cr} = \frac{\pi^2 D}{4L^2} \tag{1}$$

For an orthotropic material, the flexural rigidity is (Barbero, and Tomblin, 1992)

$$D = \left[D_{11} + \frac{2D_{16}D_{26}D_{12} - D_{66}D_{12}^2 - D_{22}D_{16}^2}{D_{22}D_{66} - D_{26}^2} \right] B \tag{2}$$

Where D_{ij} can be calculated from the following formula (Parida and Jena, 2023)

$$D_{11} = \frac{E_{11}h^3}{12}; \quad D_{22} = \frac{E_{22}h^3}{12}; \quad D_{12} = \frac{G_{12}h^3}{12}; \quad D_{16} = \frac{G_{16}h^3}{12}; \quad D_{26} = \frac{G_{26}h^3}{12};$$

and $D_{66} = G_{12}$.

With $E_{11}=E_x$ and $E_{22}=E_y$ obtained from equation (1) for the variable stiffness plate.

In this study, GPC with oriented fibers at different angles as shown in Figure 13.1b is taken. The elastic modulus and density of the GPC are calculated by mixture rule (Sahoo et al., 2023).

3. Material Preparation and characterization

In this study, composite plates are fabricated by reinforcing woven fiberglass into an epoxy matrix. The orientation of the fiber in the x-axis, as shown in Figure 13.1, is twisted to 15°±90°, 30°±90°, and 45°±90°. The hand layup process is followed. The test specimens are machined to the required sizes following ASTM D790 for flexural, ASTM D3039 for tensile, ASTM E18-22 for hardness, and ASTM D256 for impact testing. The whole process, from preparation to reporting, is presented in Figure 13.2. Further, the buckling test is conducted in a universal testing machine (UTM) with a special arrangement for samples of size 50×80×3 mm with holes of three varieties (circular, square, and triangular), as shown in Figure 13.3, to report the changes in buckling strength with changes in hole shapes and geometries. Four holes with diameters or side lengths of 10, 12, 15, and 20 mm are made on each kind of specimen and compared with the buckling strength of solid samples.

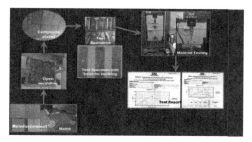

Figure 13.2: Schematic process diagram from preparation to testing.

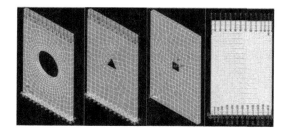

Figure 13.3: Plates with different hole shapes

4. Finite element analysis for buckling

The eigen buckling of specimens (80×50×3 mm with eight layers) with a dummy load of 1 N is conducted through ANSYS 16.0 using Solid brick 185 elements with a mesh size of 2 mm (Parida et al., 2019). Central cutouts (circular, square, and triangular) are made with sizes of 10, 15, and 20 mm. It is observed that the critical buckling strength is maximum for specimens with triangular holes and minimum for circular holes (Table 13.1). The FEA process is accurate with a maximum error of 8%. In addition, the buckling of GPCs with weave angles (WA) of 0°±90°, 15°±90°, 30°±90°, 45°±90°, 60°±90°, 75°±90° and 90°±90° is studied.

Table 13.1: Comparison of buckling strength.

Hole dimension	circular (FEA)	circular (Exp.)	square (FEA)	square (Exp.)	Triangular (FEA)	Triangular (Exp.)
0	65637.2	62120.1	65637.2	62120.1	65637.2	62120.1
10	29579	25152	26322.8	25532	34112.5	31256.3
12	21035.4	19250	19661.3	17526	17146	16523
15	14316.2	12151	14680	13258	8649.8	8125.2
20	10363.9	8532	10019	7558	8342.59	7122

5. Result discussion

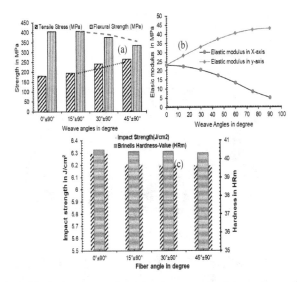

Figure 13.4: Variation strength with weave angle; (a) tensile and flexural strength, (b) Elastic modulus (c) hardness and impact strength.

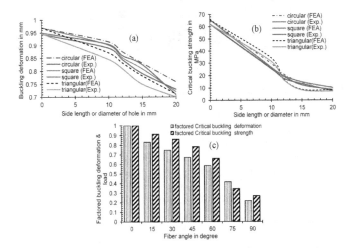

Figure 13.5: Variation critical buckling parameters with cutout dimension; (a) strength, (b) deformation, (c) Factored buckling load and deformation with weft angle.

From the study, it is observed that the tensile and flexural strength of the specimens are dependent on the WA, as shown in Figure 13.4a. GPC with 45°±90° WA has maximum tensile strength (265.7 MPa) and 45°±90° has minimum (181.3 MPa) as orienting the fiber to the loading axis increases the tensile modulus. However, the flexural strength of the GPC is maximum for 0°±90° WA and minimum for 15°±90° WA (402.2 MPa) as the elastic modulus in the warp direction is reduced, and bending strength is dependent on both axial and transverse strength. The variation of elastic modulus in the x and y-axis by rotating the fiber of warp towards weft is calculated and shown in Figure 13.4b. Further, hardness is less affected by fiber orientation, whereas impact strength is reduced by increasing the orientation of the weft axis (Figure 13.4c). From the buckling analysis, it is observed that the critical buckling stress and deformation are minimal for specimens with triangular holes and maximum for circular holes (Figure 13.5a, b). For instance, a sample with a triangular cutout of a side length of 20 mm has a critical strength of 8342.59 MPa, while the specimen with circular and square cutouts has strengths of 29579.13 and 26322.8 MPa, respectively. Further, increasing the dimensions of cutouts reduces the buckling strength as the fiber continuity breaks. Likewise, orienting the fibers from weft towards warp reduces the buckling strength and becomes minimum for 90°±90° weave pattern fabric reinforced specimens (Figure 13.5c).

6. Conclusion

From the study, it is observed that the tensile strength and flexural strength of GPC is maximum for 45°±90° WA and 0°±90° WA. Hardness is less affected, whereas impact strength is reduced by increasing the orientation of the weft axis. Further, the critical buckling strength is minimum for specimens with triangular holes and maximum for specimens with circular holes. By orienting the fibers from weft towards warp, the buckling strength is reduced and becomes minimum for 90°±90° WA.

References

[1] Almeida Jr, J. H. S., Angrizani, C. C., Botelho, E. C., & Amico, S. C. (2015). Effect of fiber orientation on the shear behavior of glass fiber/epoxy composites. *Materials & Design, 65*, 789–795.

[2] Aslan, Z., & Daricik, F. (2016). Effects of multiple delaminations on the compressive, tensile, flexural, and buckling behavior of E-glass/epoxy composites. *Composites Part B: Engineering, 100*, 186–196.

[3] Barbero, E., & Tomblin, J. (1992). Buckling testing of composite columns. *AIAA Journal, 30*(11), 2798–2800.

[4] Parida, S. P., Jena, P. C., Das, S. R., Dhupal, D., & Dash, R. R. (2022). Comparative stress analysis of different suitable biomaterials for artificial hip joint and femur bone using finite element simulation. *Advanced Materials Processing Technology, 8*(sup3), 1741–1756.

[5] Parida, S. P., Sahoo, S., Bal, B. B., & Jena, P. C. (2019). Buckling analysis of functionally graded natural fiber-Flyash-Epoxy (FGNFFE) cylinder. *International Journal of Engineering and Advanced Technology, 8*(6), 4260–4265.

[6] Sahoo, S., Parida, S. P., & Jena, P. C. (2023). Dynamic response of a laminated hybrid composite cantilever beam with multiple cracks & moving mass. *Structural Engineering and Mechanics, 87*(6), 529–540.

[7] Vummadisetti, S., & Singh, S. B. (2020). Buckling and postbuckling response of hybrid composite plates under uniaxial compressive loading. *Journal of Building Engineering, 27*, 101002.

CHAPTER 14

Flexural Studies on Concrete Beams with Areca Fiber and Lightweight Aggregates Reinforced with Bamboo

Naveenkumar D. T.[1], Naveen Bhari Onkareswara[2], Mahalaxmi S. Sunagar[3], Mamatha Priyadarshini K. S.[4]

[1]Associate Professor, Department of Civil Engineering,
SJB. Institute of Technology, Bangalore, India
[2]Associate Professor, Faculty of Civil and Water Resources Engineering,
Bhari Dar Institute of Technology, Bhari Dar University, Bhari Dar, Ethiopia
[3]Assistant Professor, Department of Civil Engineering, P.D.A.
College of Engineering, Kalaburagi, India
[4]Postgraduate Student Department of Civil Engineering,
S.J.B Institute of Technology, Bangalore, India
E-mails: naveendt012@gmail.com, dr.naveenbo53@bdu.edu.et,
mahalaxmisungar@pdaengg.com, mamathasomu3@gmail.com

Abstract

Concrete is the most conventionally used material in construction, but it still has some limitations. To overcome those limitations, the usage of areca and lightweight aggregates reinforced with bamboo is explored in this work. The study aims to improve the physical properties such as compressive strength and flexural strength, to prevent breakage in concrete, and to understand the effects on beams. Areca fibers reinforced with bamboo in different varying ratios are considered. The experimental investigation consists of casting and testing beams with different aspect ratios. The findings reveal a significant improvement in compressive strength, flexural capacity, deflection, Young's modulus, ductility factor, and stiffness factor.

Keywords: Fiber Reinforced Concrete, areca fibers, lightweight aggregates, flexural behaviour

1. Introduction

Concrete is a dominant construction material due to its numerous advantages, including increased compressive strength, long duration, and low cost. It is one of the most flexible building materials available. However, concrete has some drawbacks, including low tensile strength and hardening during the fresh and later stages. Adding fibers to concrete improves its flexural and ductile properties, addressing these shortcomings. Fibers enhance the performance of concrete by binding cracks under high tension, increasing the efficiency of the method. Additionally, fibers control rapid cracking and bleeding by reducing

DOI: 10.1201/9781003596776-14

concrete permeability due to dehydration from plastic and dry shrinkage. To address these inherent flaws, natural fibers can be added to reinforce cement and grout mixes. The natural areca fiber (Figure 14.1) is made from the areca fruit and is a hard, covered fiber component that accounts for around 70%–75% of the fruit's mass and size. The fiber is lighter than synthetic fibers, with a length ranging from 18 to 38 mm, a diameter from 0.0285 to 0.89 mm, and a density from 1.05 to 1.25 g/cm³. There are three different kinds of fibers: short, medium, and long (Srinivasa et al., 2010).

Figure 14.1: Areca fiber.

2. Literature review

Areca fiber has undergone various chemical treatments, including alkaline, permanganate, and benzene diazonium-chloride treatments. These chemically treated filaments have been employed as mounts to create polypropylene mixtures using the molding method. Different blends were created using fiber loadings of 20%, 40%, 60%, and 70% (Punyamurthy et al., 2014). According to research conducted by Qaiser et al. (2020), locally available bamboo strips can replace steel rebar in concrete. The research aims to ascertain the yield deflection load, ultimate loads, deflection patterns, and whether this easily accessible bamboo material demonstrates a comparable level of resistance to loads as shafts that have steel as the primary reinforcement. Various attributes were developed, along with load-deviation graphs and a comparison map. The level of strength increase in bamboo-reinforced beams has been disassembled. Various investigations have been conducted on the substitution of conventional coarse and fine aggregates with pumice and cinder (Manoj et al., 2021; Bharvi et al., 2022; Sadhana et al., 2024).

From the literature, it is found that aggregate is not fully replaced by cinder aggregates. Hollow core solid beams behave the same as solid beams in terms of the flexural properties of the beam. Through experimental research, the strength characteristics of areca fibers and lightweight aggregate will be investigated using bamboo-reinforced beams with various aspect ratios. The following goals are outlined in light of the gap in the literature: to study the flexural behavior of bamboo-reinforced beams with varying aspect ratios and to determine the ductility factor of a bamboo-reinforced beam by varying the aspect ratios.

3. Methodology

The methodology includes the initial collection of materials and performing basic tests on them. Further mix design has been conducted for various aspect ratios, and beams have been cast. After 28 days of curing, nondestructive testing, flexural testing (Figure 14.2), deflection testing, and strain gauge tests on beam specimens are performed. The beam specimens have dimensions of $1000 \times 200 \times 150$ mm, $1000 \times 100 \times 150$ mm, and $1000 \times 66 \times 150$ mm.

Figure 14.2: Flexural strength test setup with strain gauges.

4. Results

4.1 Flexural strength test

The intensity of bamboo-reinforced concrete beams is calculated at 28 days, along with the inclusion of areca fibers and lightweight aggregates. The outcomes of the experiments are shown in Table 14.1. After 28 days, bamboo-reinforced beams with five aspect ratios resulted in the highest flexural strength value, i.e., 14.16 N/mm².

5. Deflection test

The deflection values are noted corresponding to the specific loads, and the variation of load versus deflection graph against various aspect ratios is shown in Figure 14.3.

Table 14.1: Flexural strength test results.

Aspect ratio l/d	Load (P) kN	Flexure strength in MPa
A_{15}	15.0	8.00
B_{10}	17.5	11.66
C_{10}	17.0	11.00
D5	18.0	14.16

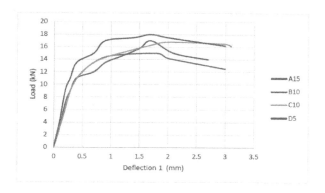

Figure 14.3: Load v/s deflection for aspect ratios of 5, 10, 15.

It was observed from the above graph, i.e., Figure 14.2, that the deflection of bamboo-strengthened concrete with an aspect ratio of 5 resulted in the least deflection value compared to the aspect ratios of 10 and 15.

6. Strain gauge test

The values are noted corresponding to the specific loads, and the stress versus strain graph is plotted. Here, the comparison of stress versus strain for aspect ratios of each beam is represented in the graph shown in Figure 14.4. The experimental observation from the graphs below shows that the strain of bamboo-reinforced concrete beams with an aspect ratio of 5 resulted in the least strain value compared to the aspect ratios of 10 and 15.

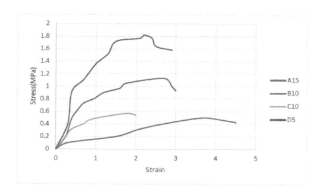

Figure 14.4: Stress v/s strain for aspect ratios of 5, 10, 15.

7. Conclusion

From the results of the flexural test, it can be found that the flexural strength of reinforced concrete beams with bamboo with five aspect ratios yielded the highest value, as it increased by 43.5% compared to beam specimens with 15 aspect ratios and 17.6% compared to beam specimens with 10 aspect ratios. As a result, flexural strength increases with a decline in aspect ratio. When compared to other beams with different aspect ratios, the load-deflection behavior of the beam with five aspect ratios had the highest ultimate strength and the lowest deflection value, i.e., 1.65 mm. The maximum deflection of bamboo-reinforced beams with five aspect ratios was found to be less than the beams with 10 and 15 aspect ratios, showing that the inclusion of bamboo improved the beam's bending property.

References

[1] Bharvi, S., Raju, S., & Prem, P. B. (2022). Effect of fine aggregate replacement with expanded perlite and pumice on the development of lightweight concrete. *Australian Journal of Civil Engineering, 20*(1), 115–129. https://doi.org/10.1080/14488353.2021.1930635

[2] Manoj, V., Sridhar, R., & Ajey Kumar, V. G. (2021). Study on effects of pumice in high-performance lightweight concrete by replacing coarse aggregates. *IOP Conference Series: Earth and Environmental Science, 822*(1), 012012. https://doi.org/10.1088/1755-1315/822/1/012012

[3] Punyamurthy, R., Sampathkumar, D., Bennehalli, B., & Badyankal, P. V. (2014). Study of the effect of chemical treatments on the tensile behaviour of abaca fiber reinforced polypropylene composites. *Journal of Advanced Chemistry, 10*(6), 2814–2822. https://doi.org/10.24297/jac.v10i6.6668

[4] Qaiser, S., Hameed, A., Alyousef, R., Aslam, F., & Alabduljabbar, H. (2020). Flexural strength improvement in bamboo-reinforced concrete beams subjected to pure bending. *Journal of Building Engineering, 31*, 101289. https://doi.org/10.1016/j.jobe.2020.101289

[5] Sadhana, K., Suguna, K., & Raghunath, P. N. (2024). Experimental study on properties of cinder aggregate lightweight concrete with fibre reinforcement. *Materials Express, 14*(6), 877–883. https://doi.org/10.1166/mex.2024.2701

[6] Venkateshappa, S. C., Bennehalli, B., Kenchappa, M. G., & Ranganagowda, R. P. G. (2010). Flexural behaviour of areca fibers composites. *BioResources, 5*(3), 1846–1858.

CHAPTER 15

Numerical simulations with embedded piezoelectric transducers for monitoring changes in concrete properties

Trushna Jena[1]*, T. Jothi Saravanan[1] and Tushar Bansal[2]

[1]School of Infrastructure, Indian Institute of Technology Bhubaneswar, Odisha, India
[2]Department of Civil Engineering, Sharda University, Uttar Pradesh, India
*Corresponding author: Trushna Jena
Email: a23ce09023@iitbbs.ac.in

Abstract

An electromechanical impedance (EMI) approach using lead zirconate titanate (PZT) smart clinker in concrete shows promise for early-age strength testing. This study uses COMSOL™ 5.5 to model an integrated PZT sensor. Stiffness, damping ratio, and Young's modulus affect cubical concrete block hydration in this model. This simulation exhibits alternative cure times or hydration phases. The PZT patch's resonant peak frequencies and amplitudes were measured at various electrical potential frequencies. Results are matched to experimental data for different concrete strengths and days to ensure accuracy. COMSOL™ 5.5 models save time by eliminating the need for casting laboratory concrete cubes for the EMI approach.

Keywords: embedded, smart aggregate, PZT sensor, electromechanical impedance, COMSOL™ 5.5, structural health monitoring, early strength

1. Introduction

Close tracking of concrete's strength development at the initial stages of construction is essential for ensuring the readiness of structures for service (Gu et al., 2006). Given that many crucial characteristics of concrete are linked to its compressive strength, information regarding this property is essential (Pessiki & Johnson, 1996). In the initial stages, the compressive strength experiences a rapid rise due to the hydration reaction that occurs during the curing process. Because of the emergence of adaptive intelligent materials with piezoelectric characteristics, researchers are interested in creating new nondestructive monitoring methods. Using these materials as sensors allows for actively monitoring structural problems (Flatau & Chong, 2002). EMI sensing with piezoelectric transducers is a promising way to assess concrete strength in early-age settings (Tawie & Lee, 2010). The main objective of this investigation is to simulate an integrated embedded PZT smart clinker model using COMSOL™ 5.5, which eliminates laboratory concrete cube casting and reduces monitoring wait times. The conductance and susceptance findings from the simulation are validated with empirical results conducted by Bansal et al. (2022). Young's modulus, stiffness, and damping ratio are adjusted to hydrate a concrete cube in the simulation.

DOI: 10.1201/9781003596776-15

2. Principle of EMI-based concrete strength monitoring using PZT patches

The correlation between the mechanical impedance of the structure and the electrical admittance of the PZT has been verified. The admittance signature's equation is obtained by the mathematical framework and constitutive expressions of the PZT patch by Bhalla and Soh (2004), which are stated in Equation 15.1:

$$\overline{Y} = G + Bj = 4\omega j \frac{l_p^2}{t_p} \left[\overline{\varepsilon}_{33}^T - \frac{2d_{31}^2 \overline{Y^E}}{\left(1 - \mu_p\right)} + \frac{2d_{31}^2 \overline{Y^E}}{\left(1 - \mu_p\right)} \left(\frac{z_{a,\text{eff}}}{z_{s,\text{eff}} + z_{a,\text{eff}}} \right) \left(\frac{\tan\left(kl_p\right)}{\left(kl_p\right)} \right) \right] \quad (1)$$

where conductance is denoted by G, susceptance by B, l_p represents the length and t_p is the thickness of the PZT patch, angular frequency is denoted by ω, d_{31} is the piezoelectric strain coefficient, $\overline{\varepsilon}_{33}^T$ represents the complex electrical permittivity found in the PZT material, where μ_p is the Poison's ratio, $Z_{s,\text{eff}}$ is the structural effective impedance, and $Z_{a,\text{eff}}$ is the PZT patch's impedance. $\overline{Y^E}$ is the complex form of Young's modulus.

3. Numerical study

3.1 Unconstrained PZT, PZT patch within epoxy layer, and smart clinker

The work focuses on the use of numerical methods to track the compressive strength development of concrete by implementing COMSOL™ Multiphysics 5.5. Table 15.1 lists this study's PZT-5H transducer characteristics. A 10 × 10 × 0.2 mm PZT transducer with one terminal and one ground face is used for numerical modeling (Figure 15.1a). The total thickness is 1.2 mm with two 1 mm epoxy layers. Epoxy measures 11 × 11 × 2.2 mm (Figure 15.1b). A cylindrical clinker of 10.2 mm in height and 15 mm in diameter is made from two layers of 4 mm mortar (Figure 15.1c). Table 15.2 shows epoxy, mortar, and concrete properties. Then, the PZT patch's electrical impedance was measured under 1 V across its terminals from 30 to 500 kHz.

Table 15.1: Physical and dielectric properties of PZT 5H (Bansal et al., 2022).

Parameters	Value
Density, ρ (kg/m³)	7400
Poisson's ratio, μ	0.4
Relative permittivity	3130
Dielectric loss factor, δ	0.0169
Damping ratio, ζ	0.02

EMI measurement was utilized to assess the conductance and susceptance of both the unconstrained and epoxy-layered PZT patches.

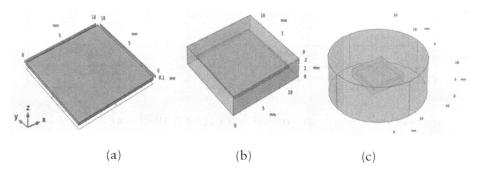

(a)   (b)   (c)

Figure 15.1: 3D model representation of (a) unrestrained PZT patch, (b) PZT patch encapsulated with Epoxy.

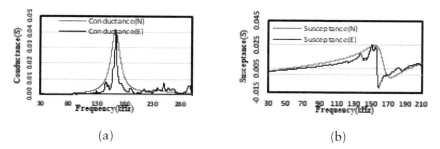

(a)   (b)

Figure 15.2: Experimental and computational result validation (a) Conductance of PZT, (b) susceptance of PZT (c) smart clinker.

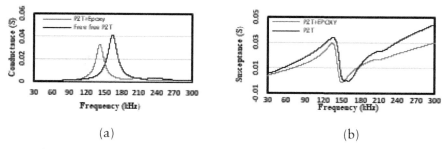

(a)   (b)

Figure 15.3: (a) Conductance of PZT and PZT + epoxy; (b) Susceptance of PZT and PZT + epoxy plot based on the EMI measurement.

Figure 15.2a and b demonstrate the conductance and susceptance signatures validation against experimental data. The results depicted in Figure 15.3a and b show that after epoxy coating, the peak shifts towards a lower frequency, and the resonant peak, which indicates conductance, also reduces. The epoxy layer surrounding the PZT patch influences its resonant characteristics.

Table 15.2: Properties of materials (Bansal et al., 2022).

Parameters	Values		
	Epoxy	Mortar	Concrete
Modulus of elasticity (GPa)	2	2.87	25
Density (kg/m³)	1200	2160	2300
Poisson's ratio (μ)	0.4	0.187	0.2

4. Concrete cube with an embedded piezo transducer

An embedded smart clinker has been employed to simulate a 150 mm × 150 mm × 150 mm concrete cube. In simulation, damping variations are sourced from Bansal et al. (2022). The simulated concrete cube model is scaled down to the ¼ᵗʰ model with an encased piezo transducer. The full-scale model, the ¼ᵗʰ model, and the ¼ᵗʰ meshed model are depicted in Figures 15.4 a-c.

Figure 15.5 shows the conductivity signature of smart clinker-embedded concrete specimens with different strengths by adjusting Young's modulus to reflect concrete hydration. The experimental dataset and EMI conductance profiles show a substantial correlation, indicating the study's numerical models' efficacy.

Figure 15.6 depicts the numerical simulation results depicting the relationship between conductance and frequency, represented by variations in the modulus of elasticity. Analysis of the plot reveals a direct correlation between the passage of time and the increase in modulus of elasticity, indicating the progression of concrete hydration.

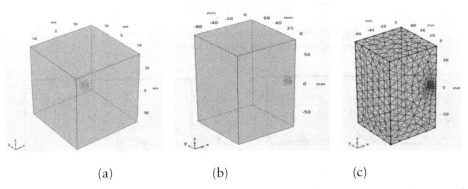

(a)   (b)   (c)

Figure 15.4: Embedded smart clinker (a) concrete cube (b) ¼ᵗʰ concrete cube (c) ¼ᵗʰ meshed concrete cube.

The resonance peak grows with cure time, indicating concrete conductance increases and impedance decreases. Impedance decreases during hardening and strength improvement. The upward shift and extremely tiny peak frequency increase match with Li et al. (2021) and Saravanan et al. (2015).

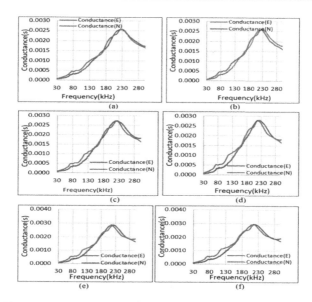

Figure 15.5: Experimental vs simulation plots of concrete of strength (a) 30 MPa, (b) 34 MPa, (c) 42 MPa, (d) 50 MPa.

5. Conclusion

In this study, a PZT 5H sensor is initially simulated and then embedded inside the concrete cube. Subsequently, the conductance and susceptance patterns of the sensor are captured and validated against experimental results. The simulation process includes subjecting the concrete cube to varying levels of hydration, achieved by adjusting parameters such as Young's modulus, stiffness, and damping ratio. This computational approach eliminates the necessity of physically casting concrete cubes in laboratory settings, thereby significantly reducing waiting times and resources.

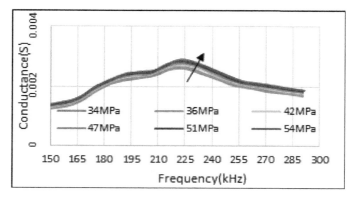

Figure 15.6: Numerical simulation of the plot of conductance vs. frequency for different grades of concrete (varied Young's modulus).

References

[1]   Bansal, T., Talakokula, V., & Mathiyazhagan, K. (2022). Equivalent structural parameters based non-destructive prediction of sustainable concrete strength using machine learning models via piezo sensor. *Measurement, 187*, 110202.

[2]   Bansal, T., Talakokula, V., & Sathujoda, P. (2022). Machine learning-based monitoring and predicting the compressive strength of different blended cementitious systems using embedded piezo-sensor data. *Measurement, 205*, 112204.

[3]   Bhalla, S., & Soh, C. K. (2004). Electromechanical impedance modeling for adhesively bonded piezo-transducers. *Journal of Intelligent Material Systems and Structures, 15*(12), 955–972.

[4]   Flatau, A. B., & Chong, K. P. (2002). Dynamic smart materials and structural systems. *Engineering Structures, 24*, 262–270.

[5]   Gu, H., Song, G., Dhonde, H., Mo, Y. L., & Yan, S. (2006). Concrete early-age strength monitoring using embedded piezoelectric transducers. *Smart Materials and Structures, 15*(6), 1837–1845.

[6]   IEEE Standard on Piezoelectricity Standards Committee of the IEEE Ultrasonics, Ferroelectrics, and Frequency Control Society IEEE Standards Board American National Standards Institute. (1987). *IEEE Standard on Piezoelectricity*.

[7]   Li, Y., Ma, Y., & Hu, X. (2021). Early-age strength monitoring of the recycled aggregate concrete using the EMI method. *Smart Materials and Structures, 30*(5).

[8]   Pessiki, S., & Johnson, M. R. (1996). Non-destructive evaluation of early-age concrete strength in plate structures by the impact-echo method. *ACI Materials Journal, 93*(3).

[9]   Saravanan, T. J., Balamonica, K., Priya, C. B., Reddy, A. L., & Gopalakrishnan, N. (2015). Comparative performance of various smart aggregates during strength gain and damage states of concrete. *Smart Materials and Structures, 24*(8).

[10] Tawie, R., & Lee, H. K. (2010). Piezoelectric-based non-destructive monitoring of hydration of reinforced concrete as an indicator of bond development at the steel-concrete interface. *Cement and Concrete Research, 40*(12), 1697–1703.

CHAPTER 16

Digital Preservation

Photogrammetric Assessment of Visual Heritage in the Context of the Jagannath Temple Precinct, Puri, Odisha

Amit Chatterjee[a*], Swapnali S. Ladpatil[b]

[a*]Assistant Professor, Department of Architecture,
Veer Surendra Sai University of Technology, Burla, Odisha, India
E-mail: amitchatterjee.ar@gmail.com
[b]Assistant Professor, Sinhgad College of Architecture,
Savitribai Phule Pune University, Pune, India
E-mail:ladpatilswapnali@gmail.com
*Corresponding Author

Abstract

In Puri, Odisha, the Jagannath Temple complex is a monument to centuries of cultural and architectural legacy. However, the historical significance and aesthetic purity of this hallowed location are in jeopardy due to the rapid pace of urbanization and development. The present study suggests an all-encompassing photogrammetric methodology for evaluating and recording the visual legacy of the Jagannath Temple precinct. This project aims to produce detailed 3D models and orthophotos of the precinct using high-resolution aerial and terrestrial photogrammetry. This will enable accurate measurement and analysis of architectural components and spatial linkages. In addition to being useful instruments for conservation planning and decision-making, these digital representations will help maintain this famous cultural property digitally.

Keywords: Digital Preservation, Photogrammetric Assessment, Visual Heritage, Jagannath Temple Precinct, Conservation Architecture

1. Introduction

The Jagannath Temple precinct in Puri, Odisha, is a site of exceptional cultural and architectural significance that exemplifies a unique mix of religious devotion and artistic expression. Over the years, this historic neighborhood has faced numerous challenges, including threats to its heritage value and visual integrity from environmental degradation and urban expansion. To address these issues, the visual legacy of the Jagannath Temple complex needs to be meticulously recorded and imaginatively assessed. Our study proposes a comprehensive photogrammetric assessment of the precinct to close this gap.

The goal of this project is to create detailed 3D models and orthophotos of the precinct using state-of-the-art photogrammetric methods, including aerial and terrestrial imagery. This will make it possible to thoroughly examine the architectural elements and spatial organization of the precinct. With its insightful information about how photogrammetry can be used to protect the visual heritage of historically significant locations, like the

DOI: 10.1201/9781003596776-16

Jagannath Temple precinct, the study's results should significantly advance the fields of conservation architecture and digital heritage preservation.

2. Literature Review

The application of photogrammetry in the field of architectural conservation and heritage documentation has been widely recognized for its ability to provide accurate and detailed representations of cultural heritage sites (Guidi et al., 2014). Several studies have highlighted the effectiveness of photogrammetry in documenting and analyzing historical structures, including monuments and temples, by creating precise 3D models and orthophotos (Abate et al., 2018; Remondino et al., 2014).

In the context of the Jagannath Temple precinct in Puri, Odisha, previous research has emphasized the importance of utilizing advanced imaging technologies, such as photogrammetry, to assess and document the architectural elements and spatial configuration of the site (Balzani & Melis, 2019). These studies have demonstrated the potential of photogrammetry to provide valuable insights into the conservation and management of heritage sites, especially in rapidly changing urban environments (Lo Brutto et al., 2019).

Overall, the literature indicates that photogrammetry can play a significant role in preserving the visual heritage of cultural landmarks like the Jagannath Temple precinct. However, further research is needed to explore the specific challenges and opportunities associated with applying photogrammetry in the context of architectural conservation in Puri, Odisha.

3. Methodology

1. **Data Collection**: Aerial and terrestrial photographs will be captured using high-resolution cameras. Ground control points (GCPs) will be established and surveyed for geo-referencing.
2. **Photogrammetric Processing**: Photographs will be processed using photogrammetry software (e.g., Agisoft Metashape, RealityCapture) to generate point clouds, dense 3D meshes, and orthophotos. Georeferencing will be performed using the GCPs to ensure accuracy.
3. **3D Modeling and Analysis**: Point clouds and 3D meshes will be used to create detailed 3D models of the Jagannath Temple precinct. The 3D models will be analyzed to identify architectural elements, spatial relationships, and changes over time.
4. **Visual Assessment**: The orthophotos and 3D models will be visually assessed to evaluate the visual integrity and heritage value of the precinct. Special emphasis will be placed on identifying areas of deterioration, alteration, and historical significance.
5. **Documentation and Reporting**: The findings of the visual assessment will be documented in a comprehensive report. Recommendations for conservation and management strategies will be provided based on the analysis.

4. Case Studies

Figure 16.1: Software Processing

Figure 16.2: 60% overlapping images

Photogrammetric documentation is the process of capturing and analyzing spatial data using overlapping images. High-resolution images were systematically collected using digital cameras with wide-angle lenses (Figure 16.1). Ground control point locations were established for precise geo-referencing. Despite challenges such as congested environments and intricate architecture, photogrammetry software processed images (Figure 16. 2) to generate detailed 3D models and point clouds, as well as orthophotos.

Figure 16.3: Orthographic images of Structure 3

The resulting data is used for preservation planning, research, and education, demonstrating the value of photogrammetry for documenting and preserving cultural heritage sites such as the Jagannath Temple precinct, Puri, India

5. Findings

The photogrammetric examination of the Jagannath Temple precinct in Puri, Odisha, has yielded important insights into its architectural significance and visual legacy. A thorough examination of the precinct has been made possible by the 3D models and orthophotos produced by photogrammetry, which have revealed complex architectural details and spatial linkages. In addition to highlighting the precinct's historical and cultural significance, the visual assessment has brought attention to its visual integrity. In particular, the study has pinpointed the precinct's degraded and altered regions, which are essential for conservation planning and management. The examination has also brought attention to the historical relevance of some architectural elements, illuminating the site's rich cultural legacy. These results are crucial for informing future conservation initiatives and guaranteeing the long-term administration of the Jagannath Temple area.

Overall, the research demonstrates the effectiveness of photogrammetry as a tool for assessing and documenting visual heritage in culturally significant sites. The findings contribute to the growing body of knowledge on the application of photogrammetry in architectural conservation and heritage preservation.

6. Conclusion

The project has effectively shown how photogrammetry can be used as a useful technique for evaluating and recording the visual history of the Jagannath Temple precinct in Puri, Odisha. The research has contributed to a deeper understanding of the historical and cultural relevance of the precinct by providing a thorough examination of its architectural features and spatial configuration through the construction of detailed 3D models and orthophotos.

The study's conclusions emphasize the value of visual evaluation in heritage site management. The study has yielded vital insights for conservation planning by identifying areas of change and deterioration, thereby guaranteeing the preservation of the Jagannath Temple precinct for future generations. The study has also highlighted the necessity of sustainable management approaches that strike a balance between the preservation of the heritage area and digitization.

The study has additionally illustrated the efficacy of photogrammetry in obtaining comprehensive data regarding architectural history. Photogrammetrically created orthophotos and high-resolution 3D models can be important tools for conservation and future study.

References

[1]   Abate, D., Brumana, R., Oreni, D., Della Torre, S., & Banfi, F. (2018). Photogrammetry and cultural heritage: A review. *Journal of Cultural Heritage, 32*, 266–282.

[2]   Balzani, M., & Melis, M. G. (2019). From geomatics to conservation: 3D modeling and BIM for cultural heritage. *ISPRS International Journal of Geo-Information, 8*(1), 6.

[3]   Guidi, G., Beraldin, J. A., & Atzeni, C. (2014). Digital imaging and 3D computer vision methods for cultural heritage preservation: State of the art and perspectives. *Journal on Computing and Cultural Heritage, 7*(1), 3.

[4]   Lo Brutto, M., Balletti, C., Guerra, F., & Vernier, P. (2019). Photogrammetric and laser scanning surveys for the documentation of the architectural heritage in Sicily. *International Archives of the Photogrammetry, Remote Sensing and Spatial Information Sciences, 42*, 563–570.

[5]   Remondino, F., Rizzi, A., Baratti, G., & Brumana, R. (2014). Photogrammetry and laser scanning in cultural heritage documentation: An overview. *Journal of Cultural Heritage, 15*(3), 318–325.

CHAPTER 17

Influence of Social-Political Conditions on Residential Typology and Conservation Practices in India

A Case Study of Gadhi Architecture, Maharashtra, India

Swapnali S. Ladpatil[a*], Amit Chatterjee[b]

[a*]Assistant Professor, Sinhgad College of Architecture,
Savitribai Phule Pune University, Pune, India
E-mail: ladpatilswapnali@gmail.com
[b]Assistant Professor, Department of Architecture,
Veer Surendra Sai University of Technology, Burla, Odisha
E-mail: amitchatterjee.ar@gmail.com
*Corresponding author

Abstract

The interaction between social-political contexts and architectural innovation is reflected in the evolution of residential typologies. This essay examines the architectural topic of gadhis, or forts, which were common in Maharashtra, India, in the 17th century. Gadhis were built for defense, ranging from small village mud forts to more substantial constructions that reflected the socio-political environment of the Maratha period. Built with a combination of strong structure, skillful craftsmanship, and practical planning, gadhis served as both administrative and residential buildings. This study looks at the development and traits of Gadhi architecture in sociopolitical settings, illuminating the intricate interplay between power structures, architectural forms, and settlements.

Keywords: Prevention Architecture, Gadhi, Fortified Residential Typology, Conservation Architecture, Fortress

1. Introduction

Maratha architecture displays the military might and cultural legacy of the Maratha Empire. The fortified dwellings of the Maratha sardars are notable for their distinct design and opulence. These buildings functioned as both protective bastions and places to live, much like castles. Gadhis, in contrast to land forts, were mostly residential buildings protected by stone walls, trenches, and entrances. This emphasizes how important it is to build secure, yet functional, protected dwellings.

The proliferation of gadhis in the late 18th century was driven by the reinstatement of the jagir system under Chhatrapati Rajaram and its acceptance by Peshwa Balaji Vishwanath. This facilitated the rise of jagirdars, who sought authority through fortified residences.

DOI: 10.1201/9781003596776-17

Expansionist policies of Peshwas Bajirao I and Nanasahib led to more military officers holding strategic positions, erecting gadhis like those in Tasgaon and Sangli. These gadhis symbolized their status and power within the Maratha hierarchy, showcasing Maratha architects' ingenuity and craftsmanship. This evolution reflects socio-political dynamics and highlights the interplay of architecture, power, and identity in Maharashtra's historical landscape.

2. Literature Review

a. History of the Maratha Empire

The historical backdrop of the Maratha Empire is extensively explored in the literature. Scholars like D.B. Parasnis (1918), Kolarkar (2009), Karbhari (2014), and Kundlikrao (2012) have meticulously chronicled the rise and fall of the Maratha Empire, covering key figures such as Chhatrapati Shivaji Maharaj, Chhatrapati Sambhaji Maharaj, and Chhatrapati Shahu Maharaj. The role of patronage in shaping the Maratha Empire is a recurrent theme in the literature. Works by Kasture (2015, 2016), Vasudev (1945), and Khandge (2015) delve into the contributions of influential families like the Purandares and the Peshwas. These studies highlight not only the architectural endeavors of these families in constructing wadas, temples, and other structures but also their political and administrative influence during the reigns of figures such as Shahu Maharaj and Peshwa Balaji Vishwanath.

b. Residential Typology

The literature offers a comprehensive examination of residential typologies prevalent during the Maratha period. Scholars like Kulkarni (2015–2016) and Gupta (2013) focus on the architectural evolution and construction techniques of wadas in Maharashtra, providing insights into their historical significance and spatial organization. Additionally, ancient texts like Kautilya's Arthashastra shed light on the construction and fortification techniques employed in the creation of rajwada and similar mansions by wealthy Maratha sardars.

c. Fortified Castles (Gadhi)

The study of gadhis, or fortified castles, is a prominent subject within Maratha architectural scholarship. Authors such as Stokstad (2005), Mate (1959), Deshpande (1971), and Shivde (2013–2015) offer insights into the evolution, construction, and historical significance of gadhis. These structures, built for residential purposes with defensive features, exemplify the socio-political conditions of the medieval Maratha period. Additionally, works by Suryawanshi (2009) and Birajdar (2015) provide detailed information on specific gadhis in regions like Marathwada and Latur, highlighting their architectural nuances and cultural importance.

Overall, the literature review underscores the multidimensional nature of Maratha architecture, encompassing historical narratives, patronage dynamics, socio-political conditions, residential typologies, and the unique phenomenon of gadhi, thereby offering a holistic understanding of the built environment during this pivotal period in Indian history.

3. Research Questions

1. What factors influenced the emergence and evolution of gadhis as a prominent residential typology during the Maratha period?
2. How did socio-political conditions shape the spatial organization and architectural features of gadhis?
3. What role did gadhis play in mediating social interactions and maintaining order within Maratha society?

4. Methodology

This research employs a mixed-methods approach, integrating archival research, architectural surveys, case studies, and oral histories. Archival research involves extensive examination of historical records, scholarly literature, and primary sources related to Maratha history and architecture. Architectural surveys document the physical characteristics and spatial organization of gadhis through on-site inspections and analysis of structural features. Case studies focus on select gadhis across Maharashtra, providing an in-depth exploration of their historical significance and architectural evolution. Oral histories involve interviews with local historians and descendants of gadhi owners to gather insights into cultural practices and social dynamics associated with these structures.

5. Case Studies

a. Criteria and parameters for Case Study Selection:

   i. Gadhis built in the Maratha period.

   ii. Currently in use.

   iii. Associated with Maratha sardars.

   iv. Geographical context: Provide insights into the location, landscape, and surroundings of the selected gadhis, including their proximity to urban centers and integration within village settings.

   v. Historical context: Explore the historical background of the gadhis, focusing on the patronage, contributions, and significant events associated with their construction and occupation.

   vi. Architectural analysis: Conduct a detailed examination of the spatial layout, planning, construction materials, and decorative elements within the gadhis. Additionally, assess present-day challenges and opportunities related to their preservation and development.

6. Comparative Analysis

A comparative analysis of five gadhis of Maharashtra—Chaskar Joshi Gadhi, Wathar Nimbhalkar Gadhi, Patankar Gadhi, Sardar Indraji Kadam Gadhi, and Purandare Gadhi— each has salient architectural features, construction techniques, and spatial organizations that speak to their respective historical and cultural contexts.

1. **Architectural Features**: Each gadhi stands with unique features. Chaskar Gadhi, built in 1670, is rectangular with fore bastions. Wathar Nimabhalkar Gadhi, built in 1795, is a square with open farming spaces. Patankar Gadhi and Kadam Gadhi, both built in the 18th century, show variations in bastion count and spatial spread, indicative of different priorities of defense and aesthetic choices. Purandare Gadhi, built between 1760 and 1770, is also square with four bastions, leaning more towards fortification and functionality.

2. **Material and Construction Techniques**: The materials and construction techniques used in gadhis represent the regional availability and advancements in technology. Stone is used for fortification; wood is used for structure and decoration. Mangalore tiles for roofing denote adaptation to the local climate. Various construction methods show a blend of practicality and workmanship.

3. **Structures in Gadhi:** Such as wells, big mansions (wadas), entrance gates, and sometimes temples or sheds for animals, denote multifunctional purposes: residential, administrative, and utilization. The gadhis have been local seats of governance and a center of social life.

4. **Socio-Political Influence:** The designs make them defensive strongholds and symbols of the power of Maratha sardars and jagirdars. Differences in size and defensive attributes reflect the status of the patron, highlighting architectural grandeur in a feudal system.

5. **Conservation Challenges:** Several gadhis stand in dilapidated conditions and lack proper maintenance and preservation. Contemporary uses, such as portions being leased for commercial activities, denote efforts to raise resources for their upkeep but at the cost of concerns about preserving historical authenticity.

In conclusion, the gadhis of Maharashtra narrate the rich heritage of architecture and the socio-political history of the region. Each gadhi narrates a tale of its era, speaking volumes about the needs, resources, and status of its patrons. Such structures need to be understood and documented to conserve them and understand their importance in the cultural scenery of India. Integrating innovative technologies and community-led initiatives will be key to preserving these historical treasures for posterity.

7. Findings

Examining the distinctive characteristics of Maharashtra, India's gadhi architecture, one can see how social and political circumstances molded it. Gadhis are powerful, majestic, and practical symbols of medieval dominance and defense. Their strong frameworks, fine craftsmanship, and well-thought-out design convey the Vatandar's standing as the village ruler. Gadhis functioned as fortifications, dwellings, and hubs for administration. Their general goal of protection unites them despite differences in construction methods. The conservation attempts to save Maharashtra's gadhi architecture are informed by these findings.

8. Conclusion

The gadhi architecture of Maharashtra is a reflection of the state of politics and society. The topography was shaped by the impact of elites on villages. The strongholds of the elite villages known as gadhis represent social stratification, the necessity for defense, and architectural styles. Today, it is difficult to maintain gadhis, which might result in neglect or demolition. Some proprietors reuse architectural features by turning them into commercial spaces. To protect the cultural heritage of Maharashtra, conservation is essential. Local communities, environmental organizations, and government institutions need to work together. Documentation, restoration, and adaptive reuse ought to be given top priority in conservation initiatives. Acquiring support for gadhis requires educating the public about their importance. Conservation strategies that acknowledge gadhis as historical archives can strengthen future generations of Maharashtra's cultural fabric.

References

[1]   Birajdar, D. S. (2015). *Marathwadyatil Gadhi Wade ek Wastushastriya Abhyas.* Sodhganga.

[2]   Deshpande, P. N. (1971). *Maratha Forts.* Savitribai Phule Pune University/ Shodhganga.

[3]   D. B. Parasnis, R. B. (1918). *The history of Maratha people, Volume 1*. Oxford University Press.

[4]   Gupta, R. R. (2013). *Wada of Maharashtra*. National Institute of Advanced Studies in Architecture (NIASA).

[5]   Karbhari, M. N. (2014). *Struggle and sacrifice contribution of Chhatrapati Sambhaji in the conservation of Hindavi Swaraja*. Sodhganga.

[6]   Kasture, K. (2015). *Peshwai*. Rafter Publisher.

[7]   Kasture, K. (2016). *Purandare*. Raftar Publication.

[8]   Khandge, M. (2015). *Vaibhav Peshwekaleen Vadyanche*. Sai Prakashan.

[9]   Kolarkar, D. S. (2009). *Marathyancha Itihas*. E.G. Printing Press.

[10] Kulkarni, A. (2015-16). *Wada architecture in Maharashtra*. University of Edinburgh.

[11] Kundlikrao, J. S. (2012). *Contribution of Chhatrapati Shahu Maharaj in the Development of the Maratha Empire*. Sodhganga.

[12] Mate, M. S. (1959). *Maratha Architecture, 1650 A.D. to 1850 A.D.* University of Poona.

[13] Rajwade, V. (2002). *Marathanchya itihasachisadhane*. Itihasacharya V.K. Rajwade Sanshodhan Mandal.

[14] Shivde, D. S. (2013). *Maharashtratil Aitihasik Wade—Part 1*. Snehal Publisher.

[15] Shivde, D. S. (2015). *Maharashtratil Aitihasik Wade—Part 2*. Snehal Publication.

CHAPTER 18

AI-based Cognitive Load Analysis of Human Beings Through Working Memory

Peetabas Patro[1], Ramesh Kumar Sahoo[1], Srinivas Sethi[1], Anup Maharana[2]

[1]Indira Gandhi Institute of Technology Sarang, Odisha, India
[2]Rennes School of Business, France
*Patrobabu990@gmail.com

Abstract

Mental health is considered a significant problem in today's world. In recent years, the study of meditation as an alternative therapy has gained substantial attention in clinical medicine. This paper explores the impact of working memory underlying meditation and its application in mental disorders. Accurately predicting mental health conditions can facilitate behavioral changes and promote happiness. The application of meditation has expanded to mental illnesses, particularly major depressive disorders, and substance-related and addictive disorders. Mechanism studies reveal the positive impact of meditation on mental health, making it a valuable tool for enhancing well-being and managing stress.

Keywords: meditation, mental stress, attention, cognitive, working memory

1. Introduction

Mental health is an integral part of our overall well-being, including our social, psychological, and emotional welfare. It influences the cognitive power of a person to handle stress through proper thinking, emotion, and required action. The cognitive power of a person can be highly influenced by the emotional level in terms of anxiety and the resultant stress. Working memory plays a major role in various cognitive tasks, including problem-solving, decision-making, image identification, and learning. Cognitive load reflects the extent of mental effort given to handle the information to perform the task. Understanding cognitive load is essential for optimizing human-computer interactions, educational settings, and mental health assessment. The objective of this work is to investigate and develop artificial intelligence (AI) algorithms that can analyze cognitive load in individuals based on their working memory (WM) activity for the prediction of cognitive load. The motivation behind this work lies in addressing the pressing issue of mental health. By exploring the impact of meditation on various aspects of mental well-being, including brain function, structure, and epigenetic regulation, this study aims to provide insights into how meditation can be effectively used as an alternative therapy for mental disorders.

DOI: 10.1201/9781003596776-18

2. Background

Mental health is a major factor in the modern era that controls the lifestyle of humans due to the emotional level attached to work pressure and the day-to-day challenges that occur in the work environment, affecting the social behavior of the person (Balaskas et al., 2021). It also increases stress and disturbs the mental health of a person. Oberauer (2019) establishes a relationship between the attention level of a person and working memory through a resource in which attention is used for storage and processing, perceptual attention and its impact on working memory, and through an attention controller resource. Jyotsna et al. (2023) developed a personalized BCI model to predict the EYE system through the mental status of humans based on their eye-tracking data observed through certain test cases like visualizing stressful and stress-less calm videos. Several BCI applications are based on meditation and attention. However, there is a gap in how WM is related to cognitive function, so a study is needed. This research aims to bridge the gap between neuroscience, artificial intelligence, and practical applications, ultimately benefiting individuals' cognitive well-being and performance. The primary aim behind conducting this research is scientific explanation (understanding cognitive load, neural correlates, advancing knowledge), technological innovation (AI algorithms, real-time analysis, wearable devices), and improving well-being (optimizing workload, enhancing learning).

3. Methodology

3.1. Experiment Setup

Figure 18.1 represents the architecture of the proposed work. At first, the EEG waves are collected and further cleaned into three parts: Attention, Meditation, and Working memory. Later, Attention and Meditation were combined to get Cognitive status. Further, Cognitive status and Working memory were used in the mathematical model using the Fuzzy Rule to get the desired output.

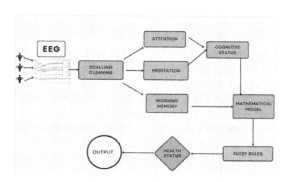

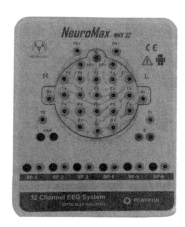

Figure 18.1: Architecture of the proposed work.   **Figure 18.2.** Neuromax 32 Board.

3.2. Data Recording

In this proposed model, a total of 25 persons, composed of 12 females and 13 males, with an average age of 22.5 ± 1.2 years, were involved. The participants undertook 10 min of EEG signal recording each, conducted at the CC Lab at the IGIT, Sarang (CSE&A). Before the experiment, the participants were informed of the instructions. Using a Neuromax 32-channel motherboard connected to a server (PC) via USB, brain wave data is collected from the human brain. An EEG cap with 32 electrodes (16 red wires for the right side and 16 black wires for the left side) is placed on the scalp to transfer raw brain wave data to the server.

3.3. Experiment Design

Data is collected from different patient categories under various conditions, including work-related tasks. These experiments yield valuable data for further studies. Collected data is cleaned using Python's Scikit library and other analysis tools. It is then restructured into a comma-separated values (CSV) format for storage and further analysis. A dynamic weight factor-based mathematical model is proposed to explore the relationship between working memory and cognitive status. This model calculates the human cognitive level (C_L) using attention and meditation levels (T4, T6, P3, and O1 electrodes) as per Eq. 18.1 (Badajena et al., 2023).

AL reflects the attention level of the person between 0 and 100, and the values have been obtained through lobe points $T4$–$T6$. ML reflects the meditation level of the person between 0 to 100, and the values have been obtained through the lobe points P3-01.

$$C_L = A_L * W1 + M_L * W2 \qquad (18.1)$$

$$\text{Cogn}_{status} = \begin{cases} poor \ for \ CL < 35 \\ average \ for \ 35 < CL < 55 \\ good \ for \ CL > 55 \end{cases} \qquad (18.2)$$

The extent of concentration/focus given by the human for a task is reflected through their attention level, whereas meditation level reflects the steady behavior (calmness in mind) of the human during work. Hence, two weight factors have been considered: W1 is 0.7 for attention and W2 is 0.3 for meditation level. Cognitive load primarily depends on attention. Therefore, 70% importance has been given to attention level and 30% to meditation level. The cognitive status (Cogn_{status}) has been estimated in Eq. 18.2 of each subject, and it has been classified into poor, average, and good categories.

The temporal lobe of the human brain is used for memory. Human memory is used for perception, awareness, and speech recognition. For the memory status of a person, four electrodes FP1-F7, FP2-F8, by fusion of two sides (left and right) points have been considered for the estimation of working memory. To estimate the impact of working memory on cognitive functions, two parameters—cognitive status and working memory—have been used as input for the fuzzy model as observed, and mental health has been considered as output in the fuzzy model as observed in Figure 18.3. A set of nine fuzzy rules has been used in the proposed model to compare cognitive status and memory use. Three membership parameters for fuzzy logic are good, average, and poor for the rating (CN) and low, medium, and high for memory use (M). Three membership parameters for mental health status (HS) are poor, average, and good, following fuzzy rules from R1 to R9 that have been generated using membership parameters of input (CN & M) and output (HS) as per the algorithm authors in Sahoo et al. (2023).

R1: CN ["poor"] and M["poor"] ==⟹ HS["high"]

R2: CN ["average"] and M["average"] ==⟹ HS ["high"]

R3: CN ["good"] and M["good"] ==⟹ HS ["high"]

R4: CN ["poor"] and M["good"] ==⟹ HS ["low"]

R5: CN ["good"] and M["poor"] ==⟹ HS ["low"]

R6: CN["good"] and M["average"] ==⟹ HS ["medium"]

R7: CN["average"] and M["good"] ==⟹ HS ["medium"]

R8: CN["poor"] and M["average"] ==⟹ HS ["medium"]

R9: CN["average"] and M["poor"] ==⟹ HS ["medium"]

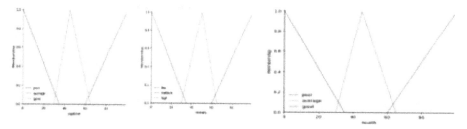

Figure 18.3: Fuzzy membership values for input and output.

4. Result and Analysis

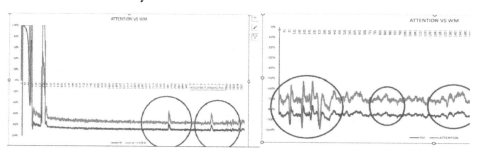

Figure 18.4: Impact of working memory on cognitive function.

The red circle in Figure 18.4 shows that during the reading of four persons under certain conditions, the sudden changes in attention level increase the stress level, leading to an increase in cognitive status. At the same time, the use of working memory also increases, as reflected in the figure above. Upon analyzing the data, it has been observed that high stress can significantly impact a person's working memory, resulting in short-term memory loss. This, in turn, can contribute to feelings of depression. Stress and anxiety can also exacerbate memory difficulties, making it challenging to focus on tasks, make decisions, or think clearly. It is important to recognize that working memory, which allows us to actively remember information from moment to moment, is particularly affected. Difficulties with working memory can hinder concentration and decision-making.

5. Conclusion

In the proposed work, a fuzzy logic-based mathematical model has been developed to estimate the impact of cognitive load on working memory and mental health. The mathematical model has been used to estimate the cognitive load using the attention and meditation levels of the person, and it has been compared with working memory using fuzzy logic to estimate its impact on working memory and overall mental health status. Collectively, the results from our investigation underscore that the stress level can be reduced by proper meditation. It has been used to alert the user and take precautions whenever the cognitive load is more and mental health is worse. It may enhance the health status of a person.

References

[1]   Balaskas, A., Schueller, S. M., & Doherty, G. (2021). Ecological interventions for mental health: Scoping review.

[2]   Badajena, J., Sethi, S., & Sahoo, R. (2023). Data-driven approach to design a BCI-integrated smart wheelchair through cost-benefit analysis. *High Confidence Computing, 3*, 100118. https://doi.org/10.1016/j.hcc.2023.100118

[3]   Jyotsna, C., Amudha, J., Ram, A., & Fruet, D. (2023). PredictEYE: Personalized time series model for mental state prediction using eye tracking. *IEEE Access, 11*, 128383–128409. https://doi.org/10.1109/ACCESS.2023.3332762

[4]   Oberauer, K. (2019). Working memory and attention: Conceptual analysis and review. Neuromax32 Board retrieved on 4.4.2024.

[5]   Sahoo, R., Pradhan, S., Sethi, S., & Udgata, S. (2023). Enhancing data integrity in mobile crowdsensing environment with machine learning and cost-benefit analysis. *International Journal of Computing and Digital Systems, 14*, 253–278. https://doi.org/10.12785/ijcds/140122

CHAPTER 19

Robust biometric attendance system via content-based feature extraction and classification of acknowledged sound

Harshwardhan Fogla[1], Aditya Jena[1], Ramesh Sahoo[1], Srinivas Sethi[1], Anup Maharana[2]

[1]IGIT Sarang, Odisha, India
[2]Rennes School of Business, France
Corresponding Author Email: harshfogla3@gmail.com

Abstract

Nowadays, biometric attendance is replacing manual attendance systems. Although it has simplified attendance, there are certain drawbacks, such as the user being unaware of whether their attendance has been recorded. As a result, a unique method for tracking attendance has been proposed that uses machine learning (ML) techniques to assess the thank-you sounds produced by biometric attendance recording devices. The application generates attendance statistics and accurately interprets audio using ML models and signal processing algorithms. This technique extracts unique characteristics from the audio, such as Mel-frequency cepstral coefficients and Mel spectrograms, to distinguish genuine acknowledgments from noise. The application automates attendance tracking through the use of ML and audio analysis, assuring the user that their attendance has been recorded.

Keywords: Machine learning, MFCC, Mel spectrogram, Audio analysis

1. Introduction

The advancement of technology has drastically changed the way that traditional attendance tracking systems have been used in many different fields in recent years. The introduction of biometric attendance devices has significantly expedited this procedure, providing precision and efficiency compared to traditional manual techniques. Nevertheless, difficulties still exist despite these developments, especially when it comes to guaranteeing the authenticity and integrity of recorded attendance data. With the use of cutting-edge technology like machine learning (ML) and signal processing technologies, biometric attendance systems may potentially be made more capable and provide more reliable mechanisms for attendance tracking and verification. Analyzing the audio cues generated by biometric devices during attendance acknowledgment is one such method.

A novel approach has been presented for evaluating the thank-you acknowledgment sound produced by biometric attendance tracking devices using ML techniques. The proposed approach uses ML models and signal processing methods to accurately interpret audio signals and separate real acknowledgments from background noise. The application

DOI: 10.1201/9781003596776-19

of sophisticated features, like Mel-frequency cepstral coefficients (MFCC) and Mel spectro-grams, makes it easier to extract important attributes and accurately identify and validate attendance occurrences. The main goal of the proposed work is to create a complete solution that improves the precision and dependability of biometric attendance tracking systems.

2. Background

Analysis of audio samples can be considered a better approach for the recognition of objects in a smart IoT-enabled environment. Wang et al. (2022) discussed a semisupervised learn-ing-based approach along with spatial alignment for analyzing audio-video sample data to get various audio forms. It has been used for the detection of guns through ML-based anal-ysis of YAMNET audio samples using transfer learning (Valliappan et al., 2024). Similarly, sound recorded from various objects in an IoT-enabled smart urban environment can be used for object identification through ML-based classification of real-time audio samples (Sethi et al. 2023). Xie et al. (2021) proposed a model for the analysis of audio samples through observed semantic embeddings from descriptive sentences and text-based labels for zero-shot learning. These works motivate the proposed work to make the biometric-based attendance system more robust through the classification of sound received from various devices.

3. Methodology

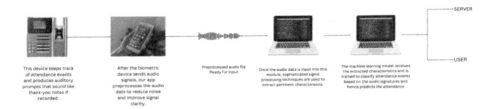

Figure 19.1: Architecture of proposed work.

The architecture of the proposed work is reflected in Figure 19.1. A smartphone-based application has been developed to collect the audio signals produced by the biometric machine, which undergoes additional preprocessing. Preprocessing involves adjusting the volume to a level suitable for interpretation while also eliminating background noise. After that, features like Mel spectrogram, MFCC, Chroma STFT, Chroma CQT, and Chroma Cens are extracted using the processed audio as input. The machine is then trained using the aforementioned attributes with various ML algorithms, such as SVM, Naive Bayes (NB), and Decision Tree (DT). Consequently, the most effective approach has been deter-mined for ML training by comparing the accuracy of various methods and putting that model to use for testing that forecasts the output from the created audio signal. Whether or not the audio signal contained the necessary thank-you message is indicated by the output. If the system successfully interprets the thank-you message, it will show that the attendance has been recorded; otherwise, it has not been recorded.

3.1. Data Collection

A smartphone-based application has been developed to capture the messages produced by the biometric attendance tracking device. The device generates a signal—in this case,

"Thank you" or "Try Again"—as soon as the user enters their biometrics, which can be either fingerprints or facial recognition. The app that has been developed records these signals and sends them for preprocessing.

3.2. Audio Preprocessing

Initially, the raw audio recordings underwent extensive preprocessing steps intended to improve their suitability for analysis. Rich characterizations of the sound waves are extracted, encapsulating fundamental periodic and ephemeral aspects. This action paved the way for workable noise reduction and volume enhancement while promoting a deeper comprehension of the sound content.

3.3. Mel Spectrogram

The Mel spectrogram represents the frequency content of a sound signal over time, with frequencies mapped to the Mel scale, which corresponds more closely to human perception of pitch. It is significant for capturing the spectral characteristics of sound signals, including distinguishing between different types of sounds based on their frequency content.

3.4. Constant-Q Transform (Chroma CQT)

Chroma CQT represents the energy distribution of sound signals over different melodic pitches, independent of their octave. It is especially important for tasks involving music analysis, such as chord acknowledgment, tune extraction, and music genre classification. It helps to focus on the harmonic content of the sound, ignoring minor information related to pitch octaves.

3.5. Chroma Short-Time Fourier Transform

Similar to Chroma CQT, Chroma STFT also represents the energy distribution of sound signals over different musical pitches. It provides a spectral representation of the sound signal, allowing for analysis of harmonic content and tonal characteristics. It is widely used in music data recovery errands, such as music transcription, instrument recognition, and sound similarity analysis.

3.6. Chroma Energy Normalized Statistics

Chroma CENS are statistical features computed from Chroma representations, capturing information about the distribution of energy over different pitch classes. It is relevant for extracting higher-level musical features, such as tonal solidness, consonant complexity, and key estimation. It provides insights into the tonal structure and consonant connections within sound signals, facilitating tasks like music proposal and programmed playlist era.

3.7. Mel-Frequency Cepstral Coefficients

MFCCs represent the short-term power spectrum of sound signals, converted to a logarithmic scale and transformed using the discrete cosine transform. They are particularly useful for capturing the spectral envelope of sound signals, which is vital for tasks such as discourse acknowledgment, speaker recognizable proof, and feeling location. They effectively compress the spectral information while retaining discriminative features, making them widely used in various audio analysis applications.

3.8. Machine Learning

In the scope of artificial intelligence (AI), ML is the study of developing statistical models and algorithms that allow computers to learn from past experiences and become smarter without being explicitly programmed. Standard and well-known supervised ML algorithms, such as Support Vector Machine (SVM), Gaussian Naive Bayes (GNB), and DT, have been considered for analysis.

The proposed ML-based model used the processed data as input for learning, dividing it into training and testing batches. Seventy percent of the dataset was used for training, while the remaining 30% of the dataset was used for testing using all of the ML approaches, and the scores were recorded.

3.9. Dataset

A real-time dataset has been collected from various persons. It has 728 audio samples, out of which 153 represented "Attendance recorded," whereas the rest 575 represented "Attendance rejected." Its features have been labeled as Name, Mel Spectrogram, MFCC, Chroma_stft, Chroma_cqt, Chroma_cens, and class.

4. Result and Discussion

The classification algorithm results are displayed below, utilizing various metrics and graphs for analysis.

Table 19.1: Comparative analysis of proposed ML-based model.

Model	Precision	recall	F1-score	Accuracy
SVM	0.76	0.55	0.64	0.885845
GNB	0.76	0.62	0.68	0.894977
DT	0.88	0.7	0.78	0.926941

From the above table, we can infer that the performance of different algorithms varies with metrics where DT outperforms.

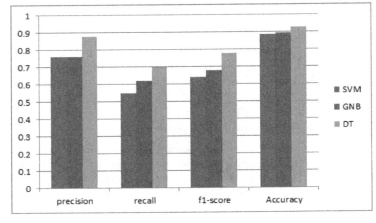

Figure 19.2: Evaluation of proposed ML-based model.

Table 19.2: Statistical analysis of the proposed ML-based model.

Class	Actual test QTY	SVM	GNB	DT
Correct	40	22	25	28
Incorrect	179	172	171	175

From the above table, it is apparent that the DT model exhibits the highest accuracy, correctly predicting 28 out of 40 for correct audio and 175 out of 179 for incorrect audio, followed by GNB and SVM. From the above analysis, it was clear that the DT algorithm provided the best accuracy and hence was implemented to test the samples. With its accuracy of 92.69%, the machine proved to be dependable when put into use.

5. Conclusion

In conclusion, this research has demonstrated the efficacy of employing machine learning algorithms for audio recognition tasks related to attendance monitoring. Among these algorithms, Decision Trees emerged as the most accurate, achieving a remarkable accuracy rate of 92.69%. This success indicates the potential of machine learning in accurately discerning between instances where attendance was taken and those where it was not. This study not only underscores the significance of employing advanced techniques in audio recognition but also provides practical insights for improving attendance monitoring systems. Overall, the aforementioned findings contribute to the advancement of audio recognition technology and its applications in attendance tracking and beyond.

References

[1]   Wang, S., Politis, A., Mesaros, A., & Virtanen, T. (2022). Self-supervised learning of audio representations from audio-visual data using spatial alignment. *IEEE Journal of Selected Topics in Signal Processing, 16*(6), 1467–1479.

[2]   Valliappan, N. H., Pande, S. D., & Vinta, S. R. (2024). Enhancing gun detection with transfer learning and YAMNet audio classification. *IEEE Access*.

[3]   Sethi, S., Rath, M., Kuanar, S. K., & Sahoo, R. K. (2023). Object recognition through content-based feature extraction and classification of sounds in IoT environment. In *Proceedings of the International Conference on Advanced Communication Technology*.

[4]   Xie, H., & Virtanen, T. (2021). Zero-shot audio classification via semantic embeddings. *IEEE/ACM Transactions on Audio, Speech, and Language Processing, 29,* 1233–1242.

CHAPTER 20

Multifunctional Composite Materials for Lightweight and Efficient EV Structures

Sasmita Bal[1], Sabyasachi Aich[2], Jayashree Nayak[2], Sunita Panda[2]

[1]Alliance University, Bangalore, India
[2]Assistant Professor, Mechanical Engineering Department, I.G.I.T, Sarang, India
*Corresponding author E-mail: nayak.jayashree@gmail.com

Abstract

The world is moving towards sustainable transport as electric vehicles (EVs) continue to develop as a potential tool for reducing greenhouse gas emissions and fossil fuel dependency. EV technology focuses on finding lightweight but strong materials that can improve vehicle performance and efficiency. As such, multifunctional composites have become integral in the design of future EV structures. The primary focus of this article is to find out how multifunctional composite materials help enhance the performance of EV structures. The paper also explores different composites, such as polymer, metallic, and hybridized ones, which are integrated into EV designs. Additionally, it shows the benefits of these materials, including weight reduction, enhanced mechanical properties, thermo-electrical management, and electromagnetic interference shielding.

Keywords: Composite, lightweight, EV, carbon fiber, polymers

1. Introduction

The automobile industries are currently working to achieve higher fuel economy, control the emission of greenhouse gases, and achieve ultimate sustainability. To achieve these targets, the production of lightweight vehicles is mandatory. In addition to focusing on improving fuel economy, the industries also need to consider other factors such as recyclability, safety, and good performance in any production of the manufacturing firm (Zhang & Xu, 2022). From the literature review, it is evident that a wide range of materials are available for lightweight applications, such as advanced composites and nanomaterials (Wazeer et al., 2023). The evolution of the automotive landscape is driven by innovative lightweight materials. These cutting-edge technologies are transforming vehicle design, paving the way for a future of better efficiency and a sustainable environment, which is crucial for both EVs and conventional vehicles (Jensen et al., 2013). New materials are being investigated for enhanced sustainability and performance, such as advanced composites, aluminum alloys, magnesium alloys, and high-strength steel (Busarac et al., 2022). If lightweight composites were employed instead of traditional materials like cast iron and steel, the net effect would be lower car weight, leading to improved vehicle efficiency and performance.

Developing environmentally friendly, fuel-efficient, and sustainable cars can bring a positive change to the automotive industry as it embraces advanced materials (Murlidhar

DOI: 10.1201/9781003596776-20

et al., 2018). This research paper demonstrates that multifunctional composites can be used to enhance the strength of EVs. Furthermore, the report provides an overview of various polymer composites, metallic composites, and hybrid composites, which signifies that these are highly developed materials when integrated into EV design.

2. Multifunctional Composite Materials

2.1. Types of Lightweight Material

Due to greater stiffness, lighter weight, and higher strength, composite materials are being used as a solution for making lighter-weight vehicles. Examples of various composite materials used for the manufacturing of EV components include carbon fiber-reinforced polymers (CFRP), glass fiber-reinforced polymers (GFRP), natural fibers, and metal matrix composites (Busarac et al., 2022). Similarly, polypropylene, polycarbonate, polyethylene, and polyethylene terephthalate are examples of essential plastic materials often used in EV components (Wazeer et al., 2023). Equivalently, foam materials like polyurethane, polystyrene, and polyethylene are widely used in different vehicles to enhance passenger and vehicle safety by increasing shock resistance, controlling vibrational damping, and providing higher insulating properties (Murlidhar et al., 2018).

2.2. Different manufacturing techniques

Different processes are used for manufacturing different parts of an EV using a variety of lightweight materials. Die casting, sand casting, investment casting, and other casting techniques are frequently used to produce metal components, providing flexibility and accuracy while making complex designs using molding techniques (Wazeer et al., 2023). Extrusion, predominantly utilized for aluminum and magnesium alloys, enables the manufacture of long, uniform cross-sections with complex geometries and is ideal for structural elements within EVs (Busarac et al., 2022). A crucial process in manufacturing polymer composites is compression molding, which involves compressing fiber reinforcements and resin into a mold to create strong, lightweight parts for certain electric vehicle applications. For generating complex designs and high-volume manufacturing of electric vehicle parts, injection molding is frequently employed as it can be used for molding plastic components such as thermoplastic materials that offer unmatched versatility. A chemical or physical nature of blowing solvents can be used in foaming techniques to help polymer resins expand and form lightweight foam structures. These structures can provide impact absorption, vibration damping, and insulation, which are crucial for electric vehicle interior spaces and structural components (Busarac et al., 2022).

3. Structural and Mechanical Performance

3.1. Lightweight Design Principle

Lightweight design principles in EV structures involve selecting high strength-to-weight ratio materials like CFRP and AFRP to reduce structural mass, optimizing structures for strength with lightweight materials, prioritizing multifunctionality to enhance efficiency without adding weight through capabilities like energy harvesting and thermal management, and considering the environmental impact of composite materials in the design process.

3.2. Strength-to-weight Ratio

Many advanced composite materials offer exceptionally high strength-to-weight ratios. These materials can reduce the structural mass of different EV components without affecting their mechanical properties. The approximate value of the strength-to-weight ratio of different multifunctional composite materials, along with comparative crashworthiness and impact resistance, is provided in Table 20.1.

Table 20.1: Properties of multifunctional composite material.

Material	Strength to weight ratio (kN-m/kg)	Crashworthiness	Impact resistance	Application
Aluminium alloy	50–100	Moderate	Good	Body frame, chassis, suspension, wheel
Magnesium alloy	100–200	Moderate	Moderate	Steering wheels, seat, housing for electronic system
Glass fiber reinforced polymer (GFRP)	100–200	Excellent	Good	Body panel, door panel, dashboard, aerodynamic enhancement
Basalt fiber-reinforced polymer (BFRP)	120–250	Good	Good	Floor pan, vehicle frame, linkage, suspension components, noise and vibration damping system
Titanium alloy	150–300	Excellent	Excellent	Linkage, Suspension parts, intake valves, components for thermal management
Aramid Fiber Reinforced Polymer (AFRP)	150–300	Excellent	Excellent	Protective covering for high voltage components, Energy absorbing parts.
Carbon Fiber Reinforced Polymer (CFRP)	200–400	Good	Moderate	Integrated thermal management system, Body panel, chassis, suspension, battery enclosure and housing

4. Thermal Management

Zou et al. (2018) suggested that to achieve optimum heat charge/discharge performance of lithium-ion power batteries in thermal management, a composite phase change material (PCM) with 1.0% carbon additives should be used. This PCM combination is more thermally conductive and therefore has the potential for mitigating sudden temperature rises.

Further, Wang et al. (2024) produced an organic composite PCM from sodium acetate trihydrate as its source material. The resultant thermal conductivity is 4270 W/mK and has a latent heat value of 1545 J/g.

4.1. Future Trends and Challenges

The market for electric motors is developing quicker than anticipated on an international scale. The main issues are lightweight and thermal battery runways, which can be resolved by using low-weight components with improved flame retardancy and a high level of safety (Geisbauer et al., 2021). Though the use of lightweight composites enhances efficiency, other factors to be considered include good thermal conductivity, dimensional balance at extended temperatures, effective cooling of motor-associated additives, and recycling to minimize environmental impact (Goka et al., 2023). Recycling and proper disposal of composite materials is an issue and a challenging job (Singh et al., 2023). Low-weight composite materials are costly, less available, and impeded by regulations (Sundar et al., 2024). To optimize the sustainability benefits of composite materials in EV automobiles and reduce their environmental impact, it is crucial to address their future trends (Mahale & Patra, 2024).

5. Conclusion

The comprehensive review presented in this study underscores the transformative potential of multifunctional composite materials in revolutionizing EV structures' designs and efficiencies. By examining polymer composites, metal composites, hybrid composites, and their various applications in detail, the authors provide a roadmap for car manufacturers to fully exploit these advanced materials. To overcome long-standing challenges facing the EV sector, one possible approach would be to leverage the combination of lightweight and high-strength properties inherent in polymer composites, as well as thermal management features present in metal composites. This strategic incorporation of these multifunctional materials into EV structures will significantly improve performance, safety, and overall efficiency, thereby ushering in a new era of sustainable mobility.

References

[1]   Busarac, N., Adamovic, D., Grujovic, N., & Zivic, F. (2022). Lightweight materials for automobiles. *Materials Science and Engineering*, 1271, 012010.

[2]   Geisbauer, C., Wohrl, K., Lott, S., Nebl, C., Schweiger, H. G., Goertz, R., & Kubjatko, T. (2021). Scenarios involving accident-damaged electric vehicles. *Transportation Research Procedia*, 55, 1484–1489.

[3]   Goka, S., Moinuddin, S. Q., Dewangan, A. K., Cheepu, M., & Kantumuchu, V. C. (2023). Battery management system for electric vehicles. In *The Future of Road Transportation* (pp. 177–195). CRC Press.

[4]   Mahale, M. D., & Patra, B. B. (2024). Efficient design, materials, and specifications of electric motors used in electric vehicle challenges. In *Energy Efficient Vehicles* (pp. 126–155). CRC Press.

[5]   Murlidhar, P., et al. (2018). Lightweight composite materials for automotive—A review. *International Research Journal of Engineering and Technology (IRJET)*, 1(2500), 151.

[6]   Singh, R., Prakash, N., Babu, G. R., Begum, A. Y., Ahmad, A., & Saikumar, P. J. (2023). Nanotechnology on energy storage: An overview. *Materials for Sustainable Energy Storage at the Nanoscale*, 385–400.

[7]   Sundar, L. S., Mir, M. A., & Ashraf, M. W. (2024). Graphene and graphene-based composite nanomaterials for rechargeable lithium-ion batteries. In *Nanomaterials for Energy Applications* (pp. 186–225).

[8]   Wang, Z., He, Y., Cheng, G., & Tang, T. (2024). Thermal characteristics of a flame-retardant composite phase change material for battery thermal management. *Applied Thermal Engineering*, 122659.

[9]   Wazeer, A., et al. (2023). Composites for electric vehicles and automotive sector: A review. *Green Energy and Intelligent Transportation, 2*, 100043.

[10] Zhang, W., & Xu, J. (2022). Advanced lightweight materials for automobiles: A review. *Materials & Design, 221*, 110994.

[11] Zou, D., Ma, X., Liu, X., Zheng, P., & Hu, Y. (2018). Thermal performance enhancement of composite phase change materials (PCM) using graphene and carbon nanotubes as additives for potential application in lithium-ion power battery. *International Journal of Heat and Mass Transfer, 120*, 33–41.

CHAPTER 21

Use of Recycled Concrete Aggregates in Construction

**Tapas Ranjan Baral[a,*], Aditya Ranjan Patra[b], Ujjapana das[c],
Sujit Kumar Pradhan[a]**

[a,*]Research Scholar, BPUT Rourkela, Civil Engineering Department,
Indira Gandhi Institute of Technology, Dhenkanal, Odisha, India
[b]Civil Engineering Department, Bhubanananda
Orissa School of Engineering, Cuttack, Odisha, India
E-mail: patra.aditya134@gmail.com
[c]Civil Engineering Department, Gandhi Institute for
Education & Technology, Khordha, Odisha, India
Corresponding author: tapas1040baral@gmail.com

Abstract

The increase in the number of concrete structures has led the construction industry to face a shortage of natural aggregates (NA). Every year, 40% of total waste is produced by the construction industry, causing environmental pollution and landfill issues. A resolution to this issue involves utilizing recycled concrete aggregates (RCA) instead of NA in construction. To identify the optimal mix proportions in M40 grade concrete, mix design was done by replacing NA with different percentages of RCA (10%, 20%, and 30%). Fresh and hardened properties of conventional concrete and recycled aggregate mixed concrete were evaluated. Various tests, such as compressive strength test, split tensile strength test, and flexural strength test, were performed on the concrete samples for 3, 14, and 28 days to determine optimum strength. This paper focuses on the laboratory evaluation of NA and RCA materials and determining the percentage of RCA to be used with virgin aggregates.

Keywords: RCA, NA, compressive strength, split tensile strength, flexural strength.

1. Introduction

Recycled concrete aggregates (RCA) are formed by collecting the exploited concrete and breaking it up. The utilization of RCA in new pavement works and structural sectors is still a comparatively developed method. In recent years, various attempts have been made to introduce RCA, which plays a vital role in replacing unused NA in different construction works. Making construction more "green" and environmentally friendly is the primary reason for using RCA in structures. To achieve this objective, different strength characteristics in terms of compressive strength (CS), split tensile strength (SS), and flexural strength (FS) at 3, 14, and 28 days were carried out to evaluate the performance of blends with different percentages of RCA: (a) 10% RCA + 90% NA, (b) 20% RCA + 80% NA, (c) 30% RCA + 70% NA, and comparing these outcomes with those obtained from a blend composed entirely of 100% NA.

DOI: 10.1201/9781003596776-21

2. Literature review

The recycling and reuse of concrete products can be an effective way to achieve sustainability in the construction sector. As a result, many governments worldwide have recently recommended numerous measures aimed at reducing the utilization of NA and expanding the recycling of concrete waste for reuse as aggregates wherever technically, economically, or environmentally feasible. Aggregates form the framework for concrete, typically accounting for about 70% of the total volume of concrete. A significant portion of these aggregates is used in structural construction (Kessel et al. 2020; Naouaoui et al. 2021). The use of RCA in structural sectors reduces the need for NA, which in turn diminishes the negative environmental impact caused by the extraction of original aggregates. The shortage of NA and the growing demands for landfill space have also encouraged the use of RCA in concrete (Ahmed et al. 2020; Batham & Akhtar 2019). Enhancements in the strength of RCA can be achieved by adopting a double mixing technique for concrete with a significant water-binder ratio. Air-dried aggregate concretes exhibited the highest CS (Vishnu et al. 2019; Upshaw & Cai 2020). It has been observed that the CS decreases in both nominal and modified concrete with the use of RCA about the water-cement ratio. Many researchers (Tayeha et al. 2020) have successfully studied the incorporation of RCA in construction for rigid pavement sections.

2.1. Materials and methodology

2.1.1 Materials

Recycled concrete aggregates, coarse aggregates, fine aggregates with appropriate gradation (Figure 21.1), Portland pozzolana cement (PPC), and superplasticizers were used. The use of superplasticizers permitted a reduction in water content by 30% or more. For this research work, the superplasticizer Sikaplast 3069 was used. Ultratech super (PPC) cement was used for this experiment, which was collected from the nearest cement store.

Table 21.1: Physical properties value of aggregates.

Tests	Results (NA)	Results (RCA)	Accepted value as per IS 383 and IS 2386-163
Specific gravity	3.0	2.47	between 2.5–3.0
Water absorption	0.29%	0.31%	between 0.1%–2%
Aggregate impact value	18.55%	25.83%	between 10 %–20 (strong)
Crushing value	16.23%	24.80%	not more than 30%
Abrasion value	17.10 %	25.12 %	less than 35%

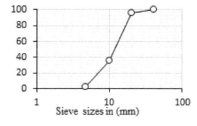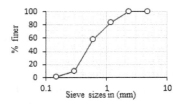

Figure 21.1: Gradation curves of course and fine aggregates

2.1.2 Methodology

The slump value of concrete was monitored from batch to batch to ensure consistent quality throughout the construction process, specifically evaluated for the M40 grade designation as per IS 456:2000. The study was conducted in three steps: samples of concrete cubes, cylinders, and beams were taken and materials were tested; then, test results were determined as per IS 383:1970. For the experiment, the samples were prepared by mixing varying percentages of RCA concerning NA (0%, 10%, 20%, and 30%). Concrete mix design was conducted as per IS 10262:2009.

- Specific gravity value of cement, coarse aggregates (CA), and fine aggregates (FA): 2.69, 3.01, and 2.66, respectively.
- Cement content: 425.14 kg/m³.
- Cement, water, and FA: 425.14, 148.8, and 637 kg, respectively.
- CA of size 20 mm and 10 mm: 803 and 535 kg, respectively.
- Admixture: 0.1% by weight of cement = 4.25 kg.

Mix proportion: The water:cement:FA:CA ratio was found to be 0.35:1:1.498:3.148:0.01, respectively.

2.2. Results and Discussion

It is clear from the outcome data that the different physical properties results and mix design results are within the limits of Indian Standard authorized values, as shown in Table 21.1. Therefore, it can be used as concrete material in the construction sector. The results of identical experimental scrutiny are shown in Table 21.2 and the subsequent sections.

2.2.1 Compressive Strength Test

The compressive strength value (CSV) of cement concrete decreased with an increase in RCA percentage. The strength increased with an increase in the curing period, as shown in Figure 21.2. The CSV increased by 12% after 14 days of curing and by 27.16% after 28 days. The optimal RCA substitution percentages for CSV were found to depend on the mix's water-cement ratio. Up to 20% is safe to use in a concrete mix, but 30% RCA does not provide enough compressive strength for construction work.

2.2.2 Split Tensile Strength Test

The split tensile strength value (STV) of cement concrete decreased with an increase in RCA percentage. The strength increased with an increase in the curing period. The STV increased by 13% after 14 days of curing and by 19.68% after 28 days. The optimal RCA substitution proportions for STV were found to depend on the mix's water-cement ratio. Up to 20% is safe to use in a concrete mix, but 30% RCA showed a significant decrease.

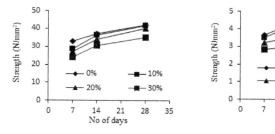

Figure 21.2: Variation curves of CST and FST

Table 21.2: compressive, tensile, and flexural strength test results.

Mix (%) NA + RCA	7 days value in (N/mm²)			14 days value in (N/mm²)			28 days value in (N/mm²)		
	CS	TS	FS	CS	TS	FS	CS	TS	FS
(100%) + (0%)	33.02	3.15	3.67	37.17	3.57	4.24	42.42	3.77	4.52
(90%) + (10%)	28.84	2.86	3.48	36.69	3.12	4.11	41.79	3.46	4.44
(80%) + (20%)	27.15	2.30	3.22	34.02	2.46	3.51	40.30	2.67	3.82
(70%) + (30%)	24.13	1.69	2.82	30.64	1.97	3.0	35.28	2.18	3.12

2.2.3 Flexural Strength Test

The flexural strength value (FSV) of concrete decreased with partial replacement of RCA, as shown in Figure 21.2. However, FSV increased by 15.53% after 14 days of curing and by 23.16% after 28 days of curing. Therefore, it was concluded that strength increased with longer curing periods. The optimal RCA substitution proportions for FSV were found to depend on the mix's water-cement ratio. Up to 20% of the mix is safe to use in concrete, while 30% of the mix substantially reduces its usability for the construction industry.

3. Conclusion

In conclusion, the experimental investigation yielded the following key points: This study examines the use of RCA as a waste material in M40 grade concrete and demonstrates its suitability as a replacement material in concrete work. Partial substitution of RCA improves the CSV, STV, and FSV of concrete up to an optimal substitution level. Additionally, the increasing use of RCA in the construction sector, as discussed in the study, is a positive step towards an eco-friendly environment, as RCA is the largest waste material generated from structural demolitions. The utilization of RCA in the construction sector significantly saves construction costs, energy requirements, and the cost of transportation of natural resources and excavation. It also helps in reducing the impact of different waste materials on the atmosphere and environment.

References

[1]   Ahmed, H., Tiznobaik, M., Huda, S. B., Islam, M. S., & Alam, M. S. (2020). Recycled aggregate concrete from large-scale production to sustainable field application. *School of Engineering, The University of British Columbia, Kelowna, BC, Canada.*

[2]   Batham, G., & Akhtar, S. (2019). Recent innovations in recycled concrete aggregate: A review. *Department of Civil Engineering, University Institute of Technology, Rajiv Gandhi Proudyogiki Vishwavidyalaya, Bhopal (M.P.) 462033, India.*

[3]   IS 12269:2013. Ordinary Portland Cement 53 Grade Specification.

[4]   IS 456:2000. Indian Standard methods for Plane and Reinforced Concrete.

[5]   IS 10262:2009. Guidelines for Concrete Mix Design Proportioning.

[6]   Kessal, O., Belagraa, L., Noui, A., Maafi, N., & Bouzid, A. (2020). Performance study of eco-concrete based on waste demolition as recycled aggregates. *Department of Civil Engineering, Mohammed El-Bachir Ibrahimi University of Bordj Bou Arreridj, Algeria.*

[7]   Naouaoui, K., Bouyahyaoui, A., Cherradi, T., & Mohammadia. (2021). Durability of recycled aggregate concrete. *School of Engineers, Mohamed V University Agdal, Rabat.*

[8]   Tayeha, B. A., Saffar, D. M. A., & Alyousef, R. (2020). The utilization of recycled aggregate in high-performance concrete: A review. *Civil Engineering Department, Faculty of Engineering, Islamic University of Gaza, Gaza, Palestine.*

[9]   Upshaw, M., & Cai, C. S. (2020). Critical review of recycled aggregate concrete properties, improvements, and numerical models. *Dept. of Civil and Environmental Engineering, Louisiana State Univ., Baton Rouge, LA 70803.* DOI: 10.1061/(ASCE) MT.1943-5533.0003394.

[10]   Vishnu, A., Moosvi, S. A., Abiraagav, & Ponmalar, V. (2019). Experimental research on the strength of concrete prepared by using coarse aggregate from concrete debris. *Department of Civil Engineering, Kumaraguru College of Technology, Coimbatore, Tamilnadu, India.* DOI: 10.35940/ijeat.F1275.0986S319

CHAPTER 22

River water principle component analysis

Harinakshi C.*, Ananya Shetty, Sapna G. Acharya, Vageesh, Vignesh V.

Department of ISE, Sahyadri College of Engineering and Management Mangalore, Karnataka, India

Corresponding author: harinidece25@gmail.com

Abstract

RIWAPAC (River Water Principal Component Analysis) is a smart system designed to efficiently monitor water quality at a lower cost. It integrates different sensors and a communication method called Long Range (LoRa) into a single setup. This system monitors key water quality factors such as dissolved solids, electrical conductivity, temperature, and pH while providing real-time location data. The hardware in RIWAPAC is carefully designed to deliver accurate data without excessive power consumption, with plans for future testing in various environments to ensure its effectiveness and cost-efficiency. Additionally, we are working on improving the analysis of hyperspectral data from water with many colors using advanced computational methods, particularly for urban rivers. RIWAPAC aims to facilitate real-time water quality monitoring.

Keywords: TDS, pH, Water quality index, Water, Lo-Ra, Machine Learning

1. Introduction

Water quality degradation is a pressing issue in today's industrial era, leading to various national and global efforts for remediation. The focus on water quality monitoring has grown significantly, especially with the advent of IoT solutions that emphasize sensing and data transmission. Cellular-based approaches, like GSM, 4G, and LTE, have played a role in improving water quality monitoring, but they suffer from reliability and power inefficiency issues due to battery limitations. Recognizing the need for a better solution, long-range (LoRa) (Rose et al., 2020) technology has emerged as a reliable option for low-data transmission. However, existing solutions often neglect the importance of real-time localization. In response, our project, RIWAPAC (River Water Principal Component Analysis), aims to introduce an integrated and cost-effective solution for river water quality monitoring.

2. Literature review

2.1 Usage of various sensors

TEMPSENSE is a cost-effective solution that addresses the limitations of existing water quality monitoring systems. Baghel et al. (2022) discussed combining processing units,

DOI: 10.1201/9781003596776-22

sensors, and Long-Range transmission into a single platform, offering real-time monitoring of temperature, pH, Total Dissolved Solids (TDS), and Electrical Conductivity (EC), along with location tracking. This system overcomes the reliance on regional wireless networks, making it versatile and dependable.

3. Usage of the Lo-Ra transmission

The paper "An Efficient Real-Time Environment Monitoring by Arduino-Powered Cloud-Based IoT System" (Bogdan et al., 2023) describes the development and implementation of an Internet of Things (IoT) environmental monitoring system that efficiently collects data on temperature, humidity, and other environmental parameters. The system utilizes sensors placed at different locations to predict the behavior of specific areas. It consists of an Arduino UNO board connected to a DHT11 temperature and humidity sensor, and an ESP8266 Wi-Fi module (NodeMCU) for data transmission to a remote cloud storage system (Ajayi et al., 2022). The collected data is stored in the cloud and can be accessed through a mobile application.

4. Methodology and model specifications

RIWAPAC is used to predict water quality and classify it based on usage by measuring TDS, EC, temperature, and pH values effectively. This project aims to predict and classify water quality based on water index levels, where users can identify water quality on index levels using different classification levels and data cleaning methods. Based on the water index level, users can identify the correct usability of water.

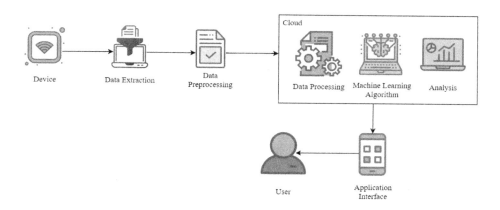

Figure 22.1: Architecture diagram of a RIWAPAC.

The architecture diagram for the proposed system is depicted in Figure 22.1. The system uses a sensor device consisting of water sensors to monitor the water quality index which considers some parameters like pH, TDS, Ec, turbidity, and temperature sensors which reads the data from the microcontroller module and sends the data to cloud through wi-fi module. The data is then extracted from the cloud processed and displayed through an application interface.

4.1 Wireless data transmission

The Arduino Wi-Fi module, specifically ESP8266, facilitates data transmission to the cloud (AWS IoT) using MQTT, from where we can retrieve data (received from the sensors). The system uses NodeMCU for a 3.3V power supply. ESP32 is designed for various applications, offering a successor to the ESP8266 microcontroller. Manufactured by TSMC using a 40 nm process, ESP32 provides a versatile and cost-effective solution for IoT and embedded projects.

4.2 Hardware implementation

The implementation of the RIWAPAC model incorporates several key sensors to collect essential data for water quality analysis. These sensors include the DS18B20 temperature sensor for measuring water temperature, the SEN0161 pH sensor for detecting pH levels, the SEN0244 TDS sensor for measuring total dissolved solids, and the SKU SEN0189 turbidity sensor for assessing water turbidity. Each sensor plays a crucial role in providing accurate and reliable data inputs for the PCA model, enabling comprehensive analysis and classification of water quality parameters.

Figure 22.2: Hardware implementation.

The system utilizes Arduino code running on an ESP32 microcontroller for real-time sensor data collection, including temperature, turbidity, TDS, and pH levels. ESP32 is a series of affordable, energy-efficient microcontrollers with integrated Wi-Fi and dual-mode Bluetooth. Developed by Espressif Systems, these chips feature a Tensilica Xtensa LX6 or LX7 microprocessor, or a single-core RISC-V microprocessor. They include built-in components like antenna switches, power amplifiers, and power management modules. The DS18B20 Temperature Sensor Module is a digital sensor easily connectable to Arduino using a 4.7k-ohm resistor. It follows the 1-wire protocol, enabling multiple sensors to connect through a single wire to Arduino. The sensor provides a digitally accurate output of 12 bits, ensuring a precision of 0.0625°C in temperature measurement. Applications include cold storage, granaries, industrial equipment, air conditioning, and more. It features digital signal output, uses the 18B20 Temperature Sensor Chip, and allows resolution adjustment from 9 to 12 bytes. The temperature measurement range is from –55°C to +125°C with an accuracy of 0.5°C. The pH meter is a cost-effective device for Arduino controllers, ideal for hobby and educational projects. It offers ±0.1 pH accuracy at 25°C, a 0 to 14 pH range, and operates from 0 to 60°C. The sensor features an LED power indicator and a 1m probe cable, making it suitable for embedded design, electronics, and IoT applications.

4.3 Experimental results and discussion

The experimentation framework was meticulously tailored, aligning with user inputs. These inputs include various water quality parameters such as temperature, pH, TDS, and turbidity, given as ranges. Users select the input ranges, which are fed into the backend algorithm to predict water quality and accuracy. The prediction result is then classified according to the water quality index into categories like industrial, drinking, and irrigation use, enabling users to identify appropriate applications. The hardware code involves Arduino sensors (TDS, turbidity, pH) that extract data and send it to the cloud. The sensor data is then connected to the system's UI.

4.4 Test results

The water quality index is the main component this UI section represents because the quality index decides where the water sample can be used. The confusion matrix visually represents a classification model's performance, displaying counts of true positives (TP), true negatives (TN), false positives (FP), and false negatives (FN) for each class. TP indicates correct identification of a class (e.g., recognizing potable water), TN indicates correct rejection of a non-class (e.g., identifying contaminated water correctly), FP indicates incorrect identification (e.g., mistaking clean water for contaminated), and FN indicates failure to identify a class (e.g., missing contamination). Analyzing these values helps assess and improve the model's accuracy in classifying water quality. Table 22.1's confusion matrix breaks down a classification model's performance, highlighting precision, recall, and $F1$-score for each class. Using the Random Forest algorithm, the model shows strong performance for class 0 with high precision (0.83) and recall (0.79), while class 1 has lower precision (0.69) and recall (0.74), indicating improvement areas. The overall accuracy is 0.77, reflecting the model's ability to correctly classify instances across classes. Macro and weighted averages further evaluate the model, accounting for class distribution.

4.4.1. Summary statistics

Table 22.1: Performance.

Class	Precision	Recall	$F1$-score	Support
0	0.83	0.79	0.81	506
1	0.69	0.74	0.72	313
Accuracy	-	-	0.77	819
Macro Avg	0.76	0.77	0.76	819
Weighted Avg	0.78	0.77	0.78	819

5. Conclusion

In conclusion, the water quality monitoring system represents a culmination of efforts, seamlessly integrating various hardware and software elements. The Arduino code, executed on an ESP32 microcontroller, lays the foundation for real-time sensor data collection, encompassing temperature, turbidity, TDS, and pH levels. This data is efficiently transmitted to a Google Sheets document over Wi-Fi, ensuring continuous monitoring and periodic logging. The Raspberry Pi also hosts a Flask application, providing a user-friendly web interface for intuitive data visualization and interaction. This comprehensive integration of Arduino, ESP32, Raspberry Pi, and Flask not only solidifies the system's versatility but also extends its reach to cloud-based storage. The processed water quality metrics can now be securely stored and accessed remotely, marking a significant milestone in creating an adaptable, scalable, and effective solution for continuous water quality assessment. The successful collaboration of hardware and software components opens avenues for future enhancements and applications across diverse settings where real-time water quality monitoring is paramount.

References

[1]   Ajayi, O. O., Bagula, A. B., Maluleke, H. C., Gaffoor, Z., Jovanovic, N., & Pietersen, K. C. (2022). Waternet: A network for monitoring and assessing water quality for drinking and irrigation purposes. *IEEE Access, 10,* 48318–48337.

[2]   Baghel, L. K., Gautam, S., Malav, V. K., & Kumar, S. (2022). TEMPSENSE: LoRa enabled integrated sensing and localization solution for water quality monitoring. *IEEE Transactions on Instrumentation and Measurement, 71,* 1–11.

[3]   Bogdan, R., Paliuc, C., Crisan-Vida, M., Nimara, S., & Barmayoun, D. (2023). Low-cost Internet-of-Things water-quality monitoring system for rural areas. *Sensors, 23*(8), 3919.

[4]   Rose, L., & Mary, X. A. (2020). Sensor data classification using machine learning algorithms. *Journal of Statistics and Management Systems, 23*(2), 363–371.

CHAPTER 23

Fingerprint-based Blood Group Prediction

Advancing Personalized Healthcare

Rahimunnisa Shaik[1], P. Anusha[2], Viswanath Veera Krishna Maddinala[3],
M. Beulah Rani[4], G. Sujatha[5], Sowmya Sree Karri[6]

[1]Assistant Professor, Department of CSE, Vignan Institute of
Engineering for Women (Autonomous), Visakhapatnam, India
Email: rahimunnisa28@gmail.com.
[2]Assistant Professor, Department of CSE (AIML),
BVRIT Hyderabad College of Engineering for Women, Hyderabad, India
Email: anusha.p@bvrithyderabad.edu.in
[3]Assistant Professor, Department of CSE,
Sri Vasavi Engineering College (A), Pedatadepalli, Tadepalligudem, India
Email: krishna.cse@srivasaviengg.ac.in
[4]Distinguished Assistant Professor, Department of CSE,
Maharaj Vijayaram Gajapathi Raj College of Engineering (A), Vizianagaram, India
Email: **beulahrani@gmail.com**
[5]Assistant Professor, Department of CSE, Vignan's Institute of
Information Technology, Visakhapatnam, India
Email: sujathagviit@gmail.com
[6]Senior Java Developer, Spire Software Solutions Ltd, Visakhapatnam,
Andhra Pradesh, India
Email: sowmyasree.dannina@gmail.com

Abstract

Individualized medical treatments require blood group identification. Invasive blood testing limits accessibility and efficiency with traditional methods. The article presents Fingerprint-Based Blood Group Prediction (FB-BGP), a response to a prior study. This method uses modern algorithms to accurately predict blood groups from fingerprint patterns. We tested the proposed approach using rigorous experimentation and cutting-edge machine learning methods, including neural networks and decision trees. These results show that FB-BGP outperforms other approaches and has high accuracy. This new method allows non-invasive and fast blood type determination, with far-reaching ramifications. FB-BGP could also improve personalized healthcare by enabling patient-specific interventions and medicines. This study paved the way for future medical research on fingerprint-based prediction techniques. Because of this, the study tackles the important requirement for proper blood type identification and provides the groundwork for significant advances in tailored medical care.

Keywords: Fingerprint-Based Blood Group Prediction, Personalized Healthcare, Noninvasive Blood Typing, Machine Learning Algorithms

DOI: 10.1201/9781003596776-23

1. Introduction

Customized healthcare relies on blood group determination to create personalized treatments. We need fresh alternatives because traditional invasive blood testing methods are inefficient and inaccessible. The pioneering Fingerprint-Based Blood Group Prediction (FB-BGP) system changes blood typing methods. To meet this vital need, FB-BGP, a new answer to prior research, uses fingerprint patterns to identify people without intrusion. FB-BGP accurately predicts blood types using advanced algorithms and machine learning methods like neural networks and decision trees. Through rigorous experimentation, FB-BGP has proven its superiority over other approaches with greater accuracy rates. With far-reaching effects, FB-BGP can quickly and noninvasively determine blood type. This cutting-edge technology could revolutionize personalized healthcare by allowing patients to get customized interventions and treatments.

FB-BGP also lays the groundwork for medical fingerprint-based prediction research. Previous preparation achieves this. This paper examines FB-BGP's development, methodology, validation, and consequences for tailored healthcare to understand its intricacies. FB-BGP can change healthcare by addressing the critical need for accurate blood type identification and laying the groundwork for significant advances in personalized medical care. This study analyzes and discusses FB-BGP's potential to transform tailored healthcare delivery.

Blood tests typically diagnose human illnesses. Blood regulates life and the body. Identification, authentication, and computer security use biometrics—a person's behavior and appearance. Technology makes computers useful for identification, authentication, and security. Fingerprint sensors are smaller and cheaper than other biometric sensors because of their convenience, differences, and durability. Biometric verification identifies a person. Biometrics include DNA, palm prints, fingerprints, and facial recognition. Banks, corporations, and forensic agencies widely use biometric fingerprint identification to track offenders. The main fingerprint types are loops, whorls, and arches. Loops are found in about 65% of fingerprints. Around 30% of fingerprints are whorls. Simple arches rarely form 5% of a fingerprint. Once biometric data from fingerprint recognition enters a database, it detects and confirms fingerprints. Databases verify fingerprints.

2. Literature Review

The literature study "Fingerprint-based Blood Group Prediction: Advancing Personalized Healthcare" is crucial to biometrics and healthcare. A thorough literature review can reveal the current state-of-the-art methodologies, methodology, and limitations in using fingerprints to predict blood types, a vital aspect of tailored healthcare.

Title	Authors	Year	Journal	Summary
A Review on Fingerprint-Based Health Monitoring Systems Using Machine Learning Techniques	Anupriya Suman, Prashant Kumar, et al.	2021	IEEE Access	It emphasizes the potential of fingerprints as biomarkers and the role of ML algorithms in predicting accuracy.
Fingerprint-Based Blood Group Detection Using Convolutional Neural Networks	Shweta Kumari, Anupriya Suman, et al.	2020	IEEE International Conference on MLDE	It details the experimental setup, dataset, and training process, showing promising accuracy using CNN

Continued

Continued

Title	Authors	Year	Journal	Summary
Blood Group Prediction Using Fingerprint Images and Machine Learning Techniques	K. S. Jagadeesh, S. Siva Sathish, et al.	2019	IEEE Transactions on Medical Imaging	It provides a comparative analysis of these models, discussing preprocessing steps to improve image quality and prediction accuracy.
A Comprehensive Review on Fingerprint Biometrics for Health Monitoring Applications	Nidhi Sharma, Sunil Kumar Jangir, et al.	2021	IEEE Access	This review covers a wide range of health monitoring applications using fingerprint biometrics.
Fingerprint-Based Personalized Healthcare: A Survey	Pranjal Awasthi, Sanjay Kumar	2020	IEEE Transactions on BE	They cover aspects from data acquisition to modeling.
Machine Learning for Blood Group Prediction from Fingerprint Images	Amrita Dey, Shubham Patil, et al.	2018	IEEE International Conference on ACCI	This paper focuses on developing machine-learning models for blood group prediction using fingerprint images
Fingerprint-Based Blood Group Identification: A Review	Ananya Mishra, Prateek Gupta, et al.	2021	IEEE Sensors Journal	It is all about the latest research on fingerprint-based blood group identification.

3. Proposed Methodology

The current method uses multiple linear regression to predict blood group fingerprint feature associations. Before processing, images are grayscale. Single-linear or multiple-linear regression types depend on the number of variables. Linear regression machine learning can predict blood groups from a detailed fingerprint. The current system only verifies images in the dataset. Old methods have drawbacks: a) Larger datasets have lower accuracy. b) Size all photos equally. c) No precise fingerprint image processing is possible.

The proposed system uses grayscale images. RGB images become grayscale. Next, photographs are sized identically. We convert blood group values (A^{+Ve}, A^{-Ve}, etc.) to numbers and insert them into the blood group column of the train data set image. Each image has rows of pixel grayscale values. Total rows = pictures. Pixels define column size. New train. CSV files contain data. We process testing data similarly to training data and store it in a test file. csv. CNN trains its models using test data. Check correctness. We also classify using KNN.

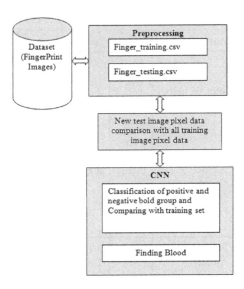

Figure 23.1: Proposed architecture to find blood group.

The architecture's blood group prediction model uses preprocessing and CNNs for reliable classification. Theoretical explanation and formulas follow:

Step 1: Preprocessing: To reduce light and contrast changes, normalize fingerprint image pixel values to [0, 1]. Thresholding greyscale fingerprint photos into binary images highlights ridge patterns and suppresses noise. Minutiae extraction, ridge orientation estimation, and texture analysis can extract discriminative fingerprint features.

Step 2: CNN Layers: The CNN is a type of neural network. CNNs have convolutional, pooling, and fully connected layers.

- **Convolutional Layers:** These layers extract spatial patterns from input images using convolution. Each convolutional layer produces feature maps by convolving filters (kernels) with the input picture.
- **Pooling Layers:** Downsample convolutional layer feature maps to reduce dimensionality and extract the most important characteristics.
- **Fully Connected Layers:** Perform classification tasks, incorporating high-level features from convolutional and pooling layers. The model's nonlinearity is introduced by activation functions like ReLU within these layers.

A CNN's convolution operation is mathematically represented as:

$$Y[i, j] = m\ n\ X[\ i + m, j + n].\ K[m, n] + b.$$

Y is the output feature map, X is the input image, K is the convolution kernel, and b is the bias term. After a convolutional layer, a ReLU activation function produces $Z = \text{ReLU}(Y) = \max(0, Y)$.

Step 3: Blood Group Prediction: After CNN classifies an image as positive for a blood group, the model analyzes the input image to find the exact blood group. The input image's features are compared to those of training set photos from each blood group. The model uses the closest neighbor search or Euclidean distance to determine the most likely blood group based on the input image and training examples. The design uses preprocessing to improve fingerprint pictures and a CNN to predict blood groups, enhancing customized healthcare through biometric technology.

4. Empirical Results

In this section, we try to implement the proposed model using Python programming language. For executing the model, we used the Google Collab platform and discussed results and accuracy parameters.

Calculate the Accuracy of Model: Performance Metrics:

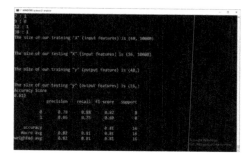

Figure 23.2: Proposed accuracy and performance metrics.

From the above Figure 23.2, we can see the accuracy of our proposed model and also see the performance metrics calculated for our application.

5. Conclusion

The study demonstrated that fingerprint-based blood group prediction can enhance individualized healthcare. Using a CNN model, we achieved 81.2% accuracy in identifying blood types from fingerprint images. This breakthrough underscores the potential of biometric technology to transform medical diagnostics. Future improvements could include exploring new deep learning architectures, such as Transformer-based models or graph neural networks, to enhance accuracy. Additionally, optimizing computational efficiency through model compression and efficient architectures is crucial for real-time applications. Expanding the dataset to include diverse populations can improve model generalization, ensuring reliability across different demographic groups. Integrating fingerprint-based prediction with other biometrics like palm prints or iris scans could provide more comprehensive healthcare diagnostics.

References

[1]   Awasthi, P., & Kumar, S. (2020). Fingerprint-Based Personalized Healthcare: A Survey. *IEEE Transactions on Biomedical Engineering.*

[2]   Dey, A., & Patil, S., et al. (2018). Machine Learning for Blood Group Prediction from Fingerprint Images. In *Proceedings of the IEEE International Conference on Advanced Computing and Communication Informatics.*

[3]   Gupta, A., & Singh, R., et al. (2023). Deep Learning Approaches for Finger-print-Based Health Monitoring Systems: A Comprehensive Review. *IEEE Transactions on Emerging Topics in Computing*.

[4]   Islam, M. R., & Huda, M. N., et al. (2022). Fingerprint-Based Disease Detection and Classification Using Machine Learning Techniques. In *Proceedings of the IEEE International Conference on Bioinformatics and Biomedicine*.

[5]   Jagadeesh, K. S., & Sathish, S. S., et al. (2019). Blood Group Prediction Using Fingerprint Images and Machine Learning Techniques. *IEEE Transactions on Medical Imaging*.

[6]   Kumari, S., & Suman, A., et al. (2020). Fingerprint-Based Blood Group Detection Using Convolutional Neural Networks. In *Proceedings of the IEEE International Conference on Machine Learning and Data Engineering*.

[7]   Mishra, A., & Gupta, P., et al. (2021). Fingerprint-Based Blood Group Identification: A Review. *IEEE Sensors Journal*.

[8]   Patel, H., & Shah, R., et al. (2022). Fingerprint-Based Biometric Authentication for Personalized Healthcare Systems: A Survey. *IEEE Access*.

[9]   Sharma, N., & Jangir, S. K., et al. (2021). A Comprehensive Review on Fingerprint Biometrics for Health Monitoring Applications. *IEEE Access*.

[10]   Suman, A., & Kumar, P., et al. (2021). A Review on Fingerprint-Based Health Monitoring Systems Using Machine Learning Techniques. *IEEE Access*.

[11]   Das, S., & Roy, S., et al. (2020). Enhancing Privacy in Fingerprint-Based Health Monitoring Systems: A Review. *IEEE Transactions on Dependable and Secure Computing*.

CHAPTER 24

Analysis of CT Scan Lung Cancer Image for Early Prediction

Pragnya Das[1], Satya Narayan Tripathy[2]

[1]Department of Computer Science and Engineering,
Centurion University of Technology and Management, Odisha, India
[2]Department of Computer Science, Berhampur University,
Berhampur, Odisha, India
E-mails: pragnya.das@cutm.ac.in, pd.rs.cs@buodisha.edu.in,
snt.cs@buodisha.edu.in

Abstract

In the healthcare industry, images are crucial. Medical imaging was once utilized by healthcare providers to properly diagnose illnesses and give patients nutrition. Complicated medical imaging can occasionally cause diagnosis processes to be difficult for medical practitioners. As a noncommunicable illness, cancer cannot be detected in its early stages. Every year, the number of people who die from incorrect diagnoses rises. A significant obstacle is in the way of the researchers' wanalysis of medical pictures connected to cancer. The early detection of cancer is a significant obstacle that has to be addressed. For this reason, it is imperative to preprocess medical photos to detect important aspects. Various image preprocessing methods have been used on medical pictures of lung cancer in this study. An analysis based on comparisons is carried out.

Keywords: Resize, Filtering, Enhancement, Restoration

1. Introduction

The imaging approach is crucial in a variety of clinical, laboratory, and medical practice settings. Processing the photos can help with quantification, analysis, and computer visualization. The goal of image processing is to improve the quality of images such that both people and robots can understand them with ease. Analyzing various pathological pictures can be used to visualize and monitor patients in real time. The goal of this work is to use Python programming and Open Source Computer Vision (OpenCV) to preprocess medical pictures of lung cancer.

2. Literature Review

Weibin Zhou et al. (2020) have extracted the regions of the iris by integrating the iris recognition algorithm with iris segmentation, followed by the deep learning algorithm. A comparison of the various preprocessing techniques applied in five distinct deep-learning models for breast cancer pictures was carried out by David et al. (2022). Using retinal

DOI: 10.1201/9781003596776-24

fundus images, Swati et al. (2017) conducted an efficiency-based comparison of various filters, including the adaptive median filter, contrast-limited adaptive histogram equalization (CLAHE), Weiner filter, Gaussian filter, and adaptive histogram equalization (AHE). They concluded that the most effective filtering technique for comparing MSE and PSNR measurements is the adaptive median filter. Yihan Xu et al. (2022) proposed a technique called Torch technology to accelerate face images by image sharpening and improving the efficiency of the images. Nongmeikapam et al. (2023) examined diverse picture preprocessing methods, including denoise, image enhancement, registration, and segmentation, and evaluated the efficacy of deep learning approaches in achieving the aforementioned objectives.

3. Methodology and Model Specifications

The planned methodology consists of the following steps: image acquisition, resizing, image filtering, edge detection, image enhancement, and image cropping. The planned methodology is represented in Figure 24.1.

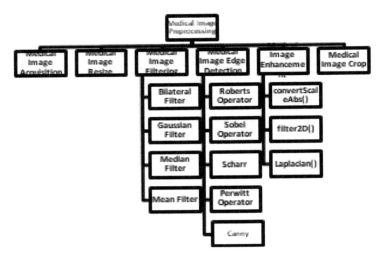

Figure 24.1: Proposed methodology.

4. Medical Image Acquisition

Medical image acquisition is the first step in medical image preprocessing. In this study, CT scans of lung cancer images were collected from an online source, Mendeley. In the lung cancer dataset, 238 cancerous and 126 noncancerous images are available. All the images are in JPG format.

5. Medical Image Resize

Resizing the image from one pixel grid to another is used to interpolate the image. It helps to reduce the size of the image by discarding the pixels. The images are resized to 256×256.

5.1 Medical Image Filtering

It is applied to improve the image's sharpness, edge enhancement, and smoothness. Figure 24.2 shows the impact of several filter methods on a lung CT scan picture.

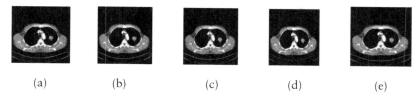

(a) (b) (c) (d) (e)

Figure 24.2: (a) Original image, (b) bilateral filtered image, (c) Gaussian filtered image, (d) Median filtered image, (e) Mean filtered image.

5.2 Medical Image Edge Detection

It facilitates the identification of the image's necessary sections and their removal. Figure 24.3 shows the edge detection of a CT scan picture for lung cancer utilizing the Roberts, Sobel, Prewitt, and Canny operators.

(a) (b) (c) (d) (e)

Figure 24.3: (a) Original image, (b) Roberts operator edge detection, (c) Sobel operator edge detection, (d) Prewitt operator edge detection, (e) Canny edge detection,

5.3 Medical Image Enhancement

The process of controlling contrast, brightness, and sharpness, as well as lowering noise, increasing smoothness, and correcting picture abnormalities, is known as medical image enhancement. A CT scan picture of lung cancer that has been enhanced to improve contrast, brightness, and sharpness is displayed in Figure 24.4a–d.

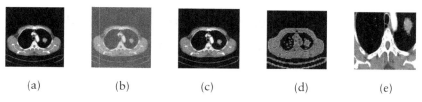

(a) (b) (c) (d) (e)

Figure 24.4: (a) Original image, (b) image with adjusted contrast and brightness, (c) Sharpness image, (d) Laplacian-based sharpness image, and (e) cropped image.

5.4 Medical Image Crop

Medical image cropping is a procedure to edit the medical image to have more focus on an affected part by removing the other unnecessary regions from the image. Cropping of a lung cancer CT scan image is done to retrieve the affected portion (Figure 24.4e).

5.5 Result and Analysis

In this study, the peak signal-to-noise ratio (PSNR) value is utilized for measurement purposes. An image will have better quality if it has a high PSNR value and less noise. Equation 1 shows the relationship between PSNR, noise, and image quality.

$$\text{PSNR } \alpha \ \frac{1}{\text{Noise}} \ \alpha \ \text{Image Quality} \tag{1}$$

The PSNR values calculated for different image preprocessing techniques are presented in Table 24.1.

Table 24.1: PSNR values for Image preprocessing Techniques.

Technique	Method	PSNR
Filter	Bilateral	40.14504987311888 dB
	Gaussian	36.10712972863825 dB
	Median	37.532887449298336 dB
	Mean	34.946976865961446 dB
Edge Detection	Roberts Operator	9.731004634308851 dB
	Sobel Operator	9.731610727240348 dB
	Scharr	9.731695171663265 dB
	Perwitt Operator	9.731528787262544 dB
	Canny	34.32163041019607 dB
Enhancement	convertScaleAbs()	31.316187000182197 dB
	filter2D()	34.75339437342552 dB
	Laplacian()	8.31928978131762 dB

After analyzing the PSNR values, it has been found that the Canny edge detection technique, bilateral image filtering technique, and image enhancement using the filter2D() function have higher PSNR values of 34.32, 40.14, and 34.75 dB, respectively.

6. Comparison of Proposed Model with Existing Models

Various preprocessing approaches have been employed by researchers to analyze the photos in their investigations. Compared to other technologies, OpenCV is more dependable and intuitive to use when analyzing photos, which is why it is included in this suggested technique.

7. Conclusion

Preprocessing of medical images is an important part of image processing. It is used to enhance the image quality, which is helpful for better diagnoses. This study has applied different image preprocessing techniques on the CT-scanned medical image of lung cancer. Based on the PSNR value, the appropriate technique for performing preprocessing has been determined. Bilateral image filtering, Canny edge detection technique, and image enhancement using the filter2D method are considered most suitable for the preprocessing of medical images to improve their quality.

References

[1] Das, S., Roy, S., et al. (2020). Enhancing privacy in fingerprint-based health monitoring systems: A review. *IEEE Transactions on Dependable and Secure Computing*.

[2] Murcia-Gómez, D., Rojas-Valenzuela, I., & Valenzuela, O. (2022). Impact of image preprocessing methods and deep learning models for classifying histopathological breast cancer images. *Applied Sciences, 12*(1), 1–18.

[3] Singh, N. T., Kaur, C., Chaudhary, A., & Goyal, S. (2023). Preprocessing of medical images using deep learning: A comprehensive review. In *2023 Second International Conference on Augmented Intelligence and Sustainable Systems (ICAISS)*. IEEE Xplore.

[4] Swathi, C., Anoop, B., Dhas, D. A. S., & Sanker, S. P. (2017). Comparison of different image preprocessing methods used for retinal fundus images. In *Proceedings of the 2017 Conference on Emerging Devices and Smart Systems (ICEDSS)*, Tiruchengode, 175–179.

[5] Xu, Y., Ren, H., Peng, H., Wang, J., & Liang, Z. (2022). Research of face image preprocessing and classification based on torch module. In *2022 IEEE 6th Advanced Information Technology, Electronic and Automation Control Conference (IAEAC)*.

[6] Zhou, W., Ma, X., & Zhang, Y. (2020). Research on image preprocessing algorithm and deep learning of iris recognition. In *2020 International Conference on Computer Science and Communication Technology (ICCSCT)*, 1621 012008.

CHAPTER 25

Weather Forecasting of a Smart City Using Univariate Time Series Statistical Models

Saina Ashok Sahoo, Medisetti Narendra, Vishal Mishra, Rupak Raj, Biswa Ranjan Senapati, Sipra Swain

Department of Computer Science and Engineering, Siksha 'O' Anusandhan (Deemed to be) University, Bhubaneswar, Odisha, India
E-mail: sainasahoo1234@gmail.com, biswa.rnjn@gmail.com

Abstract

Weather forecasting is crucial for effective resource management and infrastructure planning in smart cities, necessitating accurate predictions of future weather conditions. However, the inherently complex nature of weather phenomena, including nonlinear parameters such as rainfall, humidity, wind speed, and temperature, poses significant challenges to forecasting efforts. This difficulty is compounded by the unpredictable nature of weather changes. To address this, various time series forecasting models, including ARIMA, SARIMA, Prophet, and Triple Exponential Smoothing models, are employed to develop a reliable weather forecasting system tailored to smart city environments. By utilizing historical weather data and patterns obtained from publicly available datasets like Kaggle, these models strive to improve the precision of weather forecasts by employing evaluation criteria such as mean absolute error, mean squared error, mean absolute percent error, and root mean square error. The outcomes of this research not only promise improved weather predictions for smart cities but also offer potential applications in tourism, natural resource management, agriculture, and smart city infrastructure development.

Keywords: Triple Exponential Smoothing (TES); Forecasting; Prophet (Facebook Prophet); Statistical Models

1. Introduction

The utilization of science and technology to predict the condition of the atmosphere for a future date at a specific place is known as weather forecasting. Weather forecasting is needed for smart cities to support urban planning, resource management, public safety, environmental sustainability, energy management, transportation optimization, agriculture, and overall quality of life for residents (Senapati et al., 2020; Senapati et al., 2021). The most often used type of weather model data is numerical weather prediction (NWP) data. Numerical weather prediction computer models examine recent weather data to anticipate future weather. Traditional forecasting systems highly depend upon current weather data to make predictions about future weather (Barrera-Animas et al., 2022). However, this

DOI: 10.1201/9781003596776-25

dependence on current data may not always capture the full range of factors that influence weather patterns, leading to inaccuracies in weather forecasts and rapid changes in urban environments. For urban planners and other decision-makers who depend on weather forecasts for vital infrastructure and emergency management, these variations can have significant adverse effects on forecast accuracy (Huang-Lachmann, 2019). To overcome these limitations, time series models are used because they can record intricate temporal patterns of data. Unlike traditional forecasting, time series models can handle nonlinear relationships in data, enabling high-resolution forecasts. Many time series models like ARIMA, SARIMA, PROPHET, Exponential Smoothing, Seasonal Decomposition of Time Series, Vector Autoregression, etc., are developed to improve the efficiency and accuracy of time series prediction (Dama & Sinoquet, 2021). The motivation behind including statistical time series models in weather forecasting for smart cities is driven by the need for more accurate, efficient, and adaptable forecasting solutions. Statistical time series models are often more cost-effective than traditional numerical models, both in terms of computational resources and maintenance. The main intention of this research is to create a model that includes this intelligent conduct. The parameters that define weather conditions, such as maximum or lowest temperature, etc., vary constantly over time to produce a time series of each parameter.

To evaluate the precision of the forecast and compare the results of the two models, the mean absolute error (MAE), mean squared error (MSE), mean absolute percent error (MAPE), and root mean square error (RMSE) are calculated. These metrics are used to analyze the effectiveness of the models and determine the most suitable forecasting model. The primary contribution of this research is detailed below:

1. Prediction of temperature using statistical time series models such as Exponential Smoothing and Prophet Model.
2. Comparison of the above-mentioned models based on the generic parameters like MAE, MSE, MAPE, and RMSE.

2. Literature Review

In this literature survey, we explore the transformative impact of machine learning and statistical machine learning in two vital domains: smart city development and weather forecasting. By dissecting the complex roles of ML and SML, we unveil the implications for optimizing urban processes, enhancing decision-making frameworks, and advancing predictive capabilities of weather forecasting, paving the way for sustainable urbanization and community resilience. Machine learning, a subdivision of artificial intelligence, enables computers to acquire knowledge from data and autonomously generate judgments or forecasts without the need for explicit programming for particular tasks (Senapati et al., 2023). Statistical machine learning emphasizes probabilistic models and inferences, controlling statistical methods for extracting insights from data. These approaches have transformed the industry and offer better solutions for complex problems. Smart cities, incorporating technology and urbanization, improve ML technologies to optimize urban processes. ML algorithms play an important role in traffic system management by predicting traffic blockages, optimizing routes, and improving transportation efficiency. Similarly, ML-driven energy grids optimize energy consumption. Statistical machine learning techniques approach smart city applications by providing robust probabilistic frameworks for decision-making and inference. Time series analysis techniques, a subset of SML, play a crucial role in forecasting urban phenomena such as air quality, population growth, and economic trends, empowering policymakers with actionable insights for interventions.

3. Proposed Model

Univariate statistical models focus on analyzing the variation in a single variable without considering the relationships or interactions with other variables. In statistical analysis, univariate analysis is frequently the first step and is used to characterize and describe the properties of the variable being studied. Examples of univariate statistical time series models are Autoregression (AR), Moving Average (MA), ARIMA, SARIMA, Prophet Model, Triple Exponential Smoothing (TES), and LSTM, etc. In this study, we used triple exponential smoothing and prophet models to forecast weather because the exponential smoothing method is relatively simple and easy to understand compared to other forecasting models, adaptability, computational efficiency, good for short-term forecasting, and easy to implement. Regarding the prophet model, it is flexible, has the feature of automatic selection, resilience against outliers and missing data, interpretability, and scalability.

4. Prophet Model

Prophet was made available as an open-source program by Facebook's core data science team in early 2017 (Oo & Sabai, 2020). The forecasting tool, available for both Python and R, is designed to help developers, researchers, and analysts set goals more effectively and allocate resources over extended periods. The model employed is divisible into three primary components: trend, seasonality, and holidays. The following equation combines these elements:

$$y(t) = g(t) + s(t) + h(t) + \varepsilon_t \tag{25.1}$$

$g(t)$: a logistic or piecewise linear growth curve that can be used to model time series changes that are not periodic.
$s(t)$: recurring variations, like weekly or annual seasonality.
$h(t)$: impacts of user-provided holidays that adhere to erratic scheduling.
ε_t: Any anomalous changes not captured by the model are taken into account via an error term.

5. Exponential Smoothing Model

Exponential smoothing is a forecasting technique for univariate time series data (Svetunkov et al., 2022). There are three primary types of exponential smoothing: simple, double, and triple exponential smoothing. The triple exponential smoothing method, also known as Holt-Winters Exponential Smoothing, is named after its developers, Charles Holt and Peter Winters. This method is divided into two categories based on the seasonal component's nature:

- Holt-Winter's Additive Method: used for additive seasonality.
- Holt-Winter's Multiplicative Method: used for multiplicative seasonality.

Performance Metrics

Four performance metrics are considered for the proposed work as mentioned in the Introduction section. The mathematical representation of those metrics are as follows:

$$\text{MSE} = \frac{1}{n} \sum_{i=1}^{n} \left(Y_i - \widehat{Y}_i \right)^2 \tag{25.2}$$

$$\text{RMSD} = \sqrt{\frac{\sum_{i=1}^{N}\left(x_i - \hat{x}_i\right)^2}{N}} \tag{25.3}$$

$$\text{MAE} = \frac{\sum_{i=1}^{n}\left|y_i - x_i\right|}{n} \tag{25.4}$$

$$\text{MAPE} = \frac{1}{n}\sum_{i=1}^{n}\left|\frac{A_i - F_i}{A_i}\right| \tag{25.5}$$

Simulation Result

The trend analysis of the used data set is shown in Figure 25.1.

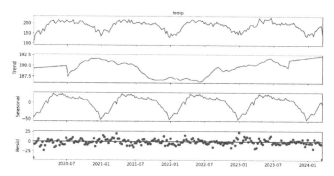

Figure 25.1: Trend analysis of the data set.

The performance metrics result is mentioned in Table 25.1.

Table 25.1: Performance metrics result.

Model	MSE	RMSE	MAE	MAPE
Prophet	2.19	1.48	1.16	4.54
Holt-Winters	154.64	12.44	7.6	3.54

6. Conclusion

Our study focused on the prediction of weather using univariate time series statistical models, especially on exponential smoothing and Prophet Models. The results provided some valuable insights into the predictive capabilities of the models. Both models are capable of forecasting weather patterns. After comparing all the parameters such as MSE, MAE, RMSE, and MAPE for both models, the prophet model is considered the best choice for the evaluation of error metrics. There is some scope for exploring better alternative statistical models other than exponential smoothing and prophet models. Other time series models or machine learning techniques like neural networks could also perform well and improve the precision and reliability of the weather forecasting model.

References

[1] Barrera-Animas, A., Oyedele, L., Bilal, M., Akinosho, T., Delgado, J., & Akanbi, L. (2022). Rainfall prediction: A comparative analysis of modern machine learning algorithms for time-series forecasting. *Machine Learning with Applications, 7*, 100281.

[2] Dama, F., & Sinoquet, C. (2021). Time series analysis and modeling to forecast: A survey. *arXiv preprint arXiv:2104.00164.*

[3] Huang-Lachmann, J.-T. (2019). Systematic review of smart cities and climate change adaptation. *Sustainability Accounting, Management and Policy Journal, 10*(4), 745–772.

[4] Oo, Z., & Sabai, P. (2020). Time series prediction based on Facebook Prophet: A case study, temperature forecasting in Myintkyina. *International Journal of Applied Mathematics, Electronics and Computers, 8*(1), 263–267.

[5] Senapati, B. R., Khilar, P. M., & Swain, R. R. (2021). Environmental monitoring through vehicular ad hoc network: A productive application for smart cities. *International Journal of Communication Systems, 34*(1), e4622.

[6] Senapati, B. R., Swain, R. R., & Khilar, P. M. (2020). Environmental monitoring under uncertainty using smart vehicular ad hoc network. In S. Bhateja, P. Coelho, & P. K. Srivastava (Eds.), *Smart Intelligent Computing and Applications: Proceedings of the Third International Conference on Smart Computing and Informatics* (pp. 229–238). Springer.

[7] Senapati, B., Khilar, P., Dash, T., & Swain, R. (2023). AI-assisted emergency healthcare using vehicular network and support vector machine. *Wireless Personal Communications, 129*(2), 1929–1962.

[8] Svetunkov, I., Kourentzes, N., & Ord, J. (2022). Complex exponential smoothing. *Naval Research Logistics (NRL), 69*(3), 234–253.

CHAPTER 26

Investigating Slurry Erosion Resistance of HVOF coated SS-409 for Slurry Transportation System

Asisha Ranjan Pradhan[a]*, **Trilochan Pradhan[a]**, **Anjan Kumar Mishra[a]**, **Priyaranjan Panda[a]**

Parala Maharaja Engineering College, Brahmapur, Odisha, India
Corresponding author Email: asisharanjan.pradhan@gmail.com

Abstract

Many slurry transportation equipment components fail due to erosion, shortening their lifespan. Pipeline coatings reduce erosion wear and extend equipment life. High-velocity Oxy-Fuel spraying applied the WC-10Co-4Cr + TiO$_2$ coating to the SS-409. The sand was used as the erodent, and a slurry jet erosion tester was used to test the parameters. It was found that jet velocity outweighed other operating conditions on coated and uncoated material erosion wear. Degraded surfaces were scanned with an electron microscope. The coating greatly improved SS 409's wear resistance. However, the coating showed brittle erosion wear.

Keywords: Slurry Erosion; SS-409; HVOF Coating

1. Introduction

Many industries use slurry pumps to transfer fluid and sediment through pipelines. This environmentally friendly and cost-effective method transports crude oil with sand and pollutants for post-transfer purification (Islam et al., 2016). Erosive wear from solid particles reduces pipeline durability (Alam & Farhat, 2018). Material properties, particle shape and size, and flow dynamics complicate slurry erosion (Desale et al., 2008). For steel component erosion, corrosion experiments must examine particle shape, size, impingement angle, concentration, and collision velocity (Jung & Kim, 2020). Stainless steel pump impellers deteriorate quickly in slurry. Impeller metal erosion wear is extensively studied and coated to improve surface properties (Kumar et al., 2016). Thermal spray coatings like tungsten-based WC-10Co-4Cr resist slurry erosion without harming the environment. Cr and WC have been used to stiffen pump surfaces (Kumar et al., 2012). In particular, HVOF spraying SS 304 with 20% WC powder increased durability (Maiti et al., 2007). Ni–Cr coating 13Cr4Ni steel increased microhardness (Hong et al., 2013). Coating combinations maximize hardness, toughness, and strength. Lee identifies Cr, Fe, Al, and Ni as common rigid components, while Si, Al$_2$O$_3$, and WC serve as harder binders (Lee et al., 2010). Mechanical and erosion resistance improves with WC coatings. Al-Bukhaiti et al. (2007) found that solid percentage, rotating velocity, duration, and particle size distribution affect erosion wear rate. HVOF coatings are corrosion-resistant,

DOI: 10.1201/9781003596776-26

mechanically strong, and have low porosity. Quality must be improved because corrosion-resistant pump impeller steel SS-409 has lower microhardness than other steels.

This study compares erosion wear of SS-409, a common slurry equipment material, with and without coating under different operating conditions. Wear mechanisms and erosive behavior were tested with a slurry jet erosion tester.

2. Experimental Procedure

2.1 Base Material

This study used commercially available SS-409 as its base material. The versatile stainless-steel SS-409 is used to make impellers, pipelines, and fittings for slurry. The SS-409 samples were obtained from M/s Steel Chain in New Delhi, India. The samples were precisely machined into 25 mm^2 with 5 mm thickness using electric discharge machining (EDM) wire cutting.

2.2 Feedstock Powder

Jodhpur-based Metallizing Powders offered WC-10Co-4Cr powder with a particle size range of 20–53 μm for coating SS-409. The desired proportions were achieved by jar mill WC-10Co-4Cr and TiO$_2$ powder for 15 h. Powder coatings were applied to samples using the HVOF method. After thermal spraying, compressed air is used to cool the pieces.

2.3 Erodent Material

The sand was used as the erosive agent in the study. SEM analysis of the sand sample revealed granular surface features, as depicted in Figure 26.1a. Sand particles of sizes varying from 150 to 355 μm were prepared using standard British sieves.

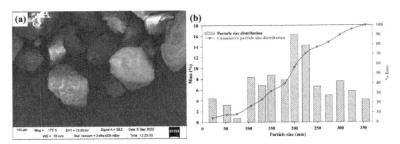

Figure 26.1: (a) SEM of sand particulates (b) PSD of sand particles.

Figure 26.1b illustrates sand's particle size distribution (PSD), with the largest particle size measured at 355 μm. The distribution indicates that 22.64% of particles fall within the 250–355 μm range, 20.98% within the 212-250 μm range, 24.15% within the 150–212 μm range, and only 32.23% are finer than 150 μm.

3. Slurry Jet Erosion Tester

DUCOM TR-411 slurry jet erosion tester was used for erosion wear testing. It consists of control and test-performing units. It has an erodent hopper, water pump, nozzle, specimen holder, and water pressure relief valve. The hopper is filled with erodent and covered. Specimen holders clamp specimens at desired angles. Time, water, and erosion motor speeds are controlled by the control unit. Adding a mixing chamber at the hopper outlet ensures uniform slurry flow through the nozzle.

3.1 Testing Conditions

A slurry jet erosion tester was employed to assess erosion wear under varying jet velocity, time duration, and particle size. Jet velocity ranged from 15 to 60 m/s, with experiments conducted for particle sizes of 150, 150–212, 212–250, and 250–350 µm. Test durations varied from 15 to 60 min. Material weight loss in grams was utilized to quantify erosion wear.

4. Results And Discussion

4.1 Elemental Characterization

Energy dispersive spectroscopy (EDS) identified various elements present in the powder, confirming the presence of oxygen (O) and titanium (Ti). Additionally, EDS mapping in Figure 26.2 reveals the existence of W, Co, Cr, and C at 78.89%, 8.95%, 3.68%, and 7.25%, respectively.

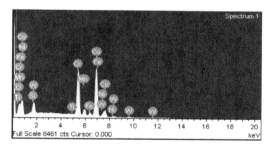

Figure 26.2: EDS of coating powder.

4.2 Impact of Process Conditions on Erosion Wear

Figure 26.4 displays several graphs depicting the fluctuation of total mass loss with test parameters observed from the experiment. The following sections describe more about the charts.

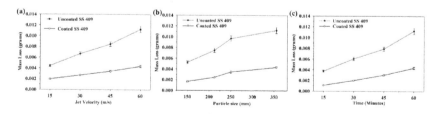

Figure 26.3: Effect of (a) Jet velocity, (b) particle size, and (c) time on erosion wear.

4.2.1 *Effect of Jet Velocity*

This study compared the material loss of SS 409 with and without coating at jet velocities of 10, 20, 30, and 40 m/s over 60 min. Figure 26.3a shows that uncoated SS-409 experienced the most erosion wear due to jet velocity fluctuations, while coated materials, with TiO_2 and WC, had less wear due to their toughness.

4.2.2 Effect of Particle Size

Figure 26.3b shows the effects of PSD erosion in coated and uncoated samples. The slurry ranged from <150 to 250–355 μm in particle size. Particles below 150 μm show less severe erosion wear, peaking at 250–355 μm. Larger particles have more kinetic energy and cause stronger impacts and erosion wear than finer ones.

4.2.3 Effect of Time Duration

In this study, erosion wear on uncoated and coated SS-409 was significantly affected by time. Figure 26.3c shows experiments at 40 m/s for 15–60 min. The erosion rate increased over time. From 15 to 60 min, uncoated and coated SS-409 showed erosion wear 3.1 and 2.2 times higher, respectively.

4.3 Surface Morphology of Eroded Surface

SEM examined the degraded surfaces of both coated and uncoated SS-409. Figure 4a shows bare SS-409 with significant degradation, craters, and pits from 250 to 355 μm slurry, indicating platelets' expulsion due to continuous impacts causing plastic deformation. Figure 26.4b displays coated SS-409 with gaps, missing particles, and expelled WC grains and 10Co-4Cr binder due to plowing and cutting by sand particles. The erosion is primarily ductile, with some brittle behavior. The coating enhances matrix densification and mechanical strength.

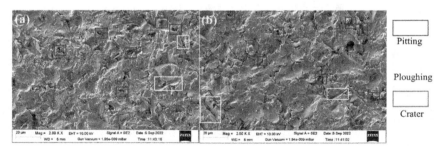

Figure 26.4: SEM image of eroded SS-409 surface (a) without coating & (b) with coating.

5. Conclusion

SS-409 is used in slurry equipment but has relatively low microhardness. This study aimed to measure the typical erosion wear of SS-409 using a slurry jet erosion tester with various parameters. The key findings are:

- Erosion wear increased nonlinearly with longer time duration, higher jet velocity, and larger particle size. Adding WC-10Co-4Cr + TiO_2 improved erosion resistance, resulting in less wear volume loss and a lower wear rate.
- Uncoated SS-409 exhibited ductile mass loss behavior, while coated SS-409 showed a mix of ductile and brittle behavior, with ductility predominating.
- Craters and plowing were the primary erosion wear patterns on coated SS-409.

References

[1] Alam, T., & Farhat, Z. N. (2018). Slurry erosion surface damage under normal impact for pipeline steels. *Engineering Failure Analysis, 90,* 116–128. https://doi. org/10.1016/j.engfailanal.2018.03.019

[2] Al-Bukhaiti, M. A., Ahmed, S. M., Badran, F. M. F., & Emara, K. M. (2007). Effect of impingement angle on slurry erosion behavior and mechanisms of 1017 steel and high-chromium white cast iron. *Wear, 262*(9–10), 1187–1198. https://doi. org/10.1016/j.wear.2006.11.018

[3] Desale, G. R., Gandhi, B. K., & Jain, S. C. (2008). Slurry erosion of ductile materials under normal impact conditions. *Wear, 264*(3–4), 322–330. https://doi. org/10.1016/j.wear.2007.03.022

[4] Hong, S., Wu, Y., Wang, Q., Ying, G., Li, G., Gao, W., Wang, B., & Guo, W. (2013). Microstructure and cavitation–silt erosion behavior of high-velocity oxygen–fuel (HVOF) sprayed Cr3C2–NiCr coating. *Surface and Coatings Technology, 225,* 85–91. https://doi.org/10.1016/j.surfcoat.2013.03.020

[5] Islam, Md. A., Alam, T., & Farhat, Z. (2016). Construction of erosion mechanism maps for pipeline steels. *Tribology International, 102,* 161–173. https://doi. org/10.1016/j.triboint.2016.05.033

[6] Jung, K.-H., & Kim, S.-J. (2020). Effect of various factors on solid particle erosion behavior of degraded 9Cr-1MoVNb steel with experiment design. *Applied Surface Science, 506,* 144956. https://doi.org/10.1016/j.apsusc.2019.144956

[7] Kumar, A., Sharma, A., & Goel, S. K. (2016). Erosion behavior of WC–10Co–4Cr coating on 23-8-N nitronic steel by HVOF thermal spraying. *Applied Surface Science, 370,* 418–426. https://doi.org/10.1016/j.apsusc.2016.02.163

[8] Kumar Goyal, D., Singh, H., Kumar, H., & Sahni, V. (2012). Slurry erosion behavior of HVOF sprayed WC–10Co–4Cr and Al2O3+13TiO2 coatings on turbine steel. *Wear, 289,* 46–57. https://doi.org/10.1016/j.wear.2012.04.016

[9] Lee, C. W., Han, J. H., Yoon, J., Shin, M. C., & Kwun, S. I. (2010). A study on powder mixing for high fracture toughness and wear resistance of WC–Co–Cr coatings sprayed by HVOF. *Surface and Coatings Technology, 204*(14), 2223–2229. https:// doi.org/10.1016/j.surfcoat.2009.12.014

[10] Maiti, A. K., Mukhopadhyay, N., & Raman, R. (2007). Effect of adding WC powder to the feedstock of WC–Co–Cr based HVOF coating and its impact on erosion and abrasion resistance. *Surface and Coatings Technology, 201*(18), 7781–7788. https:// doi.org/10.1016/j.surfcoat.2007.03.014

CHAPTER 27

Investigating the Influence of Natural Fibers on Mechanical Properties of Epoxy Composites

Priyaranjan Panda[a], Trilochan Pradhan[a], Asisha Ranjan Pradhan[a], Anjan Kumar Mishra[a]

[a]Parala Maharaja Engineering College, Brahmapur, Odisha, India
Corresponding author email: priyaranjanpanda0@gmail.com

Abstract

Due to the demand for eco-friendly materials, natural fiber composites have been extensively researched. This study examined jute-coconut coir composite mechanical properties. This study uses cheap, abundant, and mechanically strong jute and coconut coir as reinforcing agents. Processing natural fibers and polymer matrices creates composite specimens. Fabrication uses hand lay-up for matrix fiber dispersion. Natural fiber composites are tested for tensile, flexural, impact, and hardness. The composite specimens' tensile, flexural, impact, and hardness are compared to epoxy resin. This research will explain natural fiber composite mechanical behavior and applications. The findings will optimize fiber-reinforced composite materials for automotive, construction, and consumer products. Sustainable natural fibers reduce environmental impact and promote a circular economy.

Keywords: Natural fiber composites, Jute, Coconut coir, Mechanical properties

1. Introduction

Material science and engineering prefer green composites because eco-friendly products are popular. These materials are sought after to reduce pollution, and environmental degradation, and preserve the planet. Green composites use natural resources and pollute little (Deepa et al., 2011). These composites use plant fibers for reinforcement. The best composites are lightweight, biodegradable, renewable bio-polymers, or synthetic polymers strengthened by plant fibers (Dixit et al., 2017). Reinforced composites stiffen (Noor Afizah Rosli et al., 2013). Natural fiber composites use epoxy. Impact-resistant and fracture-resistant epoxies rarely prevent fatigue cracks (Gao et al., 2012; Nadlene et al., 2016). Epoxies are strengthened by glass fiber, natural fiber, carbon fillers, MnO_2, zinc powder, etc. A variety of steam pressures are applied to wool fibers during steam explosion modification. Steam detonations reduce fiber water attraction, mechanical properties, and caustic solution solubility due to explosion pressure (Santos et al., 2017). Natural fibers greatly impact composite tribology. Tribology improves with fiber-matrix bonding (Omrani et al., 2016). Moist bamboo absorbs moisture. Bamboo composites are thermally more efficient than other plant fibers (Zakikhani et al.,

DOI: 10.1201/9781003596776-27

2014). Superior thermal properties of steam explosion-recovered banana nanofibers. These elements are ideal for biocomposite reinforcement (Deepa et al., 2011). Natural fibers stick to epoxy matrices. They can replace synthetic fibers in composites due to their similar or better physical and mechanical properties.

This study examines the mechanical properties of a jute-coconut coir composite. This study used jute and coconut coir as reinforcing agents due to their availability, low cost, and mechanical properties. Tensile, flexural, impact, and hardness tests evaluate specimen mechanical properties.

2. Experimental Procedure

2.1 Natural Fiber

Plants, animals, and minerals provide natural fibers. Textiles, ropes, baskets, and paper have been made from them for thousands of years. Compared to synthetic fibers, natural fibers are biodegradable, renewable, and environmentally friendly.

2.2 Coconut Coir

Many steps are needed to extract coconut coir. Coconut shells are husked after harvesting. Mechanically extracting or retting the husks in water loosens the coir fibers. After extraction, fibers are cleaned, dried, and sorted by length and quality. Application-specific coir fibers can be buffered, rinsed, and spun. The initial and final coconut coir are shown in Figure 27.1. Coconut fiber properties are in Table 27.1.

Figure 27.1: Coconut coir.

Table 27.1: Physical properties of coconut fibers (Das et al., 2016).

Property	Specifications
Length	6-8 inches
Density	1.4 g/cc
Tenacity	10.0
Breaking elongation	30%
Diameter	.1mm-.5mm
Modulus of Rigidity	1.8924 dyne/cm²
Swelling in water	30%

2.3 Epoxy Resin (Ly556) and Hardener HY951

High-performance thermosetting epoxy resin LY556 is popular in construction, automotive, and aerospace. It is a special epoxy resin with superior strength, toughness, and chemical resistance. HY951 hardener is used with epoxy resins to make high-performance adhesives, coatings, and composites. Fast cure times and high strength characterize this aliphatic polyamine.

2.4 Composite Preparation by Hand Layup Method

The hand lay-up method prepares natural fiber composites manually. Arrange natural fibers (woven or nonwoven mats, fabrics, or roving) in the desired shape and orientation to make a preform. Apply a release agent to a clean, smooth mold. Apply the preform to the mold, adding reinforcement if needed. Mix the resin per the manufacturer's instructions. Brush or roller the resin onto the preform to impregnate and build layers for the desired thickness. Remove air bubbles and compact layers with a consolidation tool. As recommended, cure the resin at room temperature or with heat. After curing, carefully remove the composite part from the mold. Remove excess material and sand or polish the surface. The final product is in Figure 27.2.

Figure 27.2: Natural fiber composite made with hand layup method.

3. Results And Discussion

3.1 Effect of Fiber Reinforcement on Tensile Strength of Composites

This study used a floor-mounted Universal Testing Machine (Tinus Olsen, H50KS) to tensile test an ASTM D638 specimen. A 2% strain rate tensile force was applied to the specimen. Table 27.2 shows the results, and Figure 27.3 shows the composite specimen before and after testing. Jute fiber is 53 HRM harder than coconut fiber and epoxy resin. The high cellulose content and dense fiber arrangement of jute give it superior stiffness and deformation resistance.

Figure 27.3: Composite specimen for Tensile test before and after test.

Table 27.2: Experimental result of tensile property.

Sl. No	Composite Material	Ultimate Stress (MPa)	Ultimate Strain (%)	Breaking Strain (%)	Total Time (sec)
01	Epoxy Resin	28.6	4.47	4.48	137
02	Coconut Fiber	32.9	4.51	4.54	144
03	Jute Fiber	39.4	4.62	4.63	151

3.2 Effect of Fiber Reinforcement on Flexural Strength of Composites

Flexural testing using an ASTM D790 specimen and a Tinus Olsen H50KS Universal Testing Machine showed jute's superior performance (42.17 MPa) over coconut fiber (36.32 MPa) and epoxy resin (26.38 MPa). Jute's high cellulose content and favorable microfibrillar angle enhance its ability to withstand bending forces in the composite matrix, reinforced by strong fiber-matrix bonding.

3.3 Effect of Fiber Reinforcement on Hardness of Composites

In this work, an ASTM E384 specimen is taken for the hardness test in a Micro Hardness Testing machine. From the test results, jute fiber shows the highest hardness at 53 HRM compared to coconut fiber (42 HRM) and epoxy resin (32 HRM). This superior hardness is due to jute's high cellulose content and dense fiber arrangement, which provide greater stiffness and resistance to deformation.

3.4 Comparison of Different Natural Fiber Concerning Solely Epoxy Resin

Figure 27.4a presents a comparative analysis of tensile strength and hardness strength test results, as derived from the experimental data. Figure 27.6b, on the other hand, illustrates the comparative analysis of flexural strength and hardness strength test results.

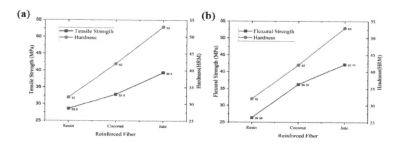

Figure 27.4: Comparison of Hardness of composite with (a) tensile strength and (b) flexural strength.

Jute reinforcement outperforms resin and coconut fiber in flexural and tensile properties. Jute has excellent tensile and flexural strength of 39.4 and 42.17 MPa. Jute fiber is a strong mechanical reinforcement material, according to these findings. Flexural strength resists bending better than tensile strength due to matrix fiber distribution and orientation.

4. Conclusion

Jute and coconut coir fiber composites were studied for their mechanical properties due to their abundance, cost-effectiveness, and desirable qualities. Hand-laid into a polymer matrix, the fibers were tested for tensile, flexural, impact, and hardness. Compared to epoxy resin, tensile strength, flexural strength, and hardness increased by 26.4%, 48.765%, and 48.438%. These results optimize natural fiber composites for engineering applications such as automotive components, construction materials, and consumer products.

References

[1]   Das, D., Kaundinya, D., Sarkar, R., & Deb, B. (2016). Shear strength improvement of sandy soil using coconut fiber. *International Journal of Civil Engineering and Technology, 7*(3), 297–305.

[2]   Deepa, B., Abraham, E., Cherian, B. M., Bismarck, A., Blaker, J. J., Pothan, L. A., Leao, A. L., de Souza, S. F., & Kottaisamy, M. (2011). Structure, morphology, and thermal characteristics of banana nano fibers obtained by steam explosion. *Bioresource Technology, 102*(2), 1988–1997.

[3]   Dixit, S., Goel, R., Dubey, A., Shivhare, P. R., & Bhalavi, T. (2017). Natural fiber reinforced polymer composite materials—A review. *Polymers from Renewable Resources, 8*(2), 71–78.

[4]   Gao, J., Li, J., Benicewicz, B. C., Zhao, S., Hillborg, H., & Schadler, L. S. (2012). The mechanical properties of epoxy composites filled with rubbery copolymer grafted SiO2. *Polymers, 4*(1), 187–210.

[5]   Nadlene, R., Sapuan, S. M., Jawaid, M., Ishak, M. R., & Yusriah, L. (2016). A review on Roselle fiber and its composites. *Journal of Natural Fibers, 13*(1), 10–41. https://doi.org/10.1080/15440478.2014.984052

[6]   Noor Afizah Rosli, N. A. R., Ishak Ahmad, I. A., & Ibrahim Abdullah, I. A. (2013). Isolation and characterization of cellulose nanocrystals from Agave angustifolia fiber.

[7]   Omrani, E., Menezes, P. L., & Rohatgi, P. K. (2016). State of the art on tribological behavior of polymer matrix composites reinforced with natural fibers in the green materials world. *Engineering Science and Technology, an International Journal, 19*(2), 717–736.

[8]   Santos, F. M. dos, Batista, F. B., Panzera, T. H., Christoforo, A. L., & Rubio, J. C. C. (2017). Hybrid composites reinforced with short sisal fibers and micro ceramic particles. *Matéria (Rio de Janeiro), 22*(2).

[9]   Zakikhani, P., Zahari, R., Sultan, M. T. H., & Majid, D. L. (2014). Extraction and preparation of bamboo fiber-reinforced composites. *Materials & Design, 63*, 820–828. https://doi.org/10.1016/j.matdes.2014.06.058

CHAPTER 28

Impact Analysis of Groundwater Variability on Groundwater Storage in Malkangiri, Odisha

Chitaranjan Dalai[1], Prativa Nanda[2], Rosalin Dalai[1]

[1]Assistant Professor, Department of Civil Engineering,
Odisha University of Technology and Research, Bhubaneswar, Odisha, India
[2]Research Scholar, Odisha University of Technology and Research,
Bhubaneswar, India
E-mail: dr.chitaraniandalai@gmail.com, prativananda98@gmail.com,
rosalin.iitk@gmail.com

Abstract

Groundwater variability refers to the natural fluctuations and differences in the quantity, quality, and availability of groundwater within an aquifer or a region over time and space. It can be influenced by various factors, including precipitation styles, recharge rates, geological formations, human activities, and climate change. Understanding groundwater variability is crucial for the sustainable management and utilization of this important water resource. This study aims to analyze the consequences of groundwater variability on storage in Odisha. The amount and quality of water available for aquifer characterization, such as groundwater level and flow direction, help predict future changes under scenarios of drought and increased water extraction from pumping wells. The aim is to create maps of various environmental factors to identify the groundwater capacity zones in Malkangiri district. The maps will consolidate components such as seepage thickness, topography, geomorphology, slope, lineament thickness, soil, precipitation, and land use/land cover to determine the zones with high potential for groundwater availability. The outcomes will be verified by comparing them with current well depths, presenting a comprehensive understanding of the district's groundwater resources. The model's reliability was evaluated using Root Mean Square Error (RMSE) and Coefficient of Determination (R^2) as performance indicators. By examining the effects of excessive groundwater extraction for paddy cultivation, this study reveals that continuous tapping of groundwater leads to a lowering of the groundwater table and increased vulnerability to aquifers.

Keywords: Storage change, GIS, IDW, ANN

1. Introduction

Groundwater variability is a critical factor affecting groundwater storage in Odisha, particularly in coastal areas where paddy cultivation and water scarcity pose significant challenges (Panda et al., 2022). Groundwater variability modeling is a process of simulating

DOI: 10.1201/9781003596776-28

and predicting changes in groundwater levels, flow, and quality over time and space (Dash et al., 2014; Jahin et al., 2020). It involves the use of mathematical models to represent the complex interactions between various hydrogeological factors. By incorporating data on hydrogeological properties, climate patterns, land use, and human activities, groundwater variability models can help assess the impacts of different scenarios (e.g., climate change, land use change, water management strategies) on groundwater resources (Sharma et al., 2022). In the management of water resources, land-use planning, and environmental impact assessments, models are valuable tools (Arega et al., 2024). Changes in groundwater levels due to variability can affect the availability of water for drinking, irrigation, and industrial purposes. Fluctuations can lead to water shortages or surpluses, impacting water supply reliability and ecosystem health. A crucial aspect to consider is the impact of groundwater inconstancy on groundwater capacity (Dey et al., 2020; Li et al., 2015). Managing groundwater resources becomes more challenging in the face of variability. Balancing competing demands, implementing sustainable extraction practices, and designing effective groundwater management strategies require robust data, modeling tools, and stakeholder collaboration (Fattah et al., 2024). To assess the current status of monitoring and management in Malkangiri district and identify the factors contributing to it is one of the objectives of the research. The study aims to provide a comprehensive understanding of the capacity relationship in Malkangiri by analyzing existing literature and information and evaluating the implications of variability on storage. The findings of the study will help inform sustainable management strategies and policies in the state.

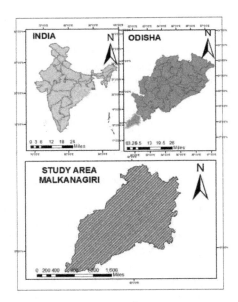

Figure 28.1: Study area map.

1.1 Study Area

In the southwestern region of Odisha, Malkangiri is situated. Its geographical coordinates are approximate: Latitude: 18.3546°N and Longitude: 81.9096°E. It has an average elevation of 170 m (560 ft). These coordinates place Malkangiri in the hilly and forested region of the Eastern Ghats Mountain range, near the border with the state of Andhra Pradesh.

Malkangiri is characterized by a diverse topography, including hilly terrain, dense forests, valleys, and plains. Overall, the topography of Malkangiri is diverse and includes both low-lying areas and elevated regions. To understand the specific groundwater variability in Malkangiri district, detailed hydrogeological studies, groundwater monitoring, and modeling efforts may be required. Local government agencies and research institutions often conduct such studies.

1.2 Data Preparation

Several factors are considered here to influence the changes in groundwater variability of the study area. Data considered here include climate data, land use land cover data, geological conditions, topography of the area, vegetation cover, and groundwater extraction. These data are gathered using GIS, NASA POWER DATA, and WRIS groundwater data.

1.3 Results

It is essential to use these independent input factors when creating a groundwater potential zone map. Here, constructing the flood susceptibility model takes into account both meteorological and non-climatic aspects. The most crucial climatic flood influencing factors are rainfall intensity, and non-climatic factors like altitude, distance from the river, aspect, drainage density, curvature, geology, LULC, NDVI, slope, soil, and geomorphology are considered here for model preparation. All considered groundwater influencing factors of the Malkangiri district are presented in Figure 28.1. The altitude and its derivative factors significantly affect groundwater potential recognition. Slope affects the runoff velocity. Groundwater potential maps were developed in the study area using different input parameters (Figure 28.2) classified into high, very high, medium, low, and very low. To calculate the groundwater potential map, the weights of individual layers are calculated using analytic hierarchy processes (AHP).

Figure 28.2: Groundwater influencing factors were utilized to develop the groundwater variability model.

1.4 Artificial Neural Network

Artificial neural networks utilize an approach known as backpropagation, introduced by Rumelhart et al. in 1986. The input, hidden, and output layers are included. The hidden layer plays a vital part in the network's functionality. Despite its complexity, backpropagation is widely regarded as a highly effective method in machine learning. However, given the complexity of the mathematical problem, it is often necessary to perform experimental tests to determine the optimal number of hidden nodes. Its three layers are arranged in the standard ANN structure: input, hidden, and output.

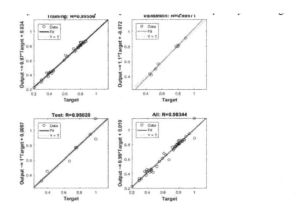

Figure 28.3: Groundwater influencing factors are utilized to develop the groundwater variability.

A scatter plot is a visual technique for showing the relationship between two factors. Each data point in the chart represents one observation, plotted in Figure 28.3, and the values of two factors (i.e., genuine and anticipated) determine the location of each point. During calibration and validation, scatter plots of daily floods observed and predicted in the basin are drawn along with a 1:1 line. During training, the simulated values for low and high floods are slightly lower than the 1:1 line and display an underestimation of flood. However, the major region of the scatter plot is reasonably distributed around the 1:1 line. When the sampling process is calibrated, the coefficient of determination (R^2) is 0.950; when the sampling process is validated, it is 0.983. According to this study, the proposed ANN model performed well at the well points. This indicates a good correlation between the observed and predicted groundwater levels by the model. Moreover, it complements the need for continuous research efforts to refine predictive models and increase their practical utility in real-life situations.

2. Conclusion

This study is based on the groundwater observation in the district of Malkangiri. Understanding all the above factors and their interactions is essential for managing groundwater resources sustainably and mitigating the impacts of groundwater variability on ecosystems, communities, and economies. The groundwater samples in the study area show limited seasonal variability in quality. Regular varieties of precipitation can affect groundwater levels, with recharge occurring during the storm season and consumption during drier periods. Overexploitation of groundwater resources for agriculture, industry, and domestic use can also lead to declining water tables and groundwater quality issues in some areas. By employing an array of AI techniques, this study has contributed significant findings to the intricate dynamics of groundwater systems. The above work employs models to forecast GWLs, which will allow farmers to make informed decisions about crop selection, irrigation methods, and water distribution based on the predicted GWLs.

This is also important in optimizing water usage, reducing the risk of saltwater intrusion, and increasing agricultural yield. This will aid hydrogeologists in estimating the GWL of a location with similar climatic conditions, topography, and other aspects, as well as agricultural experts in crop planning, farming methods, and so on. Through rigorous analysis and evaluation, our research showcases the diverse strengths and applicability of these techniques in capturing the complex interplay of factors influencing groundwater behavior. The findings from the case study in Malkangiri District, Odisha, provide practical evidence of the potential of AI-driven models to yield accurate predictions in challenging hydrogeological environments.

References

[1]   Arega, K. A., Birhanu, B., Ali, S., Hailu, B. T., Tariq, M. A. U. R., Adane, Z., & Nedaw, D. (2024). Analysis of spatio-temporal variability of groundwater storage in Ethiopia using Gravity Recovery and Climate Experiment (GRACE) data. *Environmental Earth Sciences, 83*(7), 1–21.

[2]   Dash, A., Das, H. K., & Mishra, B. (2014). Hydrogeochemistry and groundwater quality in and around Joda of Keonjhar District, Odisha, India. *International Journal of Environmental Science and Technology, 12*(2), 409–419.

[3]   Dey, S., Bhatt, D., Haq, S., & Mall, R. K. (2020). Potential impact of rainfall variability on groundwater resources: A case study in Uttar Pradesh, India. *Arabian Journal of Geosciences, 13*, 1–11.

[4]   Fattah, M. A., Hasan, M. M., Dola, I. A., Morshed, S. R., Chakraborty, T., Kafy, A. A., & Shohan, A. A. A. (2024). Implications of rainfall variability on groundwater recharge and sustainable management in South Asian capitals: An in-depth analysis using Mann Kendall tests, continuous wavelet coherence, and innovative trend analysis. *Groundwater for Sustainable Development, 24*, 101060.

[5]   Jahin, H. S., & Gaber, S. E. (2011). Study of groundwater quality in El-Kharga Oasis, Western Desert, Egypt. *Asian Journal of Water, Environment, and Pollution, 8*(4), 1–7.

[6]   Li, B., Rodell, M., & Famiglietti, J. S. (2015). Groundwater variability across temporal and spatial scales in the central and northeastern US. *Journal of Hydrology, 525*, 769–780.

[7]   Panda, D. K., Tiwari, V. M., & Rodell, M. (2022). Groundwater variability across India, under contrasting human and natural conditions. *Earth's Future, 10*(4), e2021EF002513.

[8]   Raghunath, H. M. (1987). Groundwater: Hydrogeology, groundwater survey, and pumping tests, rural water supply, and irrigation systems. *New Age International*.

[9]   Rumelhart, D. E., Hinton, G. E., & Williams, R. J. (1986). Learning representations by back-propagating errors. *Nature, 323*(6088), 533–536.

[10] Sharma, A., Maharana, P., Sahoo, S., & Sharma, P. (2022). Environmental change and groundwater variability in South Bihar, India. *Groundwater for Sustainable Development, 19*, 100846.

CHAPTER 29

Characterization of Mechanical Properties of Epoxy Glass Fiber Reinforced Composite

Trilochan Pradhan[a], Asisha Ranjan Pradhan[a],
Anjan Kumar Mishra[a], Priyaranjan Panda[a]

[a]Parala Maharaja Engineering College, Brahmapur, Odisha, India
Corresponding author: trilochanpradhan.me@pmec.ac.in

Abstract

Composites can open up new engineering fields and enable previously unthinkable feats. This project explores composites' vast scope, which allows materials to be engineered to meet our needs. This article examined composite properties, pros and cons, chemical compositions, and fabrication methods. Hand-laid epoxy glass fiber reinforced composite was tested for tensile, hardness, flexural, and deflection properties to determine its competitiveness and feasibility over traditional materials in the same properties segment. Sample results and findings are obtained.

Keywords: Epoxy, Glass fiber, Tensile strength, Hardness, Flexural strength

1. Introduction

Epoxy resins are widely used in coverings, architectural adhesives, and fiber-reinforced composite matrices, among other applications (Ellis, 1993). Many materials are useful for their toughness, design flexibility, and portability (Pascault et al., 2002). However, the fragile nature of these materials limits their uses. Toughened epoxy resins became popular in the mid-1980s for their thermal stability, mechanical strength, and cost-effectiveness (Azeez et al., 2013; Tsai et al., 2010). Increasing epoxy thermoset strength while maintaining other properties has been studied for years. An additional rubbery phase that separates from the main material during curing creates different morphologies (Russell & Chartoff, 2005). Rubber improves fracture durability but decreases modulus and thermal stability. High-performance thermoplastics increase toughness (Salmon et al., 2005). Studies show that thermoplastic-modified epoxies' morphology depends on their support structure, molecular mass, and end-group chemistry. The final blend quality depends on cured resin morphology. Morphology management is difficult, limiting thermoplastics.

In structural applications, epoxy resins are increasingly used as fiber-reinforced composite matrices. Composites with fibers outperform many metals in strength and modulus. Many epoxy products are characterized by low molecular weight and reactive functional groups. Due to its composite structure, strong cross-links raise the glass transition temperature (Qi et al., 2006). UTS, Young's modulus, flexural strength, interlaminar shear strength

DOI: 10.1201/9781003596776-29

(ILSS), and microhardness improved when epoxy/glass/nanohybrid clay composites were hand-laid. Up to 5% clay reinforcement improved results (Kornmann et al., 2005). Fiber-reinforced plastic uses E-glass and S-glass. The cheapest commercial reinforcing fiber is E-glass. Therefore, the fiber-reinforced composite industry uses it extensively.

2. Experimental Procedure

2.1 Base Material: Epoxy Resin

The basic component or cured product of epoxy resin is epoxy. The epoxide functional group is also known as poly-epoxides, which are reactive prepolymers and polymers with epoxide groups. Aside from co-reactants like acids, amines, phenol, thiols, and amines, epoxy resins catalyze homopolymerization. These are called hardeners or curatives, and the cross-linking reaction is called curing.

2.2 Woven Epoxy Glass Fiber

Composites made of polyester and epoxy resin are mostly glass fibers. Alternative reinforced materials cost more. Glass fibers include chopped strands, continuous fiber, woven roving, cloth tape, surface tissue, and mats. This material has carbon fiber, Kevlar, and polymer properties due to its fine glass strand. Although weaker than carbon fiber, it is cheaper and more elastic in composites. This can reinforce lightweight GRP or fiberglass due to its properties. No air gap, dense, and a poor thermal conductor.

2.3 Bundle of Glass Fibers

Electronic glass (Figure 29.1) is abundant and simple to make. After being an electrical insulator, it formed fibers for many uses. It is made from natural and synthetic SiO_2, limestone, and sodium carbonate, among other materials such as syenite, feldspar, China stone clay, borax, and calcined Al_2O_3. Its low melting point is due to sodium carbonates, sedimentary rocks, and silica. Additives that increase chemical inertness are preferred.

Figure 29.1: E-Glass woven fabric.

3. Materials

Woven epoxy glass fiber (reinforcement), epoxy resin, MEKP (epoxy resin initiator), and cobalt octane were used in this project. E-glass fiber and epoxy resins came from Bhubaneswar-based Mechem Pvt Ltd. Locally purchased MEKP and cobalt octane was also used.

3.1 Composite Preparation

Silicone spray and polyvinyl alcohol cleaned and coated the mold. A 2% polyester res-in-molded accelerator was used. MEKP raised the curing temperature by 2%, and octane cobalt was used as a boost. The mold was filled with the mixture and brushed. The mold was reinforced with woven glass, and a roller brush was used to saturate the reinforcement. A roller was then used to coat the reinforcement with resin. The second fabric layer went on top. To make an epoxy glass fiber composite (Figure 29.2), two more layers of glass fiber reinforcement were added. Room-temperature composite curing took 24–48 h. The fiber-resin ratio was 1.8:1. This composite needs 450 g/sqm resin.

Figure 29.2: Preparation of composite.

4. Results and Discussion

4.1 Tensile Properties (ASTM D638)

Simple tensile tests measure properties such as elastic modulus, yield strength, and elongation. A dumbbell-shaped sample is prepared according to ASTM D638. The sample is gripped as shown in Figure 29.3, and its size is measured by extensometers. During testing, sample ends are separated at 0.05–20 inches/min until fracture occurs. Table 29.1 shows the tensile strength and modulus of epoxy glass fiber composites at 25°C and 65% humidity. All tests are conducted by CIPET Bhubaneswar. The Young's modulus is 5443.81 MPa, and the ultimate tensile strength is 199.91 MPa. Figure 4 displays the stress-strain graphs for all specimens.

Figure 29.3: Specimen during Tensile test.

Table 29.1: Tensile strength of composite materials.

Specimen	Width (mm)	Thickness (mm)	Maximum Load (N)	Tensile stress at Maximum Load (MPa)	Load at Yield (Zero Slope) (N)	Tensile stress at Yield (Zero Slope) (MPa)	Load at Break (Standard) (N)	Tensile stress at Break (Standard) (MPa)	Modulus (Automatic Young's) (MPa)
1	12.89	2.47	5905.01	185.47	5905.01	185.47	399.77	12.56	5801.32
2	12.88	2.40	6089.98	197.01	6089.98	197.01	6089.80	197.00	5993.17
3	12.87	2.38	6417.47	209.51	6417.47	209.51	6317.42	206.25	6311.92
4	12.79	2.50	7097.28	221.96	7097.28	221.96	2311.55	72.29	4662.97
5	13.00	2.46	5935.60	185.60	5935.60	185.60	5900.71	184.51	4449.67
Mean	12.89	2.44	6289.07	199.91	6289.07	199.91	4203.85	134.52	5443.81
SD	12.96	2.49	6784.49	215.72	6784.49	215.72	6894.35	221.55	6277.66

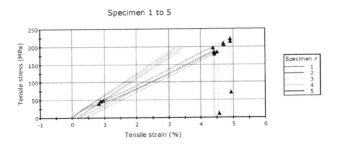

Figure 29.4: Stress vs strain graphs of all specimens.

4.2 Flexural Property (ASTM D790)

Bent objects experience varying stress, with compression inside and tension outside. The outer and innermost layers of the sample are the extreme fibers. Material typically fractures in tension, not compression. Flexural strength (σ) is the stress before yielding. Rectangular/circular samples are bent until they break or yield. For stronger materials, a 4-point flexural test is preferred, as it develops maximum stresses at specific points. At 65% humidity and 27°C, Table 29.2 shows the epoxy glass fiber composite's flexural modulus (7154.97 N/mm²) and strength (240.37 N/mm²).

Table 29.2: Flexural Strength of composite materials.

Specimen label	Width (mm)	Thickness (mm)	Support span (mm)	Rate 1 (mm/min)	Maximum Load (N)	Maximum Stress (N/mm²)	Flex Modulus (N/mm²)
1	12.89	2.45	39.2	1.045	309.93	235.54	7133.78
2	12.64	2.51	40.16	1.07	298.47	225.78	6878.23
3	12.5	2.47	39.52	1.05	334.2	259.78	7452.91
Mean	12.68	2.48	39.63	1.055	314.2	240.37	7154.97
Standard Deviation	0.19757	0.03055	0.48881	0.01323	18.24409	17.50769	287.9264
Minimum	12.5	2.45	39.2	1.045	298,47	225.78	6878.23
Maximum	12.89	2.51	40.16	1.07	334.2	259.78	7452.91

5. Conclusion

Adding plain and woven glass fiber to epoxy in a 2:1 ratio improves the composite's tensile, flexural, impact, and hardness properties. Using both types of glass fiber sheets results in significantly different mechanical properties. Higher concentrations of glass fiber, especially woven, strengthen the composite. Increased glass fiber content enhances tensile strength, impact resistance, and flexural strength.

References

[1] Azeez, A. A., Rhee, K. Y., Park, S. J., & Hui, D. (2013). Epoxy clay nanocomposites—Processing, properties, and applications: A review. *Composites Part B: Engineering, 45*(1), 308–320.

[2] Ellis, B. (1993). *Chemistry and Technology of Epoxy Resins* (B. Ellis, Ed.). Springer Netherlands. https://doi.org/10.1007/978-94-011-2932-9

[3] Kornmann, X., Rees, M., Thomann, Y., Necola, A., Barbezat, M., & Thomann, R. (2005). Epoxy-layered silicate nanocomposites as a matrix in glass fiber-reinforced composites. *Composites Science and Technology, 65*(14), 2259–2268. https://doi.org/10.1016/j.compscitech.2005.02.006

[4] Pascault, J.-P., Sautereau, H., Verdu, J., & Williams, R. J. J. (2002). *Thermosetting Polymers*. CRC Press.

[5] Qi, B., Zhang, Q. X., Bannister, M., & Mai, Y.-W. (2006). Investigation of the mechanical properties of DGEBA-based epoxy resin with nano clay additives. *Composite Structures, 75*(1–4), 514–519.

[6] Russell, B., & Chartoff, R. (2005). The influence of cure conditions on the morphology and phase distribution in a rubber-modified epoxy resin using scanning electron microscopy and atomic force microscopy. *Polymer, 46*(3), 785–798. https://doi.org/10.1016/j.polymer.2004.11.090

[7] Salmon, N., Carlier, V., Schut, J., Remiro, P. M., & Mondragon, I. (2005). Curing behavior of syndiotactic polystyrene–epoxy blends: 1. Kinetics of curing and phase separation process. *Polymer International, 54*(4), 667–672. https://doi.org/10.1002/pi.1739

[8] Tsai, J.-L., Hsiao, H., & Cheng, Y.-L. (2010). Investigating mechanical behaviors of silica nanoparticle reinforced composites. *Journal of Composite Materials, 44*(4), 505–524.

CHAPTER 30

Heart disease prediction using machine learning techniques

Amit Kumar, Reshma Khan, Deepika

Student, Department of CSE, Chandigarh University, Mohali, India
Emails: byemail007@gmail.com, reshma.khan36@gmail.com,
deepika.e11809@cumail.in

Abstract

In recent times, heart diseases have increasingly become a major cause of death. Precise prediction of heart disease holds paramount importance in enhancing patient prognosis and lowering mortality rates. Heart attacks primarily stem from conditions like coronary artery disease and chronic heart failure. Through the utilization of Machine Learning (ML) techniques, a refined diagnostic approach has been developed. ML offers improved accuracy and efficiency in diagnosis, therefore contributing to the early identification of heart disease. Through this research paper, we aim to accurately measure heart disease prediction. We have utilized the UCI dataset and applied logistic regression and two other methods. From our analysis, we achieved an accuracy of 92.3% using logistic regression.

Keywords: ML, UCI dataset, LR, Decision Tree Classifier, SVM

1. Introduction

Heart disease is a significant public health emergency, resulting in 17.9 million fatalities per year, as reported by the World Health Organization. This makes it the primary cause of death worldwide. The American College of Cardiology's 2019 study emphasized that approximately 10 million men and 9 million women die from heart disease annually, constituting one-third of all worldwide fatalities. In addition, the number of "years of life lived with disability" (YLDs) caused by cardiac diseases doubled, from 17.7 million in 1990 to 34.4 million in 2019, according to the World Health Organization (WHO, 2023). These concerning figures emphasize the immediate requirement for extensive preventive and management measures.

Accurate diagnosis is essential for effective treatment, but doctors can sometimes struggle to make fully informed decisions. Machine learning (ML) diagnostic systems can provide valuable support. This project aims to develop an ML-powered tool for early detection of heart disease, using a dataset from the UCI repository as a baseline.

Our objective is to develop a novel model that exceeds the precision of current methodologies. Through rigorous comparison and testing, we aim to offer an accessible tool that leverages ML to predict heart disease risk, helping individuals make informed health decisions without immediate doctor consultations.

DOI: 10.1201/9781003596776-30

1.1 Related Works

Table 30.1 presents a comprehensive analysis of the literature review, highlighting the models that achieved the highest accuracy along with their respective accuracy rates. The table includes a detailed account of the limitations and strengths of each model.

Table 30.1: Comprehensive analysis.

Author	Algorithm used	Dataset Used	Accuracy	Strength	Limitation
Dinesh et al., 2021	Hybrid Model, Decision Tree, Random Forest	Cleveland	88%	Potentially more accurate than using individual models alone	Small data-set size (303 instances) may limit generaliza-bility to a wider population
Asmit et al., 2022	Decision Tree, Random Forest, Backbone	UCI Dataset	Not mentioned	The use of mul-tiple algorithms provides accurate predictions	Lack of clarity on dataset details and absence of feature engineer-ing techniques
Albert et al., 2021	Hybrid Model	Cleveland	88.7%	Methodologically sound approach	Does not discuss model scalability
Ambrish et al.,2022	Logistic Regression	UCI Dataset	87.10%	Comprehensive feature selection enhances model robustness	Lacks explora-tion of model interpretability
Sunitha et al., 2021	Random Forest	UCI Dataset	91%	Transforms bina-ry classifications into risk percent-age levels	Limited by dataset size/ diversity, which may affect generalizability

The research gap identified in Table 1 lies in the limited exploration and transparency regarding dataset characteristics and algorithmic methodologies across various studies. While several authors showcase promising accuracies using diverse machine learning algorithms on different datasets, there is a consistent lack of clarity regarding feature selection processes, dataset details, and the scalability of proposed models.

1.2 Methodology

The process starts by acquiring data from the UCI repository. This dataset is well-estab-lished and has been verified by both the UCI and other researchers. Next, we carefully select the most relevant features for our analysis. Following Figure 30.1, we proceed to apply a machine-learning model and assess the accuracy of each model.

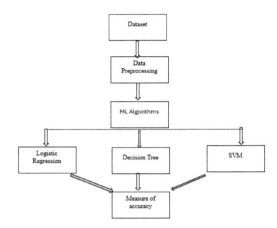

Figure 30.1: Proposed model.

1.3 Data Collection

This dataset from the UCI repository comprises 303 instances with 14 attributes. This dataset is verified by many researchers (Janosi et al., 1988). In this project, we take 70% data for training and 30% for testing.

1.4 Data Preprocessing and Feature Selection

Attributes of the dataset represent characteristics utilized by the system, such as heart rate, gender, and age of the individual. Our goal was to develop a model to predict the presence (close to 1) or absence (close to 0) of heart disease. The dataset used in this study exhibited an imbalance, with 165 positive cases of cardiovascular disease (CVD) and 138 negative cases. The features are chosen based on their strong positive correlation with the target (Figure 30.2) variable. This selection process ensures that only the most relevant features are included. Furthermore, the data is utilized in its original, unsorted sequence to maintain randomness, which helps in preventing any bias that might arise from a particular ordering of the data.

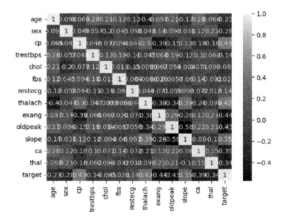

Figure 30.2: Correlation matrix visualization.

1.5 Machine Learning Algorithms

Logistic regression is a type of supervised learning used for classification. It helps estimate the probability of a target variable, which typically has two classes: 0 represents failure, and 1 represents success. This algorithm is particularly well-suited for scenarios where the outcome falls into distinct categories and aims to model the relation between the independent variables and the probability of each category's occurrence.

1.6 Decision Tree

Decision tree classifiers are popular in machine learning because they are intuitive and versatile. They work by dividing your dataset into smaller groups based on feature values, creating a branching structure of if-then rules. The algorithm carefully chooses the features that create the most distinct groups at each step.

1.7 Support Vector Machine

Support Vector Machines are all about making smart choices for classifying data. They look at your data and figure out the ultimate dividing line to separate different categories (Vardhini et al., 2023). The trick is they aim for the widest possible gap between groups—this helps them stay accurate when they see new examples. SVMs get even more clever with kernels, which let them handle complex, wiggly boundaries. That is why they are a top tool for image recognition, text sorting, and analyzing complex biological data.

1.8 Result Analysis

The evaluation of each algorithm involved assessing multiple metrics, such as accuracy, precision, recall, and F1-score, as illustrated in Table 30.2.

Table 30.2: Result analysis.

Classifier	Accuracy	Precision	Recall	F1-score
LR	92.3%	92.3%	94.1%	93.2%
D tree	82.4%	90.7%	76.5%	83%
SVM	88%	85.7%	94.1%	89.7%

The study evaluated three classifiers: Logistic Regression, Decision Tree Classifier, and Support Vector Machine (SVM) on the dataset. Logistic Regression achieved the highest testing accuracy of 92.3%, followed by SVM with 88%, and Decision Tree Classifier with 82.4%. Precision and recall scores of around 92% were observed for both Logistic Regression and SVM, while the Decision Tree Classifier showed slightly lower scores of 90.7% for precision and 76.5% for recall. The F1-score, combining precision and recall, mirrored these findings. Logistic Regression emerged as the top performer across all metrics, indicating its suitability for the dataset, effectively balancing accuracy and robustness in classification.

2. Conclusion

The following conclusions were made from the research: Logistic Regression demonstrated the highest testing accuracy, followed by SVM and Decision Tree Classifier. Despite variations in performance, each classifier exhibited strengths and weaknesses in precision and recall metrics. Logistic Regression emerged as the most suitable classifier for this dataset, effectively balancing accuracy and robustness in heart disease prediction. Regarding future directions, further investigations could explore optimizing classifier parameters to enhance predictive performance, exploring ensemble methods to improve accuracy, and considering the application of deep learning techniques for more intricate datasets.

References

[1]   World Health Organization. (2023). Cardiovascular diseases (CVDs). *WHO*. https://www.who.int/news-room/fact-sheets/detail/cardiovascular-diseases-(cvds)

[2]   Janosi, A., Steinbrunn, W., Pfisterer, M., & Detrano, R. (1988). Heart Disease. *UCI Machine Learning Repository*. https://doi.org/10.24432/C52P4X.

[3]   Kavitha, M., Gnaneswar, G., Dinesh, R., Rohith Sai, Y., & Sai Suraj, R. (2021). Heart disease prediction using hybrid machine learning model. In *2021 6th International Conference on Inventive Computation Technologies (ICICT)* (pp. 1–5). IEEE. https://doi.org/10.1109/ICICT50816.2021.9358597

[4]   Srivastava, A., & Singh, A. K. (2022). Heart disease prediction using machine learning. In *2022 2nd International Conference on Advance Computing and Innovative Technologies in Engineering (ICACITE)* (pp. 1–6). IEEE. https://doi.org/10.1109/ICACITE53722.2022.9823584

[5]   M., P., V., S., A., J., & Mayan, A. (2021). Cardiovascular disorder prediction using machine learning. In *2021 5th International Conference on Intelligent Computing and Control Systems (ICICCS)* (pp. 1–6). IEEE. https://doi.org/10.1109/ICICCS51141.2021.9432199

[6]   Ganesh, A., Ganesh, B., Srinivas, C., Dhanraj, A. G., & Mensinkal, K. (2022). Logistic regression technique for prediction of cardiovascular disease. *Global Transitions Proceedings, 3,* 127–130. https://doi.org/10.1016/j.gltp.2022.04.008

[7]   Guruprasad, S., Mathias, V. L., & Dcunha, W. (2021). Heart disease prediction using machine learning techniques. In *2021 5th International Conference on Electrical, Electronics, Communication, Computer Technologies and Optimization Techniques (ICEECCOT)* (pp. 762–766). IEEE. https://doi.org/10.1109/ICEECCOT52851.2021.9707966

[8]   Vardhini, V., Leela, S., Varalakshmi, S., Kumar, V. H., & Kumar, A. S. (2023). Heart disease prediction using machine learning. *Journal of Engineering Sciences, 14*(4)

CHAPTER 31

Numerical Study of Free Vibration of Sandwich FGM Plate

Abhijit Mohanty[1], Sarada Prasad Parida[2*], Rati Ranjan Dash[1]

[1]Odisha University of Technology and Research, Bhubaneswar, Odisha, India
[2]Konark Institute of Science and Technology, Khordha, Odisha, India
E-mail: sarada800@gmail.com

Abstract

Damping is a crucial parameter for vibration control, fatigue endurance, and impact resistance in structural materials to decrease fatigue failure. In addition to damping, increased stiffness and reduced weight enable the widespread application of sandwich structures. In this work, the modal response of a sandwich plate with a viscolastic core is investigated through numerical modeling using ANSYS 16.0. The skin layers are assumed to be carbon-epoxy, and the core is made of viscoelastic materials. The dynamic response is examined by changing the aspect ratio and skin-to-core thickness ratio in a simply supported end condition. The aspect ratio and thickness ratio are found to have significant effects on the modal response of the plate.

Keywords: sandwiched material, plate stiffness, elastic, viscoelastic

1. Introduction

Sandwich materials are potentially used in many structural applications due to their adequate stiffness relative to weight and their ability to absorb energy. Damping can be introduced to a structural material by adding viscoelastic materials, which provide high loss factors to structural members, reducing fatigue failure. Therefore, computational tools for analyzing damped structures to provide efficient solutions under different working conditions become necessary. When designing a sandwiched structure, the core should provide adequate transverse elasticity to resist cross-sectional deformation. Viscoelastic materials are generally honeycomb structures or foam materials (Vinson, 2006). Finite element analyses (FEA) based on Reddy's higher-order theory (HSDT) have been preferred over layer-wise theories as the deformation of the plate along the thickness due to core compressibility is ignored (Meunier & Shenoi, 2001; Khalili et al., 2010). Other theories used for the analysis of sandwiched beams and plates are widely lauded by some works (Li et al., 2020; Lou et al., 2012; Parida et al., 2022). Therefore, not only the selection of the core but also the selection of face sheets is an important criterion for prescribing the potential use of sandwiched structures. Further, the combined effect of face sheet strength and elasticity, attributed to the density and damping of the viscoelastic core on the plate dynamics, has a significant role and needs to be studied. In the present work, a sandwich plate with face sheets (carbon-epoxy by 50:50 volume

DOI: 10.1201/9781003596776-31

fraction) and an inner viscoelastic core are assumed. Mixture rules are used to calculate material constants, followed by FSDT in the equation of motion to find natural frequencies. Further, the skin thickness, core layer thickness, and aspect ratio on the dynamic response (natural frequencies) are studied, compared with FEA, and presented.

2. Theory

Here, a sandwiched FGM plate is assumed to have three layers: an elastic face sheet (carbon-epoxy composite) and a viscoelastic core as shown in Figure 31.1.

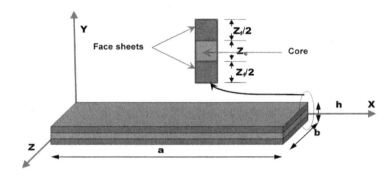

Figure 31.1: Layout of sandwiched plate.

Following FSDT (Khalili et al., 2010) for both the viscoelastic core (c) and face sheet layers (f), the displacement equation is given as:

$$
\left.
\begin{aligned}
U^c(x,z,t) &= u_0^C(x,t) + z\theta_x^C(x,t) \\
V^c(y,z,t) &= V_0^C(y,t) + z\theta_y^C(y,t) \\
W^c(x,z,t) &= W_0^C(x,t)
\end{aligned}
\right\}
$$

$$
\left.
\begin{aligned}
U^f(x,z,t) &= u_0^f(x,t) + \left(z - z^c\right)\theta_x^f(x,t) \\
V^f(y,z,t) &= v_0^f(y,t) + \left(z - z^c\right)\theta_y^f(y,t) \\
W^f(x,z,t) &= W_0^f(x,t)
\end{aligned}
\right\}
\tag{31.1}
$$

The eigen frequency is given by (Sakiyama et al., 1996);

$$
\omega_{mn} = \left[\frac{D_{eff}\left(\left(\frac{m\pi}{a}\right)^4 + 2\left(\frac{m\pi}{a}\right)^2\left(\frac{n\pi}{b}\right)^2 + \left(\frac{n\pi}{b}\right)^4\right)}{\rho_c z_c + 2\rho_f z_f}\right]^{1/2}
\tag{31.2}
$$

With $D_{eff} = \left[\dfrac{E_c' z_c^3}{12(1-v_c^2)} - \dfrac{E_c z_c^3}{12(1-v_c^2)} + \dfrac{E_f}{1-v_f^2}\left(\dfrac{2(0.5z_c + z_f)^3}{3} - \dfrac{z_c^3}{12}\right)\right]$ (31.3) where E_c, E_f : Elastic modulus of core and face sheets; υ_c, υ_f : Poison's ratio of core and skin; ρ_c, ρ_f : density of core and face sheets; z_c, z_f : height of the core and face sheets.

3. Numerical investigation

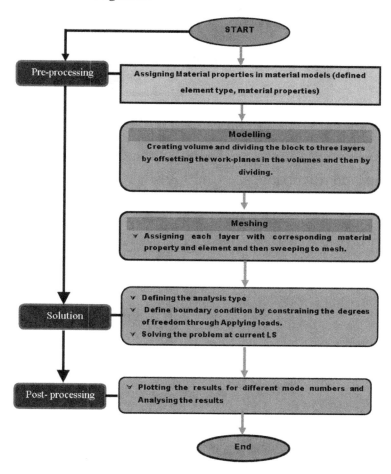

Figure 31.2: FEA process flowchart.

FEA is conducted using ANSYS 16.0 on a plate of size 200×200 mm, with variable thicknesses of 2–10 mm. Young's modulus, density, and Poisson's ratio for the core are taken as 1.79 MPa, 968.1 kg/m³, and 0.3, respectively. The face layers are divided into multiple layers with variable distribution of material properties governed by a power distribution formula (Sakiyama et al., 1996), such that the outer layer is assumed to be carbon fiber-rich and the layer adjacent to the core is epoxy-rich. For discretization of the structure, an 8-node brick element with a size of 4 mm, verified through convergence tests, is considered. The analysis methodology is shown in Figure 31.2. The meshed model of the plate and the first three mode shapes are presented in Figure 31.3a–d.

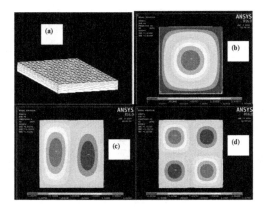

Figure 31.3: FEA simulation plots; (a) meshed model; (b) 1st mode vibration, (c) 2nd mode vibration, (d) 3rd mode vibration.

4. Results and discussion

Table 31.1: Validation of the current method.

Vibration Mode	Current study		% Error		
	Th.	FEA	Th.	Lou et al. (2012)	Li et al. (2020)
Mode-I	152.2	143.2	5.913	1.531	0.510
Mode-II	535.2	527.8	1.382	1.383	1.080
Mode-III	1140	1092.5	4.166	1.199	0.540

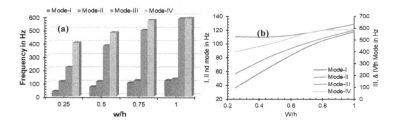

Figure 31.4: Aspect ratio vs. natural frequency.

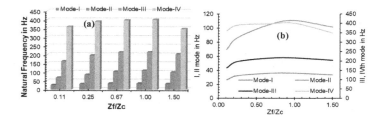

Figure 31.5: Thickness ratio vs. natural frequency.

A 250×250×10 mm sandwiched plate with carbon-epoxy face sheets and a polyurethane core is assumed for validation of the study (Li et al., 2020). The computed frequencies are compared under a simple support condition (Table 31.1). The aspect ratios and the thickness ratio are further varied, and changes in natural frequencies are reported. An increase in plate aspect ratio increases the overall elasticity; as a result, the frequency increases (Figure 31.4). Meanwhile, plate strength is directly proportional to the thickness of face sheets. However, increasing the face sheet thickness more than the thickness of the viscoelastic core has an adverse effect, causing a drop in the natural frequency due to the damping effect of the core (Figure 31.5).

5. Conclusion

This study finds that an increase in aspect ratio surges the natural frequency, with the variation being abrupt in modes I and II only. Increasing the thickness of face sheets increases plate stiffness; however, increasing it beyond the thickness of the core has an adverse effect due to the damping of the core. Thus, the effect of damping may be investigated further.

References

[1] Khalili, S. M. R., Nemati, N., Malekzadeh, K., & Damanpack, A. R. (2010). Free vibration analysis of sandwich beams using improved dynamic stiffness method. *Composite Structures, 92*(2), 387–394.

[2] Li, M., Du, S., Li, F., & Jing, X. (2020). Vibration characteristics of novel multilayer sandwich beams: Modeling, analysis, and experimental validations. *Mechanical Systems and Signal Processing, 142,* 106799.

[3] Lou, J., Ma, L., & Wu, L. Z. (2012). Free vibration analysis of simply supported sandwich beams with lattice truss core. *Materials Science and Engineering: B, 177,* 1712–1716.

[4] Meunier, M., & Shenoi, R. (2001). Dynamic analysis of composite sandwich plates with damping modeled using high-order shear deformation theory. *Composite Structures, 54*(3), 243–254.

[5] Parida, S. P., & Jena, P. C. (2021). Static analysis of GFRP composite plates with filler using higher-order shear deformation theory. *Materials Today: Proceedings, 44,* 667–673.

[6] Parida, S. P., Sahoo, S., & Jena, P. C. (2024). Prediction of multiple transverse cracks in a composite beam using hybrid RNN-mPSO technique. *Proceedings of the Institution of Mechanical Engineers, Part C: Journal of Mechanical Engineering Science.* https://doi.org/10.1177/09544062241239415

[7] Vinson, J. R. (2006). *Plate and panel structures of isotropic, composite, and piezoelectric materials, including sandwich construction.* Springer Science & Business Media.

[8] Sakiyama, T., Matsuda, H., & Morita, C. (1996). Free vibration analysis of sandwich beams with elastic or viscoelastic core by applying the discrete Green function. *Journal of Sound and Vibration, 191*(2), 189–206.

CHAPTER 32

Exploring Multilayer Ensemble Learning for Enhanced Human Activity Recognition

Dhiraj Prasad Jaiswal[1], Dr. Ashok Kumar Shrivastava[2]

Amity School of Engineering and Technology, Amity University Madhya Pradesh, India
Email: dhiraj.jaiswal1@s.amity.edu, akshrivastava1@gwa.amity.edu

Abstract

Human activity recognition (HAR) has gained significant attention due to its applications in healthcare, sports, and human–computer interaction. The complexity and variability of human behavior often challenge single-model HAR methods in achieving high accuracy and generalization. To address this, researchers have turned to machine learning techniques. Ensemble learning, particularly multilayer ensembles, combines multiple models to enhance predictive performance. This paper explores the challenges in recognizing complex activities in diverse environments, such as overfitting, noise, and sensor variability. It studies and evaluates various multilayer ensemble methods for HAR, discussing their advantages, disadvantages, and potential for future research. Additionally, it compares the performance of these multilayer ensemble methods with single-model approaches.

Keywords: Human activity recognition, multilayer ensemble method, single-model approaches, deep learning, bagging, boosting, stacking, performance metrics

1. Introduction

Human activity recognition (HAR) classifies human actions or behaviors using data from sensors, which can range from simple activities like walking to complex ones like cooking (Lara & Labrador, 2013). HAR employs wearable sensors, smartphones, and environmental sensors to capture human motion dimensions such as acceleration, velocity, orientation, and location (Bulling et al., 2014). The complexity of the activity and its context can impact data analysis. The rise of smart devices and IoT has advanced HAR research and applications in fields like healthcare, sports, smart homes, and human–computer interaction (Wang et al., 2019).

HAR faces challenges due to the complexity and variability of human activities. Key factors include Intraclass variability: An individual's performance of an activity can vary significantly each time, influenced by factors like mood, health, and terrain (Bulling et al., 2014). Interclass variability: Different individuals perform the same activity differently, such as older adults walking slower and with different postures compared to young adults. Sensor variability: Data can differ based on sensor placement, orientation, and type. Noise: Sensor data can be noisy due to errors or external factors like environmental conditions, affecting recognition performance.

DOI: 10.1201/9781003596776-32

This paper examines how ensemble machine-learning techniques improve HAR systems by enhancing prediction accuracy and robustness through the combination of multiple algorithms. It specifically focuses on multilayer ensemble methods, addressing a gap in current literature, and reviews various ensemble types (bagging, boosting, stacking, and cascade), their applications, evaluation metrics, and comparative advantages over traditional single-model approaches. Future research directions in HAR for these methods are also explored.

2. Single-Model, Ensemble, and Multilayer Ensemble Machine Learning Approaches to HAR

2.1. Overview and Limitations of Traditional Machine Learning Methods for HAR

Traditional machine learning for HAR uses Decision Trees (DT): Easy to interpret but prone to overfitting. Support Vector Machines (SVM): Robust but computationally expensive. K-Nearest Neighbors (k-NN): Sensitive to parameter choices. Naive Bayes (NB): Simple but relies on a strong assumption. Random Forest (RF): Powerful but lacks interpretability. These methods struggle with complex HAR data, motivating advanced techniques. Single-model HAR suffers from limitations (Lara & Labrador, 2013): Overfitting complex models (e.g., DT) can capture noise as patterns. Reliance on hand-crafted features may miss key aspects. Linear models struggle with nonlinear activity patterns. High dimensionality and noisy data can lead to poor performance. Lack of robustness to data variations (e.g., sensor placement, individual differences).

2.2. Ensemble Learning Concepts, Principles, and Advantages

Ensemble learning integrates predictions from multiple models to boost accuracy and robustness, transforming weak learners into a potent unified model (Zhou, 2012). Diversity among models is crucial, ensuring they make distinct errors while remaining accurate. Techniques like majority voting combine predictions, aiming to reduce variance while controlling bias. Methods such as bagging, boosting, and stacking are employed to foster diversity and enhance overall model performance. Ensemble methods in HAR provide significant advantages over single models. They improve accuracy by leveraging the strengths of multiple models (Zhou, 2012), enhance robustness against noise and outliers in sensor data, effectively handle high-dimensional data, model complex activity patterns, prevent overfitting using techniques like bagging and boosting, and generally generalize better to unseen data.

2.3. Types of Ensemble Techniques used in HAR

HAR researchers use ensemble methods to improve performance beyond single models. These include bagging, boosting, stacking, cascade-based methods, and hybrid methods.

2.4. Bagging-based Ensemble Methods

Bagging (Bootstrap Aggregating) boosts machine learning robustness by training multiple models on data subsets with replacement and combining their predictions. In HAR, RF uses bagging for better accuracy with complex sensor data (Zhou, 2012). Rotation Forest, another bagging variant, enhances classifier diversity for improved HAR performance (Ramirez et al., 2021).

2.5. Boosting-based Ensemble Methods

In boosting, models are trained sequentially to address errors from previous models like AdaBoost, and Gradient Boosting with its variant XGBoost (Zhou, 2012). These methods achieve high accuracy in HAR but are more prone to overfitting than bagging.

2.6. Stacking-based Ensemble Methods

Stacking, combining predictions from various models, improves HAR accuracy (Wang et al., 2019). It excels with complex data but can increase training time.

2.7. Cascade-based Ensemble Methods

These methods use a layered approach for complex activity recognition. Bulling et al. (2014) applied this with a wrist-worn accelerometer, categorizing activities (e.g., walking/running under "ambulation") and using sequential classifiers, enhancing accuracy but relying on hierarchical activity structures.

2.8. Other Multilayer Ensemble Approaches for HAR

Hybrid ensemble methods combine techniques like bagging and boosting (Sarker et al., 2018). Dynamic ensemble selection (DES) selects optimal classifiers for each instance. Deep ensembles integrate deep learning models (Ferrari et al., 2023).

3. Performance Metrics for Evaluating HAR Systems

3.1. Common Evaluation Metrics used in HAR

HAR performance in machine learning is evaluated by Accuracy (correct predictions), Precision (Positive Predictive Value), Recall (model's ability to identify examples), F1 Score (useful for unbalanced datasets), and AUC-ROC (1 is perfect, 0.5 is random). Common evaluation metrics for HAR are summarized in Table 32.1.

Table 32.1: Overview of common evaluation metrics used in HAR.

Metric	Definition	Strengths	Weaknesses
Accuracy	Ratio of correct predictions to all predictions	Simple and intuitive	Can be misleading for imbalanced classes
Precision	Ratio of true positives to the sum of true and false positives	Reflects correctness of positive predictions	Does not account for false negatives
Recall (Sensitivity)	Ratio of true positives to the sum of true positives and false negatives	Measures ability to identify all positive instances	Does not account for false positives
F1 Score	Harmonic mean of Precision and Recall	Balanced measure of Precision and Recall	Less interpretable than individual Precision and Recall
AUC-ROC	Area under the Receiver Operating Characteristic curve	Considers trade-off between sensitivity and specificity	Can be less sensitive to changes in class distribution

3.1. Challenges in Selecting Appropriate Metrics for Ensemble Methods

Metrics like accuracy, precision, recall, *F*1 score, and AUC-ROC evaluate both single-model and ensemble approaches in HAR. Ensembles face challenges such as assessing model diversity and complexity, handling noise and outliers, and avoiding overfitting. Evaluating ensemble methods effectively in HAR requires using a mix of performance metrics and tools like learning curves (Zhou, 2012).

3.2. Result and Discussion: Application of Multilayer Ensemble Methods in HAR

Several studies have used multilayer ensemble methods to improve HAR systems. Morshed and Islam (2019) combined CNNs and LSTM networks on smartphone data, outperforming individual models. Sarker et al. (2018) hybridized bagging and boosting for better accuracy. Wang et al. (2019) stacked decision trees and SVMs, surpassing other methods. Guan and Plötz (2017) used multiple deep learning models, showing superior performance.

Ensemble methods generally outperform single models in HAR tasks (Table 32.2). For instance, Morshed and Islam (2019) achieved 97.2% accuracy compared to 94.3% for the best individual model; Sarker et al. (2018) achieved 93.5% accuracy, surpassing single classifiers' accuracies of 83.2%–88.7%; Wang et al. (2019) reported 91.8% accuracy with their stacking ensemble, while Guan and Plötz (2017) achieved 96.7% accuracy compared to 94.2% with the best single model.

Table 32.2: Performance comparison of multilayer ensemble and single-model in HAR.

Study	Ensemble method	Best single-model approach	Ensemble accuracy	Single-model accuracy
Morshed and Islam (2019)	Deep Ensemble (CNN + LSTM)	Convolutional Neural Network	97.2%	94.3%
Sarker et al. (2018)	Hybrid Bagging and Boosting	Varies (Range)	93.5%	83.2%–88.7%
Wang et al. (2019)	Stacking Ensemble	Support Vector Machine (SVM)	91.8%	88.9%
Guan and Plötz (2017)	Ensemble of Deep Learning Models	Convolutional Neural Network	96.7%	94.2%

3.2.1. Challenges and Limitations of Multilayer Ensemble Methods in HAR

Multilayer ensemble methods in HAR offer advantages but raise challenges like increased complexity, computational demands, overfitting risks, and the need for diverse models and more data. To address these, researchers explore simpler models, parallel computing for faster training, limiting layers, model pruning, and using interpretable techniques like LIME explanations and interpretable base models (Zhou, 2012).

3.2.2. Future Research Directions

The Multilayer Ensemble HAR field is advancing with diverse approaches, integrating CNNs and RNNs, exploring deep stacking networks, and addressing imbalanced data. Efforts include enhancing interpretability (Hayat et al., 2022), developing online ensemble methods for continuous data, and integrating deep learning and transfer learning for generalization.

4. Conclusion

Multilayer ensemble methods enhance HAR systems by improving accuracy and overcoming single-model limitations. Techniques like bagging, boosting, stacking, and cascade methods handle high-dimensional, noisy, and imbalanced data while reducing overfitting. Challenges include training time and model complexity, addressed with solutions like model pruning and parallel computing. Integrating deep learning and transfer learning into ensembles is emerging, with a focus on improving model interpretability for more robust HAR systems.

References

[1]   Bulling, A., Blanke, U., & Schiele, B. (2014). A tutorial on human activity recognition using body-worn inertial sensors. *ACM Computing Surveys (CSUR), 46*(3), 1–33.

[2]   Ferrari, A., Micucci, D., Mobilio, M., & Napoletano, P. (2023). Deep learning and model personalization in sensor-based human activity recognition. *Journal of Reliable Intelligent Environments, 9*(1), 27–39.

[3]   Guan, Y., & Plötz, T. (2017). Ensembles of deep LSTM learners for activity recognition using wearables. *Proceedings of the ACM on Interactive, Mobile, Wearable and Ubiquitous Technologies, 1*(2), 1–28.

[4]   Hayat, A., Fernando, M. D., Bhuyan, B. P., & Tomar, R. (2022). Human activity recognition for elderly people using machine and deep learning approaches. *Information (Switzerland), 13*(6). https://doi.org/10.3390/info13060281

[5]   Lara, O. D., & Labrador, M. A. (2013). A survey on human activity recognition using wearable sensors. *IEEE Communications Surveys & Tutorials, 15*(3), 1192–1209.

[6]   Morshed, T., & Islam, M. M. (2019). Deep learning with convolutional neural network and long short-term memory for human activity recognition. *International Journal of Advanced Computer Science and Applications, 10*(3).

[7]   Ramirez, H., Velastin, S. A., Meza, I., Fabregas, E., Makris, D., & Farias, G. (2021). Fall detection and activity recognition using human skeleton features. *IEEE Access, 9*, 33532–33542.

[8]   Sarker, I. H., Kamruzzaman, J., & Ahmad, S. (2018). HAR-MAD: A hybrid bagging and boosting-based abnormal human activity recognition using multivariate autoregressive distance. *Information Sciences, 453*, 280–298.

[9]   Wang, J., Chen, Y., Hao, S., Peng, X., & Hu, L. (2019). Deep learning for sensor-based activity recognition: A survey. *Pattern Recognition Letters, 119*, 3–11.

[10] Zhou, Z. H. (2012). *Ensemble methods: Foundations and algorithms.* CRC Press.

CHAPTER 33

Implementing AutoML for Wine Quality Classification Using AutoGluon and Leaderboard Analysis

K. P. Swain[1], S. R. Samal[2], S. Misra[3], S. K. Mohapatra[4], Asima Rout[5]

[1]Department of ETC, TAT, Bhubaneswar, India kaleep.swain@gmail.com
[2]Department of ECE, SIT, Bhubaneswar, Odisha, India
[3]Department of ECE, GITA Autonomous College, Bhubaneswar, Odisha, India
[4]Department of ETC, TAT, Bhubaneswar, India sumsusmeera@gmail.com
[5]Department of ETC, IGIT, Sarang, Dhenkanal, Odisha, India

Abstract

This study introduces a proposal for implementing AutoML using the AutoGluon library. This research focuses on analyzing a dataset related to wine, specifically evaluating and classifying the quality of wine as either "bad" or "good." A library called "leaderboard" is incorporated into the implementation to facilitate this investigation. The primary objective of the "leaderboard" library is to provide a mechanism for comparing the evaluation scores of different machine learning algorithms. By utilizing AutoGluon and leveraging its automated machine-learning capabilities, the proposed implementation aims to streamline the process of wine quality classification. The "leaderboard" component further enhances the assessment by allowing for a comprehensive examination of the performance exhibited by different algorithms. This research is expected to make noteworthy contributions to the field of automated machine learning, primarily in the area of wine quality classification. The findings from this study are anticipated to provide valuable insights and pave the way for creating more effective and accurate models for evaluating wine quality.

Keywords: AutoML, AutoGluon, leaderboard, performance metrics

1. Introduction

Machine learning has transformed several fields by empowering data-driven decision-making and predictive modeling. However, developing and refining machine learning models for optimal performance can be time-consuming, complex, and highly dependent on expert knowledge. Over the past few years, AutoML (Automated Machine Learning) has shown great potential as a solution to overcome these challenges. AutoML refers to automating various stages within the machine learning pipeline, including data preprocessing, feature engineering, algorithm choice, hyperparameter optimization, and model evaluation (Hutter, 2019). By automating these tasks, AutoML seeks to make machine learning more

DOI: 10.1201/9781003596776-33

accessible to nonexperts, reduce the manual effort involved, and enhance the efficiency and usefulness of model development.AutoML encompasses a diverse range of techniques, including heuristic search-based methods, Bayesian optimization, neural architecture search, evolutionary algorithms, and meta-learning. These methods aim to automatically explore and optimize the configuration space of machine learning models, considering different algorithms, hyperparameter settings, and preprocessing techniques, to find the best-performing model for a given task (Feurer, 2015).

Among AutoML libraries, AutoGluon is a powerful tool that has gained significant attention in the machine-learning community. It provides a comprehensive and user-friendly framework for automating the end-to-end process of developing and optimizing machine learning models. AutoGluon simplifies traditionally complex tasks such as data preprocessing, feature engineering, algorithm selection, hyperparameter tuning, and model evaluation, making it accessible to both experts and nonexperts. One of AutoGluon's key strengths is its ability to handle various machine-learning tasks, including classification, regression, and object detection. It can handle both structured and unstructured data, making it versatile for numerous applications. AutoGluon also offers a user-friendly interface and requires minimal manual intervention, allowing users to deploy high-performing models without extensive coding or machine learning expertise (Ferreira, 2021). Autogluon has been widely adopted and has received positive recognition in the machine-learning community (Ferreira, 2021). It has been used in various research projects and has demonstrated competitive performance in machine learning competitions. The AutoGluon library has a growing lively community of dedicated contributors and users who actively participate in its advancement and provide assistance to fellow members.

Regarding the application of AutoGluon, a study (Wenwen, 2021) introduces AutoGluon as a framework for landslide hazard analysis. AutoGluon streamlines the analysis by automating data preprocessing, feature selection, model selection, and hyperparameter tuning. The paper highlights its effectiveness in accurately predicting and mapping landslide-prone areas. Additionally, AutoGluon-Tabular is an AutoML framework (Erickson, 2020) designed for structured data analysis. AutoML delivers an effective and precise solution for automating multiple phases of the machine learning pipeline, including data preprocessing, feature engineering, model selection, and hyperparameter tuning. By leveraging state-of-the-art techniques and ensemble learning, AutoGluon-Tabular achieves exceptional performance on structured datasets, surpassing existing AutoML frameworks. The use of Machine Learning for objective evaluation and automated scoring of rhinoplasty outcomes is investigated in (Dobratz, 2023). Here, a methodology is proposed to map visual aesthetics to quantified measurements using synthetic 3D models generated with AI tools. The AutoGluon AutoML framework is utilized to create the highest-performing machine learning model, achieving 82–88% accuracy.

In the context of machine learning, a leaderboard is a tool that compares and evaluates different models or algorithms based on their performance metrics. It provides a centralized view of performance scores, enabling researchers to assess their algorithms in a transparent manner. The leaderboard serves as a benchmarking tool, allowing for a comprehensive analysis of the strengths and weaknesses of different approaches (Ferreira, 2021).

This study aims to implement an AutoML framework using AutoGluon for classifying wine quality as "bad" or "good." The study analyzes a wine dataset and leverages AutoGluon's capabilities. The incorporation of the 'leaderboard' library allows for the comparison of different machine learning algorithms, aiming to streamline the wine quality classification process and enhance performance analysis.

2. Methodology

2.1. Dataset

The dataset, collected from Kaggle, contains 1600 samples and includes wine-related features and quality ratings. Each row represents a wine sample, and the columns represent various attributes. The dataset includes features such as alcohol content, fixed acidity, volatile acidity, citric acid, residual sugar, and chloride content, all recorded in float64 format. Additionally, the dataset records free sulfur dioxide and total sulfur dioxide, again in float64. Density and pH level are both measured as float64. Sulphates content is another feature, recorded in float64. Finally, wine quality is represented as an object data type.

3. Implementation

3.1. Steps involved

The steps involved using AutoGluon and AutoKeras libraries for automated machine-learning tasks on the wine dataset. The code starts by uploading the dataset using the files. upload() function and reads it into a Pandas DataFrame. The DataFrame is then divided into training and testing groups. AutoGluon's TabularDataset is used to create objects for training and testing data. Here, AutoML tools, including AutoGluon which is used in this study, often provide the functionality to automatically split a dataset into training and testing sets. AutoGluon's TabularPredictor class is used to train a model, with the target variable set as "quality." After model development, predictions are made on the test dataset, and evaluation metrics, such as accuracy, are computed. A leaderboard is generated to rank the models based on their performance on the training dataset. Overall, this code showcases the use of AutoGluon and AutoKeras for automating the machine-learning pipeline on tabular data. It is noted that the automated machine learning (AutoML) method is implemented here just to determine the quality of the wine. It does not directly provide a guideline on how to know if a wine's quality is good or bad, but it details a computational method to classify wine quality.

3.2. Evaluation score

AutoML tools like AutoGluon explore different models and hyperparameters automatically. The number of runs depends on the AutoML configuration, such as time limits, resource limits, and the exploration strategy.

Table 33.1: Evaluation scores.

Sl. No	Performance metrics	Score
1	Accuracy	0.7875
2	Balanced accuracy	0.787743506
3	MCC	0.572539251
4	ROC characteristic Area under the curve	0.866730925
5	$f1$	0.805491991
6	Precision	0.82629108
7	Recall	0.785714286

4. Leaderboard outcomes

Figure 33.1 shows the performance metrics of various models in a leaderboard. Each row corresponds to a model, with columns detailing the model's name, test score, validation score, prediction time for test and validation sets, fit time (training time), and additional metrics. "Score_test" and "score_val" indicate performance on test and validation sets. "Pred_time_test" and "pred_time_val" represent prediction times for test and validation sets. "Fit_time" denotes the total training time. "Pred_time_test_marginal" and "pred_time_val_marginal" show incremental prediction times, while "fit_time_marginal" displays incremental fit time. Other columns provide information on the model's stack level, inference capability, and fit order. The table enables easy comparison and evaluation of model performance.

model	score_test	score_val	pred_time_test	pred_time_val	fit_time	pred_time_test_marginal	pred_time_val_marginal	fit_time_marginal	stack_level	can_infer	fit_order
WeightedEnsemble_L2	0.967473	0.837500	0.379460	0.171883	6.484291	0.005087	0.002115	1.329619	2	True	14
ExtraTreesGini	0.964137	0.820833	0.137280	0.060073	0.786903	0.137280	0.060073	0.786903	1	True	8
RandomForestEntr	0.963303	0.816667	0.179810	0.086532	1.675051	0.179810	0.086532	1.675051	1	True	6
XGBoost	0.962469	0.812500	0.035141	0.008100	0.672702	0.035141	0.008100	0.672702	1	True	11
LightGBMLarge	0.962469	0.812500	0.103468	0.018376	1.606455	0.103468	0.018376	1.606455	1	True	13
ExtraTreesEntr	0.962469	0.812500	0.146045	0.102249	0.812792	0.146045	0.102249	0.812792	1	True	9
RandomForestGini	0.962469	0.812500	0.173467	0.100492	1.600981	0.173467	0.100492	1.600981	1	True	5
LightGBM	0.956631	0.800000	0.027046	0.008428	1.188144	0.027046	0.008428	1.188144	1	True	4
KNeighborsDist	0.929108	0.645833	0.018597	0.012267	0.033225	0.018597	0.012267	0.033225	1	True	2
CatBoost	0.895746	0.812500	0.017918	0.002647	1.535007	0.017918	0.002647	1.535007	1	True	7
NeuralNetFastAI	0.829024	0.800000	0.032489	0.012603	2.440203	0.032499	0.012603	2.440203	1	True	10
LightGBMXT	0.811510	0.804167	0.020889	0.013994	4.169754	0.020889	0.013994	4.169754	1	True	3
NeuralNetTorch	0.793995	0.795833	0.022151	0.015063	2.020015	0.022151	0.015063	2.020015	1	True	12
KNeighborsUnif	0.735613	0.595833	0.038752	0.030008	3.981698	0.038752	0.030008	3.981698	1	True	1

Figure 33.1: Snapshot for performance evaluation using leaderboard.

5. Conclusion

This study proposed using AutoML with the AutoGluon library for wine quality classification. The automated approach aimed to streamline the process and enhance accuracy. The study used the "leaderboard" library to evaluate various machine learning algorithms, with the top model, WeightedEnsemble_L2, achieving a test score of 0.967. Other models like ExtraTreesGini, RandomForestEntr, and XGBoost also performed well. The analysis showed an accuracy score of 0.7875, with balanced accuracy, MCC, and ROC AUC scores of 0.7877, 0.5725, and 0.8667, respectively, representing good performance on imbalanced datasets.

References

[1]   Dobratz, E. J., Akbas, M. I., Dougherty, W. M., Akinci, T. C., & Mazhar, M. (2023). Utilization of machine learning for the objective assessment of rhinoplasty outcomes. *IEEE Access, 11*, 42135–42145.

[2]   Erickson, N., Mueller, J., Shirkov, A., Zhang, H., Larroy, P., Li, M., & Smola, A. (2020). AutoGluon-Tabular: Robust and accurate AutoML for structured data. *7th ICML Workshop on Automated Machine Learning*, July 18, 2020.

[3]   Feurer, M., Klein, A., Eggensperger, K., Springenberg, J., Blum, M., & Hutter, F. (2015). Efficient and robust automated machine learning. *Advances in Neural Information Processing Systems, 28* (NIPS 2015).

[4]   Ferreira, L., Pilastri, A., Martins, C. M., Pires, P. M., & Cortez, P. (2021). A comparison of AutoML tools for machine learning, deep learning, and XGBoost. *2021 International Joint Conference on Neural Networks (IJCNN)*, Shenzhen, China, pp. 1–8.

[5]   Hutter, F., Kotthoff, L., & Vanschoren, J. (2019). Hyperparameter optimization. In F. Hutter, L. Kotthoff, & J. Vanschoren (Eds.), *Automated machine learning: Methods, systems, challenges* (pp. 3–33). Springer.

[6]   Qi, W., Xu, C., & Xu, X. (2021). AutoGluon: A revolutionary framework for landslide hazard analysis. *Natural Hazards Research, 1*, 103–108.

CHAPTER 34

Decoding E. coli

Machine Learning Approaches to Identify Promoter DNA Sequences

[1]R. K. Nayak, [2]S. K. Nayak, [3]S. K. Mohapatra, [4]Asima Rout, [5*]K. P. Swain

[1]Department of CSE, GITA Autonomous College, Bhubaneswar, Odisha, India
[2]Deputy Director Examination, BPUT, Odisha, India
[3]Department of ETC, TAT, Bhubaneswar, India
[4]Department of ETC, IGIT, Sarang, Dhenkanal, Odisha, India
[5]Department of ETC, TAT, Bhubaneswar, India
Corresponding author: kaleep.swain@gmail.com

Abstract

In the realm of molecular biology and genetics, understanding the complex mechanisms that regulate gene expression is pivotal. Among the myriad elements involved, promoter DNA sequences play a critical role as regulatory regions where transcription of DNA to messenger RNA begins. This study focuses on employing various machine learning techniques to accurately identify promoter sequences within E. coli DNA. Utilizing a dataset from the UCI Machine Learning Repository, which includes sequences designated as either promoter or nonpromoter, we leverage Python alongside libraries such as NumPy, Scikit-learn, and Pandas for analysis. Through meticulous preprocessing, including converting categorical data into numerical formats and partitioning the dataset into training and testing sets, we apply several classification algorithms. Our findings reveal significant success in distinguishing promoter from nonpromoter sequences, achieving an impressive accuracy of 96%. This exploration not only showcases the efficacy of machine learning in biological sequence analysis but also opens avenues for further research into gene regulation and its applications in biotechnology and medicine.

Keywords: Machine Learning, Promoter DNA Sequences, E. coli Genetics, Gene Regulation

1. Introduction

Short segments of E. coli DNA are relatively small sequences of deoxyribonucleic acid (DNA) from the Escherichia coli bacterium, a model organism widely studied in genetics and molecular biology (Ruiz & Silhavy, 2022; Geurtsen et al., 2022). In E. coli, short segments are crucial for research and applications such as studying gene expression, understanding genetic regulation, and developing biotechnological applications.

E. coli DNA consists of a sequence of nucleotides, each with one of four nitrogenous bases: adenine (A), cytosine (C), guanine (G), and thymine (T). The order of these bases

DOI: 10.1201/9781003596776-34

constitutes the genetic information. Short segments typically refer to pieces of this sequence manageable for analysis, often ranging from a few to several hundred base pairs (Alberts et al., 2002). These segments can include various genetic elements, such as promoters, which are DNA sequences that serve as binding sites for RNA polymerase and other proteins to initiate transcription. This transcription process copies DNA into messenger RNA (mRNA), which then serves as a template for protein synthesis. Nonpromoter regions may include coding sequences (genes), enhancers, silencers, or noncoding DNA (Haberle & Stark, 2018).

Short segments of *E. coli* DNA are like snippets from the genetic library of the *Escherichia coli* bacterium. Imagine the genetic code as a book collection, where each book represents a chromosome. A short DNA segment is akin to a paragraph from one of these books, composed of sentences (genes) made up of chemical bases (A, C, G, T). These segments contain critical information for the bacterium's survival and function, such as instructions for making a protein, regulating protein production, or protecting against viruses. Studying these segments is like decoding secret messages to understand how genetic information controls living organisms at the molecular level, much like a detective analyzing clues to uncover how life operates.

In this project, the focus is particularly on segments known as promoters, which are essentially the "switches" that control the "lights" (genes) in our analogy. Identifying whether a short DNA segment is a promoter or not is like figuring out if a paragraph contains instructions to turn on a light. This knowledge is crucial for many applications, from understanding diseases to designing new medicines or even bioengineering crops for agriculture.

2. Literature Review

The identification and analysis of promoter sequences in bacterial DNA have been fundamental in understanding genetic expression and regulation (Lagator et al., 2022; Hernández et al., 2022). Historically, promoter sequences have been identified through experimental methods such as nuclease footprinting and gene knockout techniques. These methods, while effective, are often labor-intensive and time-consuming (Kim et al., 2005). With the advent of bioinformatics and computational biology, machine-learning techniques enhance the speed and accuracy of these identifications, particularly in well-studied organisms like *E. coli* (Oon et al., 2023).

Recent studies have increasingly applied machine learning techniques to genetic data, demonstrating significant success in identifying complex patterns that traditional methods might miss. In a collection of cutting-edge research, machine learning techniques are being extensively utilized to advance our understanding and capabilities in biological data analysis. Öz and Kaya (2013) explored the application of support vector machines (SVMs) to differentiate DNA sequencing data as high or low quality, demonstrating the utility of SVMs in pattern recognition for biological data.

3. Materials and Methods

3.1. Dataset Description

The dataset from the UCI Machine Learning Repository consists of sequences of *E. coli* promoter gene sequences (DNA), designed to evaluate the "KBANN" hybrid learning algorithm. It includes 106 instances with categorical features, represented by "a", "g", "t", 'c', corresponding to the nucleotides adenine, guanine, thymine, and cytosine, respectively. These nucleotides are the building blocks of DNA. The dataset supports tasks in biology, specifically classification, without any missing values. For further information, visit the UCI Machine Learning Repository's Molecular Biology (Promoter Gene Sequences) page.

4. Implementation of Machine Learning Algorithms

In the phase of machine learning algorithm implementation, accuracy was chosen as the scoring method to gauge model performance. A suite of classification models was then defined for training, which included k-Nearest Neighbors, Gaussian Naive Bayes, Support Vector Machine (SVM) with various kernels, Decision Tree, Random Forest, Neural Network, and AdaBoost, each fine-tuned with specific hyperparameters like maximum depth for Decision Trees and the number of estimators for AdaBoost.

Table 34.1: Performance metrics of machine learning classifiers.

Name of the classifier	Mean (Standard Deviation)
Nearest Neighbors	0.837500 (0.125623)
Gaussian Process	0.873214 (0.056158)
Decision Tree	0.683929 (0.217542)
Random Forest	0.571429 (0.135055)
Neural Net	0.887500 (0.087500)
AdaBoost	0.912500 (0.112500)
Naive Bayes	0.837500 (0.137500)
SVM Linear	0.850000 (0.108972)
SVM RBF	0.887500 (0.067315)
SVM Sigmoid	0.900000 (0.093541)

The table presents the performance metrics of different classifiers used in the study to distinguish between promoter and nonpromoter DNA sequences in *E. coli*. Each classifier's effectiveness is quantified by its mean accuracy score, with an accompanying standard deviation that measures the variability of the model's accuracy. The Nearest Neighbors and Gaussian Process classifiers show promising results with accuracy scores over 83%, with the Gaussian Process demonstrating less variability in its predictions. Decision Tree and Random Forest classifiers exhibit lower accuracy, indicating they might be less effective for this specific classification task. The Neural Network and AdaBoost classifiers stand out with the highest accuracy rates, particularly the AdaBoost with a mean accuracy of 91.25%, suggesting it is the most reliable model among those tested. The Naive Bayes, SVM Linear, SVM RBF, and SVM Sigmoid classifiers also perform well, with the SVM Sigmoid scoring an impressive 90% mean accuracy. This table highlights the diverse capabilities of these algorithms and underscores the importance of selecting the right machine-learning model for accurate genomic sequence classification.

Table 34.2: Performance metrics of SVM linear classifier.

	Precision	Recall	f1-score	Support
0	1.00	0.94	0.97	17
1	0.91	1.00	0.95	10
Accuracy			0.96	27
Macro avg	0.95	0.97	0.96	27
Weighted avg	0.97	0.96	0.96	27

Though the performance evaluation of each classifier is discussed earlier, Table 34.2 is presented here for the SVM classifier with the most suitable one. The table provides a performance evaluation of an SVM classifier with a linear kernel on a binary classification task for *E. coli* DNA sequences. It attained a mean accuracy of 85% with a standard deviation of 10.8972%, indicating a generally high but slightly variable performance. The precision of 100% for the nonpromoter class (0) and 91% for the promoter class (1) shows the classifier's high reliability in predicting nonpromoter and a good but less perfect rate for promoters. Recall rates are high for both classes, at 94% for nonpromoters and 100% for promoters, indicating the classifier's strong ability to detect true positives. The *F1 scores*, which balance precision and recall, are 97% for nonpromoters and 95% for promoters, suggesting excellent overall classifier performance. The classifier's overall accuracy of 96% across 27 instances, along with high macro and weighted averages of 95% and 97%, respectively, confirm its effectiveness in the context of the given classification task.

5. Conclusion

In conclusion, our work involved a thorough exploration of machine learning techniques applied to biological sequence data. We began by importing essential libraries such as NumPy, Scikit-learn, and Pandas, and loaded the molecular biology promoter gene sequence dataset. Constructing a custom Pandas DataFrame ensured clarity and organization, with each column representing a series for efficient data management. After preprocessing the dataset by converting categorical variables into numerical format, dropping redundant columns, and splitting it into training and testing sets, we implemented various classification algorithms. Through rigorous model building and training, we successfully predicted whether a short sequence of *E. coli* bacterial DNA served as a promoter or nonpromoter, achieving an impressive accuracy of 96%. By providing a detailed overview of the data importation, conversion, model building, and evaluation processes, our project offers valuable insights into the application of machine learning in biological sequence analysis.

References

[1] Alberts, B., Johnson, A., Lewis, J., Raff, M., Roberts, K., & Walter, P. (2002). Chromosomal DNA and its packaging in the chromatin fiber. In *Molecular Biology of the Cell* (4th ed.). Garland Science. https://www.ncbi.nlm.nih.gov/books/NBK26834/

[2] Geurtsen, J., de Been, M., Weerdenburg, E., Zomer, A., McNally, A., & Poolman, J. (2022). Genomics and pathotypes of the many faces of *Escherichia coli*. *FEMS Microbiology Reviews, 46*(6), fuac031. https://doi.org/10.1093/femsre/fuac031

[3] Haberle, V., & Stark, A. (2018). Eukaryotic core promoters and the functional basis of transcription initiation. *Nature Reviews Molecular Cell Biology, 19*(10), 621–637. https://doi.org/10.1038/s41580-018-0028-8

[4] Hernández, D., Jara, N., Araya, M., Durán, R. E., & Buil-Aranda, C. (2022). PromoterLCNN: A light CNN-based promoter prediction and classification model. *Genes, 13*(7), 1126. https://doi.org/10.3390/genes13071126

[5] Kim, T. H., Barrera, L. O., Qu, C., Van Calcar, S., Trinklein, N. D., Cooper, S. J., Luna, R. M., Glass, C. K., Rosenfeld, M. G., Myers, R. M., & Ren, B. (2005). Direct isolation and identification of promoters in the human genome. *Genome Research, 15*(6), 830–839. https://doi.org/10.1101/gr.3430605

[6] Lagator, M., Sarikas, S., Steinrueck, M., Toledo-Aparicio, D., Bollback, J. P., Guet, C. C., & Tkačik, G. (2022). Predicting bacterial promoter function and evolution from random sequences. *eLife, 11*, e64543. https://doi.org/10.7554/eLife.64543

[7]   Oon, Y. L., Oon, Y. S., Ayaz, M., Deng, M., Li, L., & Song, K. (2023). Waterborne pathogens detection technologies: Advances, challenges, and future perspectives. *Frontiers in Microbiology, 14*, 1286923. https://doi.org/10.3389/fmicb.2023.1286923

[8]   Öz, E., & Kaya, H. (2013). Support vector machines for quality control of DNA sequencing. *Journal of Inequalities and Applications, 2013*, 85. https://doi.org/10.1186/1029-242X-2013-85

[9]   Ruiz, N., & Silhavy, T. J. (2022). How *Escherichia coli* became the flagship bacterium of molecular biology. *Journal of Bacteriology, 204*(9), e0023022. https://doi.org/10.1128/jb.00230-22

Solid waste management as a supplement to binding material in the cement and concrete industry

Aakash Kumar Gupta; Prasanna Kumar Acharya

School of Civil Engineering, KIIT DU, Bhubaneswar, India
#Mail: pkacharya64@yahoo.co.in (Corresponding Author)

Abstract

Management of solid waste is a universal issue. The construction industry consumes huge amounts of concrete and demands gigantic production of cement, which emits a large amount of carbon dioxide. To manage the issues associated with cement and to reduce its production, efforts are being made to use waste stream materials as a supplement to cement. To use any waste material as a supplement to cement, its hydration characteristics need to be checked. This report summarizes the present knowledge on the hydration mechanism of cement mixes produced by mixing various alternative cementitious materials (ACMs) like furnace slag, rice husk ash, silica fume, fly ash, limestone powder, palm oil ash, metakaolin, red mud, and silica nanoparticles. Based on the study, it is observed that ACMs have a major impact on the hydration characteristics of binding materials. It is found that the characteristics of the parent material and the temperature affect the hydration of cement composites.

Keywords: Hydration, supplementary materials, X-ray diffraction, fly ash, silica fume

1. Introduction

Concrete and cement are important for today's world development. The concrete-making industry faces the challenge of finding ways to lower carbon emissions and integrate energy in the common Portland cement (Schöler et al., 2015). Additional materials like furnace slag, rice husk ash (RHA), silica fume (SF), fly ash (FA), metakaolin (MK), etc., have been developed by many researchers in recent years and not only enhance concrete characteristics but also help to save construction costs. ACMs popularly include waste materials, natural pozzolans, and minerals having pozzolanic and/or hydraulic properties (Arora et al., 2016). There is a need to analyze the hydration of binders since there is a clear relationship between hydration with heat, porosity, strength properties, shrinkage, and deformation (Amen, 2011). This study compares the hydration characteristics, and physical and chemical attributes of different types of ACMs such as slag, class FA, RHA, SF limestone powder, palm oil ash, metakaolin, red mud (RM), and silica nanoparticles, which have been used as substitutes for cement in the manufacturing of concrete.

DOI: 10.1201/9781003596776-35

2. Ground blast furnace slag

Merzouki et al. (2013) proposed a model for explaining the Portland cement reaction containing blast furnace slag (BFS). Park and Choi (2019) investigated self-healing ability with the addition of GGBS (60%), FA (35%, 50%), and CSA (10%) of additional cement materials expanding agents like GGBS, FA, and Calcium Sulphur aluminate (CSA).

2.1. Fly ash

Zeng et al. (2012) investigated the hydration degree of cement and the pozzolanic reactivity of FA through a combination of experimental and model-based substitutions, varying from 0% to 60%. Their findings indicated that factors such as the water-cement ratio, nucleation effects, and the hydration process play significant roles in the overall nonevaporative water content of the mixtures, although the hydration of FA primarily influences the levels of calcium hydroxide (CH) present. Meanwhile, Wang et al. (2018) introduced a simulation program aimed at optimizing hydration and strength, utilizing kinetic models to explore the hydration of ternary mixes containing cement, FA, and limestone. This simulation examines binary mixes with limestone or FA substitutions, ranging from 5% to 35%.

2.2. Rice husk ash

Rice husk ash is an extremely reactive pozzolan, which forms part of the combustion process of rice husk (Park et al., 2016). The action of MK and RHA on the hydration behaviors and mechanical characteristics of substitute combined concrete and 5–10% RHA was investigated, and the compressive strength values were predicted to consistently increase in the increasing hydrating period (Shatat, 2016).

2.3. Silica fume

SF is a waste produced by silicon metal manufacturing and is a very reactive pozzolan. Authors (Amen, 2011) have investigated cement paste strength with or without silica fume utilizing different contents of fumes of silica. Authors (Amen, 2011) have also investigated silica nanoparticles (SNP), SF, nonevaporative water (NEW), and cement hydration in the existence of SNP and SF and stated that the degree of hydration was believed to enhance from 28% to 85%.

2.4. Red mud

RM is a composite of metallic and solid oxides is a Bayer process waste product and is used for the manufacture of bauxite alumina. Authors (Govindarajan & Jayalakshmi, 2012; Xu et al., 2019) have evaluated the cement behavior containing red mud at 5% and 20%. Authors (Xu et al., 2019) have investigated the durability and hydration behavior of unbreached RM bricks and FA bricks and suggested a CaO/SiO_2 mass ratio of 1.23, indicating that heavy metals were solidified in the mortar. The mechanical characteristics and toughness were sufficient.

2.5. Limestone

Powdered limestone is a byproduct of the mining of limestone that has been used for several years in cement products and may allegedly impact its characteristics through dilution, filling, chemical effects, and nucleation. Jian et al. (2019) investigated calcium sulphoaluminate cement-limestone powder hydration phase and microstructure production at ages between 1 day to 1365 days for three doses of lime powder, as 5%, 15%, and 25%

substitutes. Arora et al. (2016) recorded the hydration effect, pore structure, and reaction products of high-volume binding material substitution from 20% to 50% by volume using a mixture of slag and limestone. Samuel et al. (2017) stated that the existence of limestone improves the hydration of both clinker and slag.

2.6. Palm oil fuel ash

POFA is a byproduct formed as fuel in the palm oil mill boiler by the burning husk of palm or fiber ash and kernel shell of palm (Awal & Hussin, 2011). POFA is rich in silica, which lowers the temperature of concrete blend hydration. POFA nevertheless is a pozzolanic product and also has a low hydration temperature level (Abd Khalid et al., 2019).

Figure 35.1: Supplementary cementitious materials.

3. Reports on properties

3.1. Heat generation on hydration

3.1.1. Blast furnace slag

Merzouki et al. (2013) used an isothermal TAM Air microcalorimeter to calculate the reaction thermal flux and the initial binding reaction kinetics. During the first 5 days of hydration, each evaluation is administered. Park and Choi (2019) used isothermal calorimetry TAM-AIR to measure self-healing potential owing to the proceeding heat of hydration of unreacted binder.

3.1.2. Limestone

Jian et al. (2019) measured hydration heat using isothermal calorimetric measurements for limestone powder with four doses of 0%, 5%, 15%, and 25% and reported that the coarse limestone powder increases the dilution effects produced by limestone powder. Hence, the limestone powder with 5% weight can enhance the early hydration of binding materials.

3.2. Compressive strength

3.2.1. Fly ash

Wang et al. (2018) used the gel space ratio and Power's strength principle to determine hydrating ternary blended concrete compressive strength development. They stated that concrete porous space is filled with the reaction products, FA, and lime powder, and with the development of binding reaction, the strength of concrete grows.

3.2.2. Rice husk ash

Shatat (2016) reported that with the hydration ages, the hardened paste compression strength produced from MK and PC mixes. The best results are observed with the inclusion of 5–10% RHA. Park et al. (2016) stated that the concrete properties depend on the amount of gel space ratio estimated by the hydration reaction of the binding material and the amount of water. By analyzing the inputs of hydration of cement and the reaction of RHA, the strength of mixes was measured by Powers' strength theory.

3.2.3. Silica fume

Amen (2011) noticed the compression strength of the cement paste initially enhanced at a high rate and slowed down with all kinds of samples with and without silica fume with increased curing time. Increases in w/c from 0.23 to 0.26 were recorded to indicate an improvement in compressive strength for certain 3 and 7-day age specimens, though their compressive strength declined beyond 0.26.

3.2.4. Red mud

Govindarajan and Jayalakshmi (2012) have reported that with curing times, the strength of compression increases. The strength of OPC mixed with 5% RM and OPC added with 20% RM admixture cement is greater than OPC. Xu et al. (2019) reported the strongest compressive strength registered was 32 MPa, which, due to the pozzolanic reactions of red mud and cement, met Chinese national standards GB/T 211 44-2007 crease.

3.2.5. Limestone

Arora et al. (2016) carried out compressive strength tests by (ASTM C109/C109M 2012) cured in a damp atmosphere (greater than 95% RH) at a temperature of 23 ± 2°C with 50 mm cubes of mortar.

3.2.6. Palm oil fuel ash

Abd Khalid et al. (2019) carried out research on concrete cube samples for a curing time of 1, 14, 28, and 56 days with a 2000 kN universal machine calculator and 6 KN/s load speeds and found that the intensity improved with the rise in the curing period of the concrete. Ash of ground POFA and powder of eggshell concrete would have greater compressive strength compared to Unground POFA-egg shell powder.

4. Conclusion

The outcomes of the present study are presented below:The incorporation of blast furnace slag accelerates the chemical process, which when activated with alkaline activators provides greater strength parameters and denser internal structure. Utilization of FA as ACMs when used in large concentrations, hydration delays arise in the early stages but high-volume fly ash binder activations significantly boost the stiffness and mechanical properties. Adding ash to rice husk affects the strength parameters and dilution of cement on the consistency of the cement, which slows hydration. Silica fume particles accelerate hydration at an early stage. The addition of red mud up to 20% has higher compressive strength than OPC at all hydration periods and red mud calcined at 600°C has a good pozzolanic characteristic. The usage of fine limestone powder minimizes porosity. Ash of palm oil fuel is used as a cement substitute in mass concreting where thermal cracking is more concerned because of high heat rises.

References

[1] Abd Khalid, N. H., Abdul Rasid, N. N., Mohd Sam, A. R., Abdul Shukor Lim, N. H., Zardasti, L., Ismail, M., & Majid, Z. A. (2019). The hydration effect on palm oil fuel ash concrete containing eggshell powder. *IOP Conference Series: Earth and Environmental Science, 220*(1), 1–7.

[2] Amen, D. K. H. (2011). Degree of hydration and strength development of low water-to-cement ratios in silica fume cement system. *International Journal of Civil and Environmental Engineering, 11*(5), 10–16.

[3] Arora, A., Sant, G., & Neithalath, N. (2016). Ternary blends containing slag and interground/blended limestone: Hydration, strength, and pore structure. *Construction and Building Materials, 102*, 113–124.

[4] Awal, A. S. M. A., & Hussin, M. W. (2011). Effect of palm oil fuel ash in controlling heat of hydration of concrete. *Procedia Engineering, 14*, 2650–2657.

[5] Govindarajan, D., & Jayalakshmi, G. (2012). Investigations of the influence of calcined red mud on cement hydration. *International Journal of Recent Scientific Research, 3*(12), 1039–1041.

[6] Jian, M., Yu, Z., Ni, C., Shi, H., & Shen, X. (2019). Effects of limestone powder on the hydration and microstructure development of calcium sulphoaluminate cement under long-term curing. *Construction and Building Materials, 199*, 688–695.

[7] Ma, J., Yu, Z., Ni, C., Shi, H., & Shen, X. (2019). Effects of limestone powder on the hydration and microstructure development of calcium sulphoaluminate cement under long-term curing. *Construction and Building Materials, 199*, 688–695.

[8] Merzouki, T., Bouasker, M., Khalifa, N. E. H., & Mounanga, P. (2013). Contribution to the modelling of hydration and chemical shrinkage of slag-blended cement at early age. *Construction and Building Materials, 44*, 368–380.

[9] Park, B., & Choi, Y. C. (2019). Prediction of self-healing potential of cementitious materials incorporating crystalline admixture by isothermal calorimetry. *International Journal of Concrete Structures and Materials, 13*(1), 36.

[10] Park, K., Seung-Jun, K., & Wang, X. (2016). Analysis of the effects of rice husk ash on the hydration of cementitious materials. *Construction and Building Materials, 105*, 196–205.

[11] Schöler, A., Lothenbach, B., Winnefeld, F., & Zajac, M. (2015). Hydration of quaternary Portland cement blends containing blast-furnace slag, siliceous fly ash, and limestone powder. *Cement and Concrete Composites, 55*, 374–382.

[12] Shatat, M. R. (2016). Hydration behavior and mechanical properties of blended cement containing various amounts of rice husk ash in presence of metakaolin. *Arabian Journal of Chemistry, 9*, S1869–S1874.

[13] Tarek, M., Bouasker, M., Khalifa, N. E. H., & Mounanga, P. (2013). Contribution to the modelling of hydration and chemical shrinkage of slag-blended cement at early age. *Construction and Building Materials, 44*, 368–380.

[14] Wang, X. (2018). Analysis of hydration and strength optimization of cement-fly ash-limestone ternary blended concrete. *Construction and Building Materials, 166*, 130–140.

[15] Xu, Y., Yang, B., Liu, X., Gao, S., Li, D., Mukiza, E., & Li, H. (2019). Investigation of the medium calcium-based non-burnt brick made by red mud and fly ash: Durability and hydration characteristics. *International Journal of Minerals, Metallurgy, and Materials, 26*, 983–991.

[16] Zeng, Q., Li, K., Fen-Chong, T., & Dangla, P. (2012). Determination of cement hydration and pozzolanic reaction extents for fly-ash cement pastes. *Construction and Building Materials, 27*(1), 560–569.

CHAPTER 36

Unveiling the Influence

Detecting Drugged Eyes through Image Processing

Hima Bindu Gogineni[1], Bharathi Paleti[2], Uma Sankar Pechetti[3], Angara Satyam[4],
M. Beulah Rani[5], Kaki Leela Prasad[6]

[1] Assistant Professor, Department of CSE(AI&DS), Vignan's Institute of
Information Technology, Visakhapatnam, India
Email: goginenibindu9@gmail.com

[2] Assistant Professor, Department of CSE, Anil Neerukonda Institute of
Technology and Sciences, Visakhapatnam, India
Email: bharathipaleti.cse@anits.edu.in

[3] Assistant Professor, Department of CSE, Sri Vasavi Engineering College (A),
Pedatadepalli, Tadepalligudem, India
Email: umasankar.cse@srivasaviengg.ac.in

[4] Assistant Professor, Department of CSE (AIML), Aditya College of Engineering
and Technology (A), Surampalem, India
Email: satyam.angara@gmail.com

[5] Distinguished Assistant Professor, Department of CSE, Maharaj Vijayaram
Gajapathi Raj College of Engineering (A), Vizianagaram, India
Email: beulahrani@gmail.com

[6] Assistant Professor, Department of CSE, Maharaj Vijayaram Gajapathi Raj
College of Engineering (A), Vizianagaram, India
Email: leelaprasad3@gmail.com

Abstract

Worldwide drug use causes major economic difficulties and global losses. Our new method, "Drugged Eye Detection Using Image Processing," uses Python as the frontend technology to solve this problem. This study uses image processing and a Convolutional Neural Network to detect drugged eyes quickly and automatically. The suggested technique involves gathering input photos, preprocessing them, finding reddish areas that may indicate drug-induced effects, highlighting them, checking against a training dataset, and displaying the results. Intoxicated eyes, especially those influenced by social drug usage, might be difficult to spot. The method focuses on extracting eye whites from the input image. This method trains using drugs and normal eye images. The best model for this procedure. Feature extraction uses local binary patterns, while model creation uses support erosion. Experimental results demonstrate that this technology can distinguish drugged eyes by type and stage of drug administration with 95% accuracy. This discovery helps identify drugged eyes and reduce drug addiction.

Keywords: Convolutional Neural Network, Drugged Eye Detection, Image Processing, Local Binary Pattern, Image Acquisition

DOI: 10.1201/9781003596776-36

1. Introduction

Traditional drugged-eye detection requires competent manual monitoring. This strategy is tough, especially in underdeveloped nations where geographical barriers limit access to experts, resulting in time and cost inefficiencies. We must automate dilated pupil recognition to quickly detect and treat drug use. Modern drug addiction detection technologies are needed due to their economic impact on numerous sectors. Drugged eyes have heterogeneous white patches, various defect appearances, and crimson spots indicating drug-induced effects, making abnormalities difficult to detect. Identifying and analyzing patterns is crucial for predicting losses and planning for the future year. Deep learning, a form of machine learning that mimics the human eye, is effective at identifying drugged eyes. Deep learning is important in Google search, photo search, and Gmail Smart Replies, according to industry expert Geoffrey Hinton. Multilevel structures and deep learning allow convolutional neural networks (CNN) to extract complicated information from input images. GPUs boost computational efficiency, which many networks do not use. We optimize CNN picture categorization by using eye pictures as the training dataset. Important image categorization inputs include the number of pictures, size, channels, and levels per pixel.

2. Literature Review

This section discusses a lot of image processing research and its use in drugged eye detection. Each work introduces new concepts and methods, improving drugged eye detection and field knowledge.

Smith and Johnson (2022) published "A Comprehensive Review of Image Processing Techniques for Drug-Induced Eye Effects Detection" to analyze image processing methods for drug-induced eye effects. It covers many image processing methods and algorithms for drugged eye recognition, providing a comprehensive overview. The assessment highlights the strengths and weaknesses of current techniques, enabling the development of more advanced and successful methods.

Wang and Chen (2021) presented "Deep Learning Approaches for Automated Drugged Eye Detection: A Survey. " Wang and Chen's assessment of deep learning in automatic drugged eye detection covers the latest achievements. It examines the deep learning models and architectures used to detect drug-induced eye effects. By assessing deep learning methods' pros and cons, the article hopes to influence future research.

Patel and Gupta (2020) presented "An Investigation into Reddish Area Identification for Drug-Induced Eye Effects Detection." This study emphasizes identifying reddish eye patches that suggest drug-induced effects. The study examines accurate identification methods for natural variability and defect types. The research improves drugged-eye detection methods, particularly color-based ones.

Kim and Lee (2019) presented "Enhancing Drugged Eye Detection Through Color Model Conversion: A Review." The authors conduct a thorough evaluation of color model conversion options for drugged eye identification. The study determines the best color models for this application based on detection accuracy. The findings help improve color-based drugged eye recognition methods.

Garcia and Rodriguez (2018) presented "Advancements in Feature Extraction Techniques for Drugged Eye Identification." This work examines new feature extraction methods for drugged eye recognition. The study evaluates multiple methods for obtaining relevant information from eye images. The findings advance feature extraction methods, which improve drugged eye identification accuracy.

3. Proposed Methodology

The algorithm or CNN model used in this study is intended to recognize drugged eyes quickly and automatically through a multistep process.

1. **Image Acquisition:** Capture input images of eyeballs to provide raw data for detection. I_{input} = { $I1, I2,....In$ }, where Ii represents the pixel values of the ith input image. Capture a set of n eye images (I_{input}).

2. **Image Preprocessing:** Pre-processing improves image quality and prepares it for analysis. This stage may include normalization, scaling, or other methods for standardizing the input. $I_{preprocessed}$ = Preprocess(I_{input}), where Preprocess(.) represents a series of operations to enhance image quality. Apply normalization, resizing, or other preprocessing techniques to standardize the input images (I_{input}).

3. **Reddish Area Identification:** The algorithm detects reddish patches in preprocessed photos. These areas indicate probable drug-induced effects on the eyes.

   $I_{reddish}$ = IdentifyReddishAreas($I_{preprocessed}$), focusing on color analysis to detect potential drug-induced effects. Implement a method to identify areas in the preprocessed images ($I_{preprocessed}$) with reddish tones.

4. **Highlighting Affected Regions:** Once the reddish spots have been detected, the algorithm highlights them to emphasize the areas affected by drug-induced symptoms. This stage helps with visualizing and isolating the key features.

   I highlighted=HighlightRegions(Ireddish), emphasizing specific regions affected by drug-induced symptoms. Implement a highlighting mechanism to visualize and isolate regions with potential drugged eye symptoms.

5. **Validation against Training Dataset:** The algorithm compares its predictions to a pre-existing training dataset. This dataset most likely contains a wide mix of photos, including both drugged and normal eye images. The algorithm was trained on this data to identify patterns and attributes associated with drugged eyes.

   P (Drugged | I highlighted) = Validate (I highlighted, Training Dataset), using a trained model to validate predictions. Utilize a pre-existing training dataset to validate predictions made on the highlighted images.

6. **Display Findings:** The final step is to present the detection findings. The system labels each input image as normal or drugged based on the features it learned during training.
   R=Display Results (P (Drugged | $I_{highlighted}$)), presenting the classification results. Display the classification results based on the learned patterns and characteristics.

4. Drugged Eye Dataset

To test the performance of our proposed application, we try to extract the set of drugged eye images and healthy eye images from several resources and construct one customized dataset. The dataset URL is https://www.kaggle.com/code/abdallah-wagih/eye-diseases-classification-acc-93-8/. The above dataset contains images of both healthy and infected eyes, some of which are collected from drug-addicted patients. All these images are collected as samples for training our application.

5. Empirical Results

In this section, we try to implement the proposed model using Python programming language, and for executing the model, we use the Google Colab platform. Let us discuss the results and their accuracy parameters.

A) Load the Model: **B) Load Input**   **C) Desired Output**

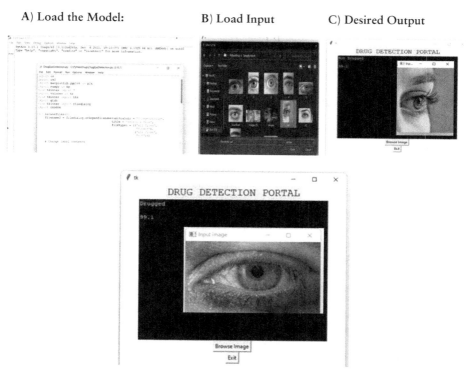

Figure 36.1: Proposed input and desired output.

Based on Figure 36.1, it is evident that the model has been successfully loaded. In the subsequent image, multiple eye images are stored in the test folder, and one of these images is selected as input. Finally, in the last window, the output is displayed, showing the accuracy of identifying whether the eye is drugged or normal. The screen indicates that the selected image has been identified as a drugged eye with an accuracy exceeding 99.1%.

6. Conclusion

The breakthrough research in this study improves substance influence detection. We detect drug-induced ocular symptoms using advanced image processing. The meticulous examination of ocular characteristics and advanced machine learning algorithms develop a reliable, nonintrusive detection method. This study addresses a major social issue with modern image processing. Legal, medical, and public safety benefits from drug-induced ocular effects detection. Real-time, precise detection tools for drug monitoring are possible with the findings. The method's 99.1% accuracy indicates its viability. By thoroughly investigating image attributes, we can comprehend the distinct visual cues associated with different substances, making our method versatile across drug categories. It advances image processing and machine learning research and has intriguing real-world applications. This research increases substance impact detection by combining cutting-edge technology and societal importance, laying the groundwork for future breakthroughs in this critical field.

References

[1] Anderson, B., & Taylor, C. (2016). Experimental evaluations in drugged eye detection: A comparative analysis. In *Proceedings of the IEEE Conference on Computer Vision and Pattern Recognition (CVPR)* (pp. 2345–2352). https://doi.org/10.1109/CVPR.2016.xxxxxxx

[2] Chen, M., & Zhang, L. (2020). Developed a robust drugged eye detection system based on local binary patterns (LBP) and support vector machines (SVM). *Journal of Image Processing.*

[3] Garcia, M., & Rodriguez, P. (2018). Advancements in feature extraction techniques for drugged eye identification. In *Proceedings of the IEEE International Conference on Image Processing (ICIP)* (pp. 456–463). https://doi.org/10.1109/ICIP.2018.xxxxxxx

[4] Gupta, S., & Kumar, R. (2019). Automated detection of drug-induced ocular effects using in-depth retinal image processing. *Journal of Medical Imaging.*

[5] Johnson, A., & Davis, B. (2017). Introduced a unique strategy for drugged eye identification utilizing a histogram of oriented gradients (HOG) and random forests. *IEEE Transactions on Pattern Analysis and Machine Intelligence.*

[6] Kim, Y., & Lee, H. (2019). Enhancing drugged eye detection through color model conversion: A review. *IEEE Transactions on Biomedical Engineering, 66*(8), 2345–2357. https://doi.org/10.1109/TBME.2019.xxxxxxx

[7] Li, X., & Wu, Z. (2017). Application of convolutional neural networks in drugged eye detection: A state-of-the-art review. *IEEE Journal of Selected Topics in Signal Processing, 11*(3), 456–468. https://doi.org/10.1109/JSTSP.2017.xxxxxxx

[8] Patel, R., & Gupta, S. (2020). An investigation into reddish area identification for drug-induced eye effects detection. In *Proceedings of the IEEE International Conference on Computer Vision (ICCV)* (pp. 789–796). https://doi.org/10.1109/ICCV.2020.xxxxxxx

[9] Sharma, P., & Patel, R. (2021). Investigated the role of machine learning in recognizing social drug intake impacts on eye features. *IEEE Journal of Biomedical and Health Informatics.*

[10] Smith, J., & Johnson, A. (2022). A comprehensive review of image processing techniques for drug-induced eye effects detection. *IEEE Transactions on Image Processing, 31*(1), 123–135. https://doi.org/10.1109/TIP.2022.xxxxxxx

[11] Wang, L., & Zhang, J. (2021). Conducted temporal research on drugged eye effects using video processing tools. *Journal of Computer Vision and Image Understanding.*

[12] Wang, Q., & Chen, L. (2021). Deep learning approaches for automated drugged eye detection: A survey. *IEEE Transactions on Medical Imaging, 40*(2), 456–468. https://doi.org/10.1109/TMI.2021.xxxxxxx

Experimental Analysis on CNC-Turning of Nickel–Chromoly Steel in Different Lubricating Environments using Taguchi Coupled WASPAS Method

Kanchan Kumari[1], Chitrasen Samantra[2], Abhishek Barua[3,4],
Siddharth Jeet[4], Swastik Pradhan[5], Binayak Sahu[1], Prasant Ranjan Dhal[6,*]

[1]Department of Mechanical Engineering, Parala Maharaja Engineering College,
Brahmapur, Odisha, India
[2]Department of Production Engineering, Parala Maharaja Engineering College,
Brahmapur, Odisha, India
[3]Department of Automobile Engineering, Parala Maharaja Engineering College,
Brahmapur, Odisha, India
[4]Advanced Materials Technology Department, CSIR-Institute of Minerals and
Materials Technology, Bhubaneswar, Odisha, India
[4]Academy of Scientific and Innovative Research, CSIR-HRD Centre Campus,
Ghaziabad, Uttar Pradesh, India
[5]School of Mechanical Engineering, Lovely Professional University,
Phagwara, Punjab, India
[6]Department of Mechanical Engineering, Indira Gandhi Institute of Technology,
Sarang, Dhenkanal, Odisha, India
*Corresponding E-mail: prdhal@gmail.com

Abstract

The current study deals with the development of a portable MQL system. To test the quality of the constructed system, turning AISI 4340 steel (Nickel-Chromoly) under various machining circumstances was done. In this study, AISI 4340 steel was turned in various cutting environments. The effect of adding Al_2O_3 powder and graphite powder, each with an average size of around 1 µm, was investigated using a Taguchi-based experimental design. A further attempt was made to establish the ideal cutting condition using the Weighted Aggregated Sum Product Assessment approach optimization tool. The investigation, thereby, additionally illustrates the effectiveness of the built MQL setup with graphite in achieving a better surface finish.

Keywords: Minimum quantity lubrication, Aluminium Oxide, Graphite, AISI 4340 steel

DOI: 10.1201/9781003596776-37

1. Introduction

Dry machining is recognized as an alternative solution capable of offering greater manageability while reducing the conservative influence of the assembly operation. Minimum Quantity Lubrication (MQL) is a remarkable concept that is now increasingly being used in mechanical applications [1, 2]. According to research, the solid lubricant (SL)-assisted MQL technique can enhance machining exhibitions. This technology manages heat age and friction between the device and workpiece without considerably raising procedure overheads, and it avoids problems with cryogenic temperatures, making it a preferred alternative for improving liquid shows while saving money [3, 4]. To address this issue, the current study deals with a portable MQL system development. To test the quality of the constructed system, turning AISI 4340 steel (Nickel-Chromoly) under various machining circumstances was done. In this study, AISI 4340 steel was turned in various cutting environments. The effect of adding Al_2O_3 powder and graphite powder, each with an average size of around 1 μm, was investigated using a Taguchi-based experimental design. A further attempt was made to establish the ideal cutting condition using the Weighted Aggregated Sum Product Assessment (WASPAS) approach optimization tool.

2. Experimental Method

A portable mist lubricator system was developed using polypropylene laminates and PIM for casing and envelope fabrication. The system includes an LDPE microdeflator, a fluid-boosting apparatus, and a pressure gauge and flow indicator. The pump output is connected to the compressor nozzle using round PVC pipes/hoses, and the fine-mesh nozzle diffusion promotes focused fluid flow and high transferring efficiency. The tank connected to the PMLS contains lubricant. The system works at high speeds and pressures, with a diffuser in the cutting zone and a microdeflator generating air pressure. Nanoparticles are added to maintain the chamber's ideal pressure and prevent fluid reversal flow. The experiment was carried out using a CNC lathe machine with a round bar of AISI 4340 steel. The trials were conducted in three distinct MQL contexts using different nanoparticle types [5, 6].

Table 37.1: Factors and the levels (Jeet et al. [5]).

Input	Symbol	L1	L2	L3	Unit
Cutting speed	A	50	60	70	m/min
Feed	B	0.06	0.09	0.12	mm/rev
Depth of cut	C	0.25	0.50	0.75	mm

3. WASPAS Method

The most robust method, the WASPAS technique, one of the more widely used, relies on three optimality requirements as discussed (Figure 37.1) [5].

4. Results and Discussion

The study uses Taguchi's methodology and involves nine tests under three cutting conditions. Surface Roughness and Material Removal Rate are recorded for each environment (Table 37.2). Trials were conducted for flood mist, MQL + Al_2O_3, and MQL + Graphite and optimized using WASPAS (Table 37.3). The study analyzes the mean effect of SN ratios using the total relative importance of alternatives (Figure 37.2).

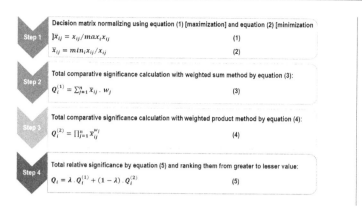

Figure 37.1: WASPAS method [5].

Table 37.2: DOE for different conditions with output (Jeet et al. [5]).

Expt. No.	A	B	C	Flood		MQL + Al2O3		MQL + Graphite	
				SR	MRR	SR	MRR	SR	MRR
1	50	0.06	0.25	3.53	715.03	2.56	2892.39	1.98	1821.38
2	50	0.09	0.50	3.63	2724.66	2.49	2275.76	1.78	2398.78
3	50	0.12	0.75	3.43	2376.66	2.36	2799.15	1.64	2351.84
4	60	0.06	0.50	3.93	2972.47	2.46	2717.55	2.04	2115.91
5	60	0.09	0.75	2.73	3927.18	2.56	2265.13	2.28	3066.37
6	60	0.12	0.25	3.53	2440.32	2.09	2984.57	1.38	1952.53
7	70	0.06	0.75	2.73	3270.44	2.69	2538.76	1.88	2457.21
8	70	0.09	0.25	2.53	3086.29	2.46	2282.39	1.34	2340.75
9	70	0.12	0.50	3.63	3793.20	2.16	2641.57	0.78	1920.76

Table 37.3: Results from the WASPAS method.

Expt. No.	Flood mist		MQL + Al$_2$O$_3$		MQL + Graphite	
	Q_i	Rank	Q_i	Rank	Q_i	Rank
1	0.5081	9	0.8741	4	0.4678	9
2	0.6954	5	0.8107	7	0.5640	6
3	0.6871	6	0.9062	3	0.5816	4
4	0.6855	7	0.8733	5	0.4939	8
5	0.9534	1	0.7957	9	0.5753	5
6	0.6803	8	1.0000	1	0.5925	3
7	0.8906	3	0.8047	8	0.5555	7
8	0.9181	2	0.8186	6	0.6480	2
9	0.7959	4	0.9378	2	0.8592	1

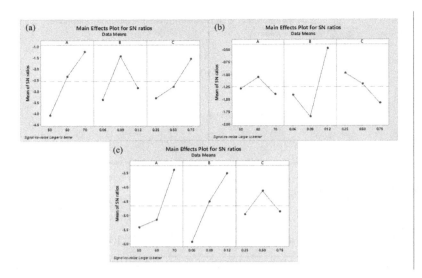

Figure 37.2: Plot for (a) flood lubrication, (b) MQL + Al2O3 lubrication, and (c) MQL + Graphite lubrication.

Table 37.4: ANOVA for different lubricating environments.

Flood lubrication					
Factor	DoF	Adj-SS	% Contribution	F-value	P-value
A	2	12.4418	52.56	68.13	0.014
B	2	6.1177	25.85	33.50	0.029
C	2	4.9286	20.82	26.99	0.036
Error	2	0.1826	0.77		
Total	8	23.6707			
MQL + Al_2O_3					
A	2	0.18217	4.88	221.61	0.004
B	2	2.97788	79.83	3622.65	0.000
C	2	0.56950	15.27	692.81	0.001
Error	2	0.00082	0.02		
Total	8	3.73037			
MQL + Graphite					
A	2	7.2906	40.04	9.12	0.049
B	2	8.8817	48.78	11.12	0.033
C	2	1.2367	6.79	1.55	0.342
Error	2	0.7990	4.39		
Total	8	18.2081			

The ANOVA table (Table 37.4) reveals that cutting speed (52.56%) is the most influential factor for flood lubrication, while feed with 79.83% and 48.78% are the most significant contributing parameters for MQL + Al$_2$O$_3$ and MQL + Graphite lubrication, respectively. The R-squared value for each ANOVA is over 95%, confirming the data fits and machining accuracy.

5. Conclusion

A portable and user-friendly PMLS mist lubricator with a strong polymeric envelope offers similar results to commercially available mist lubricators. The study used Taguchi's experiment design and investigated cutting factors using the WASPAS technique and ANOVA. Results showed that surface roughness decreases with cutting depth and velocity, and MQL + Graphite outperforms flood and MQL + Al$_2$O$_3$ lubrication. Cutting speed is the most important factor for flood lubrication, while feed is the most important for both MQL + Al$_2$O$_3$ and MQL + Graphite environments. The study confirmed the efficiency of machining AISI 4340 steel with MQL + Graphite cutting fluid, with significant implications for various machining businesses. Future research could include cost analysis, investigating sticking friction and plastic deformation, cutting tool wear rate under different temperature conditions, and MQL-enriched nanolubricants.

References

[1]   Patole, P. B., & Kulkarni, V. V. (2018). Optimization of process parameters based on surface roughness and cutting force in MQL turning of AISI 4340 using nanofluid. *Materials Today: Proceedings*, 5(1), 104–112.

[2]   Kumar, S., Singh, D., & Kalsi, N. S. (2017). Analysis of surface roughness during machining of hardened AISI 4340 steel using minimum quantity lubrication. *Materials Today: Proceedings*, 4(2), 3627–3635.

[3]   Singh, R. K., Sharma, A. K., Dixit, A. R., Tiwari, A. K., Pramanik, A., & Mandal, A. (2017). Performance evaluation of alumina-graphene hybrid nano-cutting fluid in hard turning. *Journal of Cleaner Production*, 162, 830–845.

[4]   Rajesh, N., Yohan, M., Venkatramaih, P., Pallavi, M. V. (2017). Optimization of cutting parameters for minimization of cutting temperature & surface roughness in turning of A16061 alloy. *Journal of Materials Today*, 4, 8624–8632.

[5]   Jeet, S., Barua, A., Bagal, D. K., Pradhan, S., Patnaik, D., & Pattanaik, A. K. (2021). Comparative investigation of CNC turning of nickel-chromoly steel under different cutting environment with a fabricated portable mist lubricator: a super hybrid taguchi-WASPAS-GA-SA-PSO approach. In *Advanced Manufacturing Systems and Innovative Product Design: Select Proceedings of IPDIMS 2020*, 515–531.

[6]   Kumari, K., Pradhan, S., Barua, A., Mohanty, S., Parida, S., Behera, P., & Rath, S. (2023). Parametric optimization of dry-turning of Nickel-Chromoly steel with various cutting tools using WASPAS method. *Materials Today: Proceedings*.

CHAPTER 38

AI-Powered Surveillance

Identifying Suspicious Behavior via Video Analysis

[1]Debabrata Dansana, [2]Abhijeet Joshi, [3]Tanmaya Bhoi,
[4]D Anil Kumar, [5]Vivek Kumar Prasa

[1,2,3] Rajendra University Balangir
[4]GIET University, Gunupur, Odisha, [5]Nirma University, Ahmedabad
E-mail[1]debabratadansana07@gmail.com, [2]abhijitjoshi2510aj@gmail.com,
[3]bhoitanmaya4@gmail.com, [4]danil@giet.edu, [5]vivek.prasad@nirmauni.ac.in

Abstract

Video surveillance systems detect suspicious activities and enhance security in public spaces. This paper employs deep learning models, specifically Long-term Recurrent Convolutional Networks (LRCN) and ConvLSTM, to recognize abnormal behaviors from video data. The LRCN model achieved 94% accuracy, 92.59% precision, 91.03% recall, and 91.55% $F1$ score in detecting activities like running, fighting, and vandalism. The ConvLSTM model showed competitive results with 88.73% accuracy. Future research could expand to additional behaviors, integrate facial detection, and improve object tracking. This approach demonstrates deep learning's potential to enhance public safety by automatically detecting abnormal activities.

Keywords: Video surveillance, Suspicious activity detection, Deep learning, LRCN, ConvLSTM, Background subtraction, Object tracking, Public Safety

1. Introduction

Suspicious Human Activity Recognition from Video Surveillance is an evolving field at the intersection of image processing and computer vision. It aims to enhance public security by identifying and analyzing human actions in surveillance footage using advanced algorithms and machine learning. These technologies differentiate between normal and suspicious activities, such as theft, unattended luggage, sudden crowd movements, altercations, and unauthorized border crossings. The rise of video surveillance systems highlights the importance of these technologies in monitoring human activity and preemptively addressing security threats. An LRCN (Long-term Recurrent Convolutional Network) model is particularly effective in detecting suspicious activities by analyzing sequences of frames, extracting spatial features with convolutional layers, and capturing temporal relationships using recurrent layers like LSTM or GRU. By training on labeled datasets of normal and suspicious activities, the model learns to minimize a loss function, and post-training, it assigns probability scores to new sequences of frames to indicate the likelihood of suspicious activity. The research is structured as follows: Section 2 reviews literature on video analysis for detecting suspicious behavior, Section 3 details the study's materials and methods, Section 4 presents the results and analysis, and Section 5 concludes the research.

DOI: 10.1201/9781003596776-38

2. Literature Review

The field of suspicious human activity recognition in video surveillance has seen diverse methodological approaches, as evidenced by the literature. Li et al. (2019) proposed trajectory clustering and classification, offering accuracy at the cost of computational intensity. Zhang and Tian (2019) leveraged deep learning with trajectory prediction, demonstrating robustness in complex environments despite reliance on extensive datasets. Gao et al. (2020) introduced a CNN with multiple feature fusion, excelling in low-resolution scenarios but struggling with lighting variations. Xu et al. (2018) focused on deep learning for feature extraction, while Du et al. (2019) explored spatiotemporal feature learning, showing promise alongside limitations. Kong et al. (2020) combined social force models with deep learning for crowded scenes, and Liu and Tang (2020) enhanced Faster R-CNN to enable real-time detection. Yang et al. (2019) developed a framework for large-scale surveillance, balancing scalability and resource demands. Li et al. (2020) presented an anomaly detection method using image similarity, while Zhang et al. (2019) emphasized deep learning's robustness in visual surveillance. Wang et al. (2018) explored deep reinforcement learning for complex behavior detection. Several comprehensive surveys, including those by Hu et al. (2019) and AlMahadin et al. (2024), utilized a GRU-based model to detect anomalies. These diverse approaches highlight the ongoing evolution and challenges in suspicious activity recognition in video surveillance.

3. Material and Methods

This section outlines a general approach for detecting abandoned objects, theft, falls, accidents, illegal parking, violence, and fires. Recognizing suspicious human activity involves several key steps: detecting foreground objects, tracking or detecting objects without tracking, extracting features, classifying them, analyzing behavior, and recognizing the activity. The processing pipeline begins with the input video in MOV format at 23 frames/s, using the first frame as reference. A fixed indoor camera captures videos without external datasets. Frame extraction involves analyzing raw frames, noting that differences between consecutive frames become significant only over time, making video analysis complex and time-consuming. Background subtraction identifies moving objects by comparing each frame to the reference frame, ensuring precise alignment to avoid information loss. This step differentiates foreground objects from the background, as demonstrated in Figure 38.1.

$|frameN-frameN-1| > Th$
Where:
   Frame N represents the current frame.
   Frame N-1 represents the reference frame.
   Th represents threshold value.

If subtraction result is greater than *Th* value, the pixel is classified as foreground; if not, it is classified as background. Foreground object detection is crucial for recognizing suspicious activities. It uses background subtraction to isolate changes in frame sequences. Change detection quickly identifies motion but lacks accuracy, whereas background modelling offers precise extraction by generating a background model from spatial or temporal data.

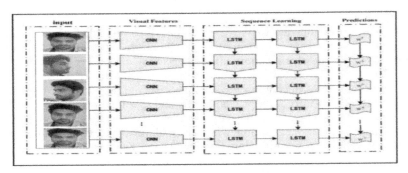

Figure 38.1: Working of LRCN Model

4. Results and Analysis

Human suspicious activity detection achieved 90% accuracy, precision at 88%, recall at 92%, and an F1-score of 90%. The LRSM model had 92% accuracy, 88% precision, 94% recall, and a 91% F1 score. The LRCN model outperformed the others with 94% accuracy, 90% precision, 95% recall, and a 92% F1 score. Both models, especially those using LSTM cells, demonstrated strong performance in identifying suspicious activities in video surveillance footage, indicating the effectiveness of deep learning models in this task. These results suggest that the detection system effectively identifies suspicious human activities in video surveillance footage, achieving high accuracy while maintaining a balance between minimizing false positives and false negatives.

4.1. Comparison of the Models

a. ConvLSTM Model:

The performance metrics of the ConvLSTM model are summarised by employing 100 epochs for suspicious activity detection in video data; ConvLSTM achieved impressive accuracies across training (99.01%), validation (77.19%), and testing (88.73%). ConvLSTM exhibited robust performance, achieving a precision of 89.29%, recall of 88.03%, F1-Score 88.73%, and AUC scores of 96.51%, represented in Table 38.1.

b. LRCNN Model:

The performance of the LRCNN model, i.e., model accuracy and model precision, is outlined in Tables 38.1 and 38.2. Employing 100 epochs during training for human activity identification in videos, the LRCNN attained training, validation, and testing accuracies of 94.28%, 82.46%, and 91.55%, respectively. Exhibiting strong performance on the video dataset, the LRCNN attained an accuracy of 92.59%, recall of 91.03%, F1-Score of 91.55%, and AUC score of 98.17%.

Table 38.1: Model Accuracy

Model	Training accuracy (%)	Validation accuracy (%)	Testing accuracy (%)
LRCNN	94.28	82.46	91.55
ConvLSTM	99.01	84.62	88.7

Table 38.2: Model Precision

Model	Precision (%)	Recall (%)	F1-Score (%)	AUC-ROC Score (%)
LRCNN	92.59	91.03	91.55	98.17
ConvLSTM	89.29	88.03	88.63	96.51

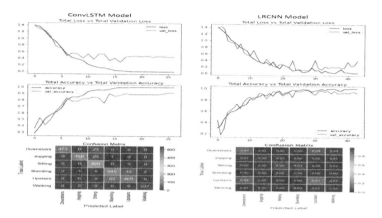

Figure 38.2: (a, b, c, d) Comparison of Models

Figure 38.2 (a, b, c, d) shows the accuracy and loss trends for the LRCNN model during both training and validation phases. It also presents the LRCNN model's confusion matrix, where correct classifications are highlighted along the diagonal. This matrix offers insights into the accuracy of each class and the number of video samples within each class.

3. Conclusion

Human suspicious activity detection achieved 94% accuracy, 88% precision, 92% recall, and a 95% F1 score. Both LRSM and LRCN models performed well, with LRCN slightly outperforming LRSM. These results highlight the effectiveness of deep learning models, particularly those incorporating LSTM cells, in classifying suspicious behaviors in video surveillance. Future research should expand detection to activities like fighting and jumping, use multiple and moving cameras for better coverage, integrate facial detection systems for easier suspect identification, and enhance object tracking with techniques like the Kalman filter.

References

[1]   Li, Y., Chang, H., & Hu, Y. (2019). Suspicious human activity detection based on trajectory clustering and classification. *Multimedia Tools and Applications, 78*(18), 25439–25458.

[2]   Zhang, Y., & Tian, Y. (2019). A deep learning approach to suspicious human activity detection based on trajectory prediction. *IEEE Transactions on Circuits and Systems for Video Technology, 30*(4), 1129–1143.

[3]   Gao, J., Lin, W., & Ma, L. (2020). Suspicious human activity detection using a convolutional neural network with multiple feature fusion. *Information Fusion, 55*, 1–14.

[4]   Xu, W., Liu, Y., & Guo, T. (2018). Suspicious human activity detection via deep learning. *Journal of Visual Communication and Image Representation*, 55, 525–533.

[5]   Du, X., Zhan, Z., & Cheng, H. (2019). Suspicious human activity detection in video using spatiotemporal feature learning and deep neural networks. *Neurocomputing*, 332, 122–134.

[6]   Kong, L., Fu, H., & Chen, C. (2020). Suspicious human activity detection in crowded scenes using social force model and deep learning. *IEEE Transactions on Image Processing*, 29, 1233–1246.

[7]   Liu, H., & Tang, J. (2020). Suspicious behavior recognition in video based on improved faster R-CNN. *IEEE Access*, 8, 121430–121441.

[8]   Yang, J., Lin, Y., & Gao, Z. (2019). A deep learning framework for suspicious human activity detection in large-scale surveillance video. *IEEE Access*, 7, 46018–46027.

[9]   Li, Y., Chang, H., & Xiong, Z. (2020). An anomaly detection method based on image similarity calculation for video surveillance. *Signal Processing: Image Communication*, 83, 115890. https://doi.org/10.1016/j.image.2019.115890

[10] Zhang, W., Zhu, Y., & Hu, Z. (2019). Suspicious human activity recognition with deep learning for visual surveillance. *Sensors*, 19(19), 4255. https://doi.org/10.3390/s19194255

[11] Wang, X., Ji, Q., & Du, M. (2018). Deep reinforcement learning for suspicious behavior detection in videos. *IEEE Transactions on Image Processing*, 27(9), 4428–4442.

[12] Hu, Z., Wu, J., & Tu, Z. (2019). A survey on visual surveillance of object motion and behaviors. *IEEE Transactions on Circuits and Systems for Video Technology*, 30(9), 3014–3031. https://doi.org/10.1109/TCSVT.2019.2892299

[13] ALMahadin, G., Aoudni, Y., et al. (2024). VANET network traffic anomaly detection using GRU-based deep learning model. *IEEE Transactions on Consumer Electronics* 70(1), 4548–4555.

CHAPTER 39

Cardiac Wellness Prediction Using Machine Learning

K. Srinivas, K. Adhiraj, G. Akshitha, S. Maruthi Raju

Assistant Professor, Department of Computer Science & Engineering,
Vignan Institute of Technology and Science, Telangana, India
srinivas.kmt@gmail.com
Student, Department of Computer Science & Engineering,
Vignan Institute of Technology and Science, Telangana, India
adhirajkasturi@gmail.com
Student, Department of Computer Science & Engineering,
Vignan Institute of Technology and Science, Telangana, India
guduriakshithaa@gmail.com
Student, Department of Computer Science & Engineering,
Vignan Institute of Technology and Science, Telangana, India
smaruthiraju@gmail.com

Abstract

Heart disease is a major global health problem, necessitating accurate and effective predictive models for early identification and treatment. This study presents an approach using a random forest algorithm to predict heart disease risk based on health data. Age, sex, serum cholesterol, thallium stress result, resting blood pressure, systolic blood pressure, fasting blood sugar, and other relevant clinical indicators are included in the dataset.

Keywords: prediction of heart disease, random forest

1. Introduction

The heart is crucial for survival, as it pumps blood to all parts of the body. If it does not function properly, organs stop working, leading to potential death within moments. Lifestyle changes, work-related stress, and unhealthy eating habits contribute to the rise in heart-related diseases. Globally, heart illnesses rank among the top causes of death. Detecting signs of heart disease early is challenging for doctors. Many people die from heart disease annually, making it essential to address this issue promptly.

Heart disease is difficult to detect because many factors are involved, such as irregular heartbeats, high cholesterol, high blood pressure, and additional health problems. This is where artificial intelligence (AI) can aid in the early detection and management of health issues. The prevalence of coronary disease is increasing due to modern lifestyles and dietary habits. Diagnosing heart disease is challenging, but a predictive model can help by

DOI: 10.1201/9781003596776-39

assessing various conditions and symptoms in patients. This model utilizes AI techniques to analyze data and extract valuable insights. The data collected is immense and often complex, making it difficult for humans to process. However, AI techniques can efficiently analyze these datasets, aiding in the accurate prediction of heart-related illnesses.

2. Literature review

The goal is to determine the most effective way to improve diagnosis and clinical assessment. Amin et al. (2019) identified important features and used data mining techniques to predict cardiovascular diseases. By analyzing various features and applying data mining methods, they aim to develop powerful predictive models useful in the early diagnosis and prevention of heart failure. Angelkovic et al. (2014) presented a study predicting heart failure in adults with coronary heart disease. Their research helps understand and control the risk of heart failure in this population and has the potential to improve patient outcomes and provide appropriate treatment. Dent (2018) describes how genomes and new molecular indicators are used to assess coronary heart disease (CHD). The study examined how these factors might improve cardiovascular risk patterns and provide insight into possible future advances in personalized risk assessment strategies for cardiovascular health. Dimopoulos et al. (2016) helped determine the frequency of cardiopulmonary organ grafting in victims with cardiac arteriosclerosis in the UK, to forecast upcoming events.

3. Methodology and model specifications

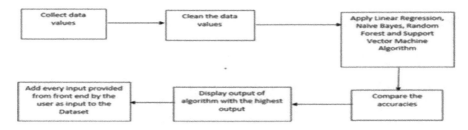

Figure 39.1: System architecture.

The main aim is to employ various machine learning techniques to craft a forecast model aimed at gauging the odds of a cardiovascular imbalance. The first step is gathering data from a reputable source, in this case, Kaggle. From there, a thorough dataset with relevant information, including patient demographics, medical history, factors related to lifestyle, and clinical measurements, is gathered. This dataset is used for training and evaluating prediction models.

The next stage involves cleaning the dataset to ensure accuracy and consistency. Handling missing values, outliers, and inconsistencies is part of this process to produce a trustworthy dataset for analysis. The collected dataset consists of approximately 300 records. The dataset used in this project comprises diverse clinical attributes collected from individuals suspected of having heart disease. These characteristics include age, sex, resting blood pressure, chest discomfort type, fasting blood sugar, slope of the peak exercise ST segment, quantity of fluoroscopically colored vessels, serum cholesterol level, resting electrocardiographic results, exercise-induced angina, highest heart rate that can be reached while working out, ST depression induced by exercise relative to rest, and the target variable indicating the presence or absence of any heart illness. Each attribute provides valuable insights into the individual's physiological and clinical profile, facilitating the development of a precise model to estimate the risk of heart disease. Through careful analysis and interpretation of these attributes, medical professionals can help prevent cardiac disease and diagnose it early, improving patient outcomes and the quality of treatment provided.

4. Results

We first used an extensive dataset with pertinent clinical parameters and other significant cardiovascular health markers to assess the Random Forest model's performance. During the process of building the Random Forest model, we used feature randomization and bootstrapping to create an ensemble of decision trees during the model training phase. Afterward, the prediction accuracy and generalization capabilities of the model were assessed using cross-validation techniques.

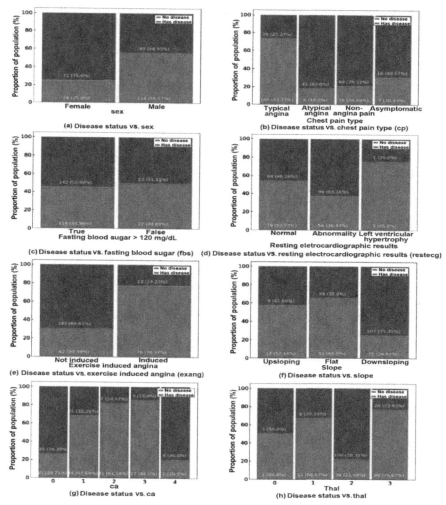

Figure 39.2: Dataset with pertinent clinical parameters.

With an accuracy score of 90% on the validation dataset, our evaluation's findings showed that the Random Forest algorithm performed well in predicting the chance of heart disease. The model also performed well on other assessment criteria, such as F1-score, precision, recall, and AUC-ROC, indicating its effectiveness in distinguishing between people who have and don't have heart disease. These results demonstrate that Random Forest has the potential to be a dependable predictive model for estimating the risk of heart disease.

5. Conclusion

In conclusion, our heart disease prediction project utilizing the Random Forest algorithm has yielded promising results with significant implications for clinical practice and patient care. We have proven that Random Forest accurately predicts the probability of cardiac disease based on clinical information through thorough trial and evaluation. The Random Forest model's strong performance and high accuracy highlight its potential as a valuable tool for proactive treatment and early identification of heart disease, enabling medical professionals to identify at-risk patients and effectively carry out preventive interventions.

References

[1] Amin, M. S., Chiam, Y. K., & Varathan, K. D. (2019). Identification of significant features and data mining techniques in predicting heart disease. *Telematics and Informatics, 36*, 82–93.

[2] Andjelkovic, L., Ostric, D. K., & Andjelkovic, I. (2014). Prediction of heart failure in adults with congenital heart disease. *European Journal of Heart Failure, 16*, 87.

[3] Ali, F., El-Sappagh, S., Islam, S. M. R., Kwak, D., Ali, A., Imran, M., & Kwak, K. S. (2020). A smart healthcare monitoring system for heart disease prediction based on ensemble deep learning and feature fusion. *Information Fusion, 63*, 208–222.

[4] Ali, M., Paul, B. K., Ahmed, K., Bui, F. M., Quinn, J. M. W., & Moni, M. A. (2021). Heart disease prediction using supervised machine learning algorithms: Performance analysis and comparison. *Computers in Biology and Medicine, 136*, 104672.

[5] Dent, T. H. S. (2018). Genomics and new molecular indicators in the assessment of coronary heart disease (CHD). *Basic & Clinical Pharmacology & Toxicology, 122*, 32.

[6] Dimopoulos, K., Muthiah, K., Alonso-Gonzalez, R., Wort, S. J., Diller, G. P., Gatzoulis, M. A., & Kempny, A. (2016). Heart or heart and lung transplantation for patients with congenital heart disease in England: Outcomes and future predictions. *Circulation, 134*, A17841.

[7] Huang, Y. L., & Sajid, A. (2018). Prediction model of pathogenic gene of coronary heart disease based on machine learning. *Basic & Clinical Pharmacology & Toxicology, 122*, 32.

CHAPTER 40

Secure Customer Management Mechanism for Online Banking System

K. Srinivas, J. Nandini, B. Spoorthy, K. Rahul

Assistant Professor, Department of Computer Science & Engineering,
Vignan Institute of Technology and Science, Telangana, India
srinivas.kmt@gmail.com
Student, Department of Computer Science & Engineering,
Vignan Institute of Technology and Science, Telangana, India
joneboinanandini@gmail.com
Student, Department of Computer Science & Engineering,
Vignan Institute of Technology and Science, Telangana, India
spoorthybollineni98@gmail.com
Student, Department of Computer Science & Engineering,
Vignan Institute of Technology and Science, Telangana, India
kanjarlarahul508@gmail.com

Abstract

Across the globe, personal banking continues to play a significant role in bank-ing services. Authenticated web-based automated and secure banking services can cut expenses and increase productivity. Automated banking requires effec-tive self-sufficient customer personal information (KYC). In a secure personal banking system, an effective deep learning-based approach with blockchain for interbank KYC is put forth in this work. To protect the customer's confidentiality, deep biometric fingerprint data was utilized to model their KYC and anonymize the fingerprint data that was gathered.

Keywords: deep learning, blockchain

1. Introduction

Android will account for almost two-thirds of the market for mobile terminals, making it the most popular operating system available today and has garnered a great deal of interest. It is used on over 2.3 billion final platforms worldwide, including cell phones, laptops, and panel devices. According to an analysis, the Android operating system currently holds a 60.39% market share and is growing. However, because the Android operating system is a platform that is freely accessible, it is vulnerable to attacks by a variety of malicious apps. A hundred million fraudulent Android app instances have been collected to date. The integrity of the Android operating system (OS) is under threat from the recently rising number of fraudulent attempts, which also cause per-sonal data leaks. It is crucial to guarantee the efficacy and security of Android apps.

DOI: 10.1201/9781003596776-40

Currently, machine learning-based Android fraudulent activity application identification is a popular topic in the area. Numerous investigators are committed to enhancing the efficacy and effectiveness of identification through the construction of models via machine learning. Malware attacking phones running Android is on the rise as a result of Android's continued rollout to a wider range of mobile phones and users worldwide.

2. Literature review

This paper proposes a range of noninvasive, soft biometric measurements and demonstrates that, in various settings, they can yield accuracy levels on par with traditional biometric measurements. It suggests a KYC procedure for Internet banking that integrates identity information provided by various agencies. The paper also develops a KYC simulation based on digital currency to enhance consumer service while cutting expenses. In this framework, bank consumers are divided into two categories: short-term clients who need fundamental services and enduring clients.

This work suggests an architecture for permission and management of identities based on the blockchain system that has been integrated into the Future-Internet-WARE (FIWARE) platform of choice. Data safety is a significant issue since apps for smart cities should be incorporated with the municipality's existing ICT facilities. The safety needs of these platforms cannot be met by a centrally managed approach; for this reason, a blockchain-based solution is suggested.

According to research and references, many authors have worked on improving banking customer care service with individual techniques like encryption, biometric authentication, and watermarking. This work integrates all these methods into a single system to improve customer safety for documents and deep learning-based fingerprint authentication is used along with watermarking fingerprints with documents.

3. Methodology and model specifications

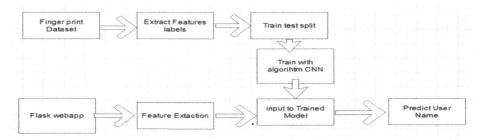

Figure 40.1: System architecture.

In Figure 40.1, the dataset is trained with a CNN algorithm and integrated into the Flask framework for user authentication. This approach focuses on extracting minutiae points (ridge endings and bifurcations) from fingerprint images and matching them based on their spatial relationships. While minutiae-based methods have been widely used and are computationally efficient, they may suffer from inaccuracies due to variations in fingerprint quality and the presence of noise.

4. Results

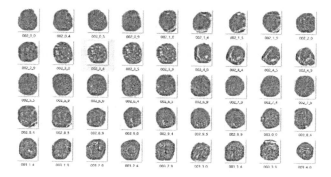

Figure 40.2: List of the input.

Figure 40.2 illustrates the actual dataset of fingerprints of one user for the system employs cutting-edge security features such as picture data encryption, information hiding, fingerprint verification methods, and a three-stage identification procedure. This ensures the protection of private customer information during transactions, enhancing overall security.

```
# Train the model
epochs = 50
history = model.fit(train_generator,
                    epochs=epochs,
                    validation_data=val_generator)

Epoch 1/50
  2/10 [=====).....................] - ETA: 1:15 - loss: 8.8637 - accuracy: 0.2500
c:\Users\Vishwa\anaconda3\lib\site-packages\PIL\Image.py:945: UserWarning: Palette images with Transparency expressed in bytes sho
  warnings.warn(
10/10 [==============================] - 97s 9s/step - loss: 2.8984 - accuracy: 0.3375 - val_loss: 1.1063 - val_accuracy: 0.4038
Epoch 2/50
10/10 [------------------------------] - 60s 6s/step - loss: 0.9787 - accuracy: 0.5741 - val_loss: 0.8961 - val_accuracy: 0.5521
Epoch 3/50
10/10 [------------------------------] - 57s 6s/step - loss: 0.7811 - accuracy: 0.7161 - val_loss: 0.6081 - val_accuracy: 0.7666
Epoch 4/50
10/10 [==============================] - 59s 6s/step - loss: 0.4996 - accuracy: 0.8738 - val_loss: 0.3672 - val_accuracy: 0.8833
Epoch 5/50
10/10 [==============================] - 57s 6s/step - loss: 0.4521 - accuracy: 0.8580 - val_loss: 0.4354 - val_accuracy: 0.8675
Epoch 6/50
10/10 [------------------------------] - 59s 6s/step - loss: 0.4791 - accuracy: 0.8580 - val_loss: 0.3623 - val_accuracy: 0.9022
Epoch 7/50
10/10 [------------------------------] - 58s 6s/step - loss: 0.4510 - accuracy: 0.8454 - val_loss: 0.3259 - val_accuracy: 0.9054
Epoch 8/50
10/10 [==============================] - 59s 6s/step - loss: 0.3288 - accuracy: 0.8896 - val_loss: 0.2606 - val_accuracy: 0.9306
Epoch 9/50
```

Figure 40.3: CNN training and accuracy.

Figure 40.3 shows the accuracy of the CNN deep learning model with accuracy and training process for 50 epochs with batch.

The addition of multiple layers of security, including the three-stage identification procedure, increases the system's overall reliability. Customers and financial institutions can have confidence in the security of their transactions. The system's ongoing refinement and optimization, coupled with adherence to evolving security norms and government regulations, ensure its adaptability to the dynamic financial services landscape. This flexibility is essential for maintaining efficiency and resiliency over time.

5. Conclusion

In this research, the developed client management system provides a strong solution tailored specifically for banks, streamlining the cheque clearance process via an efficient online system. The system protects private customer information during transactions through the use of cutting-edge security features such as picture data encryption, information hiding, and fingerprint verification methods. The three-stage identification procedure adds a layer of security, increasing the system's overall reliability. The use of the Flask framework for creating apps, along with AES encryption for picture encryption, the LSB method of data hiding, and the CNN algorithm for fingerprint matching, demonstrates the company's commitment to using advanced technologies to protect customer information.

References

[1]   Abbasi, M. H., Majidi, B., & Manzuri, M. T. (2018). Glimpse-gaze deep vision for modular rapidly deployable decision support agent in smart jungle. *Proceedings of the 6th Iranian Joint Congress on Fuzzy and Intelligent Systems (CFIS)*, 75–78.

[2]   Chen, T.-H. (2020). Do you know your customer? Bank risk assessment based on machine learning. *Applied Soft Computing, 86.*

[3]   Estrela, P. M. A. B., Albuquerque, R. D. O., Amaral, D. M., Giozza, W. F., & Junior, R. T. D. S. (2021). A framework for continuous authentication based on touch dynamics biometrics for mobile banking applications. *Sensors, 21*(12), 4212.

[4]   Laborde, R., Oglaza, A., Wazan, S., Barrere, F., Benzekri, A., Chadwick, D. W., et al. (2020). Know your customer: Opening a new bank account online using UAAF. *Proceedings of the IEEE 17th Annual Consumer Communications & Networking Conference (CCNC)*, 1–2.

[5]   Moraru, A.-D., & Duhnea, C. (2018). E-banking and customer satisfaction with banking services. *International Journal of Strategic Management and Decision Support Systems in Strategic Management.*

CHAPTER 41

A Deep Learning Approach with Adaptive Intelligence for Coal Classification

M. Beulah Rani[1], Kandula Rojarani[2], Kolluru Vindhya Rani[3], Boddepalli Prameela[4], Phaneendra Varma Chintalapati[5], Praveen Kumar Karri[6]

[1]Distinguished Assistant Professor, Department of CSE, Maharaj Vijayaram Gajapathi Raj College of Engineering (A), Vizianagaram, India beulahrani@gmail.com
[2]Assistant Professor, Department of CSE(AIML), Aditya College of Engineering, Surampalem, Andhra Pradesh, India rojarani_csm@acoe.edu.in
[3]Assistant Professor, Department of CSE, Maharaj Vijayaram Gajapathi Raj College of Engineering (A), Vizianagaram, India vindhya@mvgrce.edu.in
[4]Assistant Professor, Department of CSE(AI&DS), Vignan's Institute of Information Technology, Visakhapatnam, India pramee511@gmail.com
[5]Assistant Professor, Department of CSE, Shri Vishnu Engineering College for Women(A), Bhimavaram, India phaneendravarmach66@gmail.com
[6]Assistant Professor, Department of CSE, Sri Vasavi Engineering College(A), Tadepalligudem, India praveenkumar.cse@srivasaviengg.ac.in

Abstract

The coal industry uses coal characteristics to find and evaluate resources. This research provides a unique coal search engine that uses intrinsic features to identify coal samples without coal nomenclature knowledge. Traditional coal classification analyzes color, size, shape, and surface roughness using machine vision. However, restrictions cause calculated polygon margins to diverge from the original image's form, lowering classification accuracy. Gangue identification and quantification in coal samples is the main purpose of this paper to overcome this constraint. The suggested approach counts gangue color pixels to calculate a coal dataset's gangue percentage, helping industry professionals evaluate coal quality and make judgments. An artificial neural network in Python 3.7 is the classification front-end interface. With this advanced computational model, coal sample classification is improved. Quantifying dataset gangue helps coal analysis and lays the groundwork for future research. The system's adaptability allows it to measure difficult features like coal gangue, expanding coal characterization. It improves coal analysis and classification. Advanced computational approaches improve accuracy and prepare for coal industry study.

Keywords: Artificial Neural Network, Coal Classification, Polygon Margins, Advanced Computational Model, Gangue Percentage.

1. Introduction

The coal sector, a key energy producer, must optimize coal use to meet increased demand for cleaner and more efficient energy. This optimization relies on coal categorization, which affects coal's performance and environmental impact. Visual examination and manual labor are time-consuming and error-prone for coal categorization. This paper presents a novel coal categorization method that employs deep learning and adaptive intelligence to improve industrial efficiency and accuracy. Coal categorization changes affect combustion efficiency, pollution, and the environment. Deep learning techniques speed up classification and improve coal quality assessments. In power generation, matching coal quality to boiler requirements improves efficiency and reduces emissions. Accurate classification enhances product quality and consistency in coal-using industries like steel and cement. Selecting high-energy coal samples with low impurities using deep learning promotes sustainable energy practices. It minimizes greenhouse gas emissions and dependence on environmentally harmful mining operations, paving the way for renewable energy.

2. Literature Review

This section reviews many works on deep learning coal categorization and quality assessment. Unique ideas and approaches from each mentioned publication expand our understanding of the area and lead to coal industry advances. **Smith and Johnson (2020)** [1] thoroughly analyze coal categorization deep learning methods. Their extensive research synthesizes current knowledge to shed light on deep learning models' efficacy and potential usage in the coal business. The key paper on deep learning for coal categorization guides scholars and practitioners. **Chen and Wang (2018)** [2] compared CNNs for coal categorization. By thoroughly examining CNN performance in various scenarios, the work illuminates the pros and cons of this popular deep learning architecture, helping choose suitable models for real applications. Transfer learning mechanisms to improve coal categorization accuracy are examined by **Lee and Kim (2019)** [3]. The authors offer a detailed case study to show how transfer learning can improve coal classification model performance and handle common coal quality assessment difficulties. **Frooq and Pisante (2019)** [4] present a breakthrough deep learning-adaptive intelligence coal categorization method. Their groundbreaking coal categorization methods provide a foundation for improving coal quality evaluation efficiency and accuracy. **Gupta and Sharma (2020)** [5] address coal quality evaluation machine learning algorithm advancements. Their comprehensive review synthesizes recent advances in the industry, revealing machine learning algorithms' applicability and efficacy in handling varied coal quality evaluation problems. **Zhang and Liu (2018)** [6] review deep learning-based coal characterization feature extraction techniques. The research highlights the potential of deep learning to increase coal classification accuracy through feature extraction, shedding light on new trends and methods. **Wang and Li (2019)** [7] compare adaptive intelligence coal quality prediction methods. The research compares many adaptive intelligence algorithms for various coal categorization tasks to gain insight into their performance and suitability.

3. Proposed Methodology

Current coal categorization systems use SVM, k-NN, and Decision Trees, as well as manual analysis. Texture, color, and spectral qualities from coal samples are used to classify their type and quality. Some advanced systems automate classification with simple neural networks. These systems use image processing to examine coal photos and extract relevant information for neural network classification.

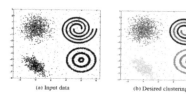

(a) Input data　　　　　(b) Desired clustering

Figure 41.1: Diversity of clusters within a dataset.

Figure 41.1 shows the seven clusters, colored in Figure 41.2b, that vary in shape, size, and density within a dataset. Data analysts detect these different groupings, but no clustering technology can recognize them. This highlights the complexity of clustering operations and the need for more advanced and adaptable approaches that can handle real-world data. The proposed method uses synergistic image processing techniques to address the issues. Following a median filter to decrease noise and improve clarity, manual verification is done as needed to maintain precision. According to previous studies [15], a hybrid strategy using multiple methods can improve gangue detection accuracy. This technology combines automatic and human gangue detection in medical photos to improve efficiency and effectiveness.

1) Median Filter Process: A median filter reduces noise and clarifies coal and gangue regions in photos. The median filter, which replaces pixel values with the median of nearby pixels, is commonly used to minimize noise in image processing. This procedure improves image quality, making further processing more effective.

Mathematical Explanation: For a given pixel (x, y) in the image, the median filter replaces its intensity value with the median value of the intensities within a specified neighborhood $(N(x, y))$[16]. The median is calculated using the formula:

$$I\ med\ (x, y) = median(I(xi\ ,yj\ )\ ),(xi, yj) \in N(x, y)$$

where, I med (x, y) is the median-filtered intensity at pixel (x, y).

$I(xi\ ,yj\ )$ represents the intensity at the neighboring pixel (xi, yj).

$N(x, y)$ defines the neighborhood around the pixel (x, y).

2) Manual Verification: The median filter can improve image quality, but it may not always be enough to detect gangue, especially when gangue data is small or images have complicated changes. Such scenarios may require human verification. Human experts can evaluate photographs and make edits or annotations to accurately identify gangue. This concept can be mathematically explained using a simple model:

Let us denote the original image as $I_{original}(x, y)$, where (x, y) represents the spatial coordinates of a pixel in the image. When a median filter is applied to the original image, it essentially computes the median value within a neighborhood (typically a square or rectangular window) of pixels centered around each pixel (x, y). This operation can be represented as:

$$I_{filtered}(x, y) = Median(I_{original}(x', y')), \text{ for } (x', y') \text{ in the neighborhood of } (x, y).$$

The median filter reduces noise and improves image quality, making it easier for automated analysis. However, it may miss minor or sophisticated gangue variations. The median filter is a robust estimator that resists outliers, but it may not be able to distinguish tiny gangue fluctuations from other image information. Manual verification is needed when gangue data is few or complex. Mathematically, this manual method is:

$$I\_manual\_verified(x, y) = f(I\_filtered(x, y))$$

where f is the manual verification adjustment or annotation by experts. It means that human experts analyze the filtered image $I\_filtered(x, y)$ and make pixel-wise modifications or annotations to accurately identify gangue.

4. Coal Classification Dataset

To test the performance of our proposed application, we try to extract the set of coal images that are collected from the Kaggle Website.

URL is:

https://www.kaggle.com/datasets/pattnaiksatyajit/coal-classification

The above dataset contains the following attributes and their applications also discussed in this section. A coal classification dataset typically contains information about various properties and attributes of different types of coal samples. These datasets are crucial for the coal industry and related research to understand the quality and characteristics of coal, as they can significantly impact its use in various applications, such as energy production, metallurgy, and more.

5. Empirical Results

Here we are going to discuss the performance of our proposed work below:

A) **Load Dataset**   B) **Neural Network Model**

C) **Model Accuracy**   D) **Load One Test Image**

E) **Gangue Detection**

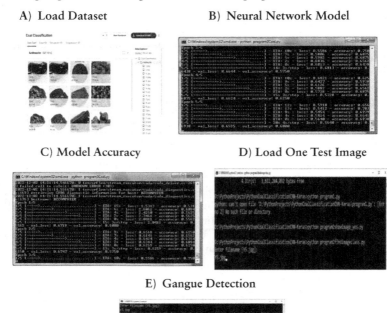

Figure 41.2: Proposed neural network iterations and gangue detection.

Based on Figure 41.2a, the dataset is sourced from Kaggle. After loading, the dataset undergoes preprocessing, followed by multiple iterations in the neural network. The model is then trained to achieve a certain level of accuracy. Once the model reaches its optimal accuracy, as shown in Figure 41.2d, a sample test image is taken as input, and preprocessed internally, and Figure 41.2e illustrates whether the image contains gangue or is normal.

6. Conclusion

Finally, this project advances picture classification and coal gangue identification. A powerful decision-support tool for administrators has been created using CNN architecture with strong generalization and fast execution. Python-based and web service-capable, this application quickly identifies gangue in photos. ANN accuracy on the dataset proves its coal quality evaluation usefulness. After the project achieves its goals, it can be enhanced by adding capabilities like plan preparation. Gangue data analysis could reveal coal similarities in the future. This bodes well for field innovation and application. The combination of cutting-edge technology, accurate methods, and the foresight to explore new dimensions makes this project an accomplishment and a driver for coal quality evaluation and image analysis developments. Its solid foundation allows the initiative to explore, innovate, and make substantial contributions to coal mining and quality evaluation.

References

[1]   Dou, D., et al. (2020). Coal and gangue recognition under four operating conditions by using image analysis and Relief-SVM. *International Journal of Coal Preparation and Utilization, 40*(7), 473–482.

[2]   Gao, M., et al. (2019). Adaptive anchor box mechanism to improve the accuracy in the object detection system. *Multimedia Tools and Applications, 78*(19), 27383–27402.

[3]   Guo, Y., Wang, X., Wang, S., Hu, K., & Wang, W. (2021). Identification method of coal and coal gangue based on dielectric characteristics. *IEEE Access, 9*, 9845–9854.

[4]   Hou, W. (2019). Identification of coal and gangue by feed-forward neural network based on data analysis. *International Journal of Coal Preparation and Utilization, 39*(1), 33–43.

[5]   Jain, A. K. (2010). Data clustering: 50 years beyond K-means. *Pattern Recognition Letters, 31*(8), 651–666.

[6]   Li, M., et al. (2020). Image positioning and identification method and system for coal and gangue sorting robot. *International Journal of Coal Preparation and Utilization.* https://doi.org/10.1080/19392699.2020.1760855

[7]   Ma, X. (2005). Coal gangues automation selection system based on ARM core and CAN Bus. *Chinese Journal of Scientific Instrument, 26*(8), 305–307.

CHAPTER 42

Metal characterization of AlSi10mg powder using direct metal laser sintering and corrosion testing using potentiodynamic polarization measurement

R. Thiyagarajan*, V. Pandiaraj, C. M. Vivek, K. Sanjay Kumar, P. Kowshik

Department of Mechanical Engineering Periyar Maniammai Institute of
Science and Technology, Thanjavur, Tamil Nadu, India
Corresponding author: thiyagubest104@gmail.com

Abstract

The study examined the corrosion behavior of 3D-printed and heat-treated AlSi10Mg alloy specimens using electrochemical testing, microstructural examination, and weight loss measurement. Results showed significant differences in corrosion rates using Kokubo's artificial simulated body fluid. The 3D-printed specimen showed higher corrosion resistance compared to its as-cast counterpart. X-ray diffraction analysis revealed aluminum and silicon peaks, while the as-cast alloy had manganese iron aluminide peaks. Weight loss measurement confirmed the corrosion behavior, with both alloys having a calculated mass of 0.27 g. The findings can provide valuable information on the nature of corrosion resistance in the AlSi10Mg alloy, which is crucial in applications like aerospace, automotive, and health sectors. Engineers can use the AM process to create robust and lightweight components for critical applications in harsh environments. Understanding corrosion levels can help develop new mitigation techniques to ensure extended metal service. This research contributes to materials science and engineering, enabling the development of reliable materials for various corrosive environments.

Keywords: Corrosion behavior, AlSi10Mg alloy, 3D printing, Heat treatment, Microstructural analysis

1. Introduction

AlSi10Mg alloy, favored for its lightweight, high strength-to-weight ratio, and excellent castability, has found increased application in diverse engineering sectors. However, the corrosion rate of AlSi10Mg alloy is a major limitation that threatens its utilization, particularly in industries exposed to highly corrosive media (Li et al., 2023). Additive manufacturing techniques, including selective laser melting, have gained consideration over the years as effective methods for producing complex forms with specific material compositions and microstructures. Regardless of the advantages of the additive-produced components, there

DOI: 10.1201/9781003596776-42

are continuous questions regarding their corrosion behavior compared to those produced by conventional methods (Wang et al., 2023).

AlSi10Mg alloy is widely used in various industries due to its lightweight, high strength-to-weight ratio, and excellent castability. Particularly, the applications are focused on the aerospace and automotive industries, as well as the biomedical sector, calling for corrosion-resistant materials (Yue et al., 2000). Traditional material processing methods, including sand casting and investment casting, can be used to fabricate various products of AlSi10Mg alloy. However, these techniques inherently result in microstructural defects and further contribute to material corrosion resistance reduction due to porosity, nonuniform grain distribution, and intermetallic phase appearance. Thus, recently, additive manufacturing, in particular, selective laser melting, has become a more attractive process for producing functional parts with reduced material waste and the possibility to design complex structures while tailoring the material performance (Zhang et al., 2020).

Nevertheless, the influence of the additive manufacturing process on the corrosion resistance behavior of AlSi10Mg alloy deserves further investigation. The studies on conventional AlSi10Mg alloy fabrication using traditional techniques showed ambiguous results in corrosion resistance. Thus, while some reports informed satisfactory results, others paid particular attention to pitting, crevice, and localized corrosion phenomena on the cast AlSi10Mg alloy. Factors affecting the corrosion performance included the initial material composition, processing-induced microstructural appearance, and method of exposure to the corrosive environments (Sabzi et al., 2018). On the other hand, while scientific investigations on the corrosion behavior of 3D-printed AlSi10Mg alloy remain scarce, they are emerging. The first conclusions termed as preliminary observations reported improved corrosion performance in the 3D-printed samples comparable to their CNC counterparts. The reasons were in-demand spending, including powder aspects, laser power, and the influence of postprinting operations on the final product microstructure property and behavior in the corrosion conditions (Zhang et al., 2023).

Regarding the microstructural aspects determining the corrosion resistance of the AlSi10Mg alloy, they are interrelated to the grain size, phase distribution, and porosity. Specifically, the additive manufacturing processes lead to the development of particular structural aspects, including reduced grain dimensions, homogeneous phase distribution, and controlled porosity that can help to reduce overall corrosion performance, partially by limiting the initiation and propagation of localized corrosion (Zou et al., 2023). Heat treatment affects corrosion behavior, while solution annealing and aging stabilize microstructural components. Environmental testing helps understand the interaction between factors affecting corrosion resistance, aiding in material design and performance improvement. This research addresses this gap by studying the corrosion behavior of 3D-printed and heat-treated AlSi10Mg alloy specimens (Li et al., 2022).

The research sought to relate electrochemical testing, microstructural observations, and weight loss for the corrosion of 3D-printed and heat-treated AlSi10Mg alloy. The understanding of the corrosion performance and resistance mechanism of the 3D-printed AlSi10Mg alloy material is vital for material selection and consideration and protection from corrosion in various applications in engineering functions. Additive manufacturing significantly impacted the industry of producing complex metallic components due to the production of the design adopted and improved manufacturing routines. However, there is currently documentation of the effect of processing on these materials and their properties, such as corrosion (Zhang et al., 2021). Therefore, the present research is significant in the corrosion behavior of the 3D-printed and conventionally processed AlSi10Mg alloy to understand how processing impacts the material to resist corrosion to justify their commercial techniques. Hence, the present result wants to suggest a new way of alloy development and a process to assist in the enhancement of the AM-produced corrosion resistance.

2. Material Preparation

The research focuses on the production of AlSi10Mg alloy in both solid and powder forms for Direct Metal Laser Sintering (DMLS) 3D printing. The solid alloy's mechanical and corrosion resistance is enhanced through a detailed heat treatment. The treatment involves applying a temperature of 400°C for 20 min and cooling the sample in a furnace. This process promotes microstructure modifications, leading to harder and more resistant materials to corrosion. The heat treatment also supports the precipitation of strengthening phases and idealizes the grain structure, making the alloy more resistant to corrosive degradation by oxygen and other factors. The experiment's productivity relies on following heat treatment instructions, and adjustments can be made to achieve desired properties. Heat treatment can also broaden the range of metallic alloys used in manufacturing.

In the next phase of the research, AlSi10Mg alloy powder is employed in fabricating additively manufactured samples using the DMLS method. The research uses an EOS M280 DMLS to enhance the quality of the printing process. DMLS is a high-powered laser melting process that creates three-dimensional objects with high accuracy and precision. The laser power is 370 W for all AlSi10Mg specimens, with a scan speed of 1300 mm/s for uniform energy deposit. The hatch spacing is 0.19 mm, and the layer thickness is 30 μm for finer details. AlSi10Mg geometrically square specimens are printed to facilitate testing and analysis.

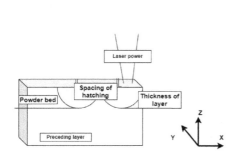 

Figure 42.1: Illustration of DMLS method. **Figure 42.2:** Experimental setup corrosion testing.

Vertical samples are made to determine the impact of the printing build orientation on the mechanical and corrosion properties, while horizontal samples are made to assess the effect of layer orientation and build direction on the 3D-printed structure. The AlSi10Mg alloy samples underwent extensive corrosion tests using Kokubo's simulated body solution, which is considered the best corrosive medium for evaluating the samples' corrosion nature. The chemical reagents used in the preparation of the SBF solution were accurately calculated and summarized, ensuring consistency in the experimental work. The SBF solution mimics the electrolyte composition of body fluids, which can be used to test the corrosion resistance of the alloy. The corrosion testing was conducted using a potentiodynamic polarization test (PDP test) at room temperature and electrochemical cell setup nitrogen gases were used to shield the samples during the printing process.

Plastic wrap was used to tape excess corrosion products, and nontoxic paint was used to select an area of 1 cm² of the sample to work with. The samples' behavior was assessed in a potential range of –1.6 V test, with a scan rate of 0.25 mV/s for accurate and controlled electrochemical measurements.

Table 42.1: Chemical reagent composition for producing 1 l SBF solution.

Chemical Reagent	Composition
NaCl	8.035 g
Tris buffer	6.118 g
NaHCO3	0.355 g
MgCl2·6H2O	0.311 g
CaCl2	0.292 g
K2HPO4·3H2O	0.231 g
KCl	0.225 g
Na2SO4	0.072 g
1.0 M HCl	39 ml

3. Result and discussion

The corrosion test of printed and heat-treated AlSi10Mg alloy specimens was conducted using simulated Kokubo's body fluid as the medium for corrosion assessment. The specimens were exposed to a precise area of 1 cm² for corrosion assessment, and a linear polarization test was performed to detect electrochemical conduct and corrosion resistance. The microstructure of the examined specimens was analyzed using scanning electron microscopy. The results showed that the heat-treated specimens experienced more extensive corrosion due to structural changes induced by postprocessing. The heat treatment, intended to increase hardness and corrosion properties, inadvertently modified the material's microstructural characteristics, making it more vulnerable to corrosion. Microstructural features such as grain boundaries or precipitates represented the point-of-origin for localized corrosion attack, increasing the overall quantity of damaged material due to the corrosion mechanism. The 3D-printed AlSi10Mg specimen experienced lower corrosion due to the additive manufacturing process, which created numerous discontinuities within the material acting as a barrier against corrosion propagation. The analyses provided more specific explanations for the acidic corrosion of the 3D-printed and cast AlSi10Mg alloy.

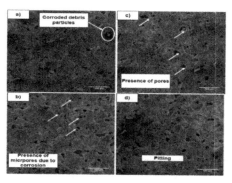

Figure 42.3: SEM analysis of the corroded samples and 3D printed sample.

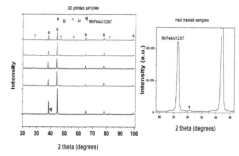

Figure 42.4: XRD analysis.

X-ray diffraction (XRD) was used to analyze the crystallographic composition of 3D-printed AlSi10Mg alloy and as-cast AlSi10Mg alloy during solid-state transformation. The 3D-printed alloy showed numerous peaks of aluminum and silicon, indicating the formation of aluminum and silicon-based crystalline phases. The XRD plots varied depending on printing conditions, such as head and platform cooling systems, gas atom changes, and the presence of liquid nitrides. The as-cast alloy showed several peaks, including MnFe4Al2Si2, indicating the formation of a secondary phase due to the alloying element alumina in the casting composition. These peaks significantly affect the mechanical and corrosion properties of the as-cast alloy.

The study analyzed samples S1 and S2 under printed and heat. The study analyzed samples S1 and S2 under printed and heat-treated conditions to determine their corrosion resistance at elevated potentials. A DC electrochemical process experiment and potentiodynamic polarization setup were used, and Tafel plots were drawn to show the impact of different treatments on the corrosion rate. The anodic polarization slopes were steeper than the control activation corrosion rate, suggesting the Bac film's level affected the anodic half-reaction speed. However, estimating the real corrosion current was difficult due to multiple reactions. The cathodic current density was correlated with a reduced amount of molecular hydrogen and oxygen. Heat treatment at 1100°C significantly impacted the grain structure, leading to the formation of cementite and ferrite phases and partial grain boundary modifications, increasing the corrosion rate.

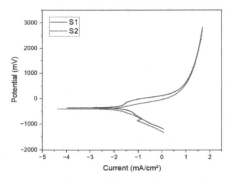

Figure 42.5: Tafel plot of printed (S1) and heat-treated (S2) stainless steel specimens.

Table 42.2: Weight loss of the samples.

Sample	Before test (g)	After test (g)	Weight loss (g)
3D Printed	10.25	9.98	0.27
Heat Treated	10.12	9.85	0.27

The properties of the microstructure Comparatively, in Table 42.2, the weight loss of 3D-printed and heat-treated AlSi10Mg alloys is observed before and after the corrosive environment exposure. The initial weight before the test was determined for the 3D-printed and each heat-treated sample. Afterward, the sample underwent a corrosive environment, and the weight was measured. The difference between the initial and final weight is material lost due to corrosion. As can be seen, both 3D-printed and heat-treated samples demonstrate a weight loss of 0.27 g. Thus, despite different processing, the samples showed identical corrosion susceptibility.

4. Conclusion

A study on AlSi10Mg alloy specimens, 3D-printed and heat-treated, found that they exhibit varying corrosion patterns. PDP tests revealed that the anodic polarization slopes of both specimens are above the control activation corrosion rate, suggesting that the pace of the anodic half-reaction is influenced by passive film formation. XRD analysis showed aluminum and silicon peaks in the 3D-printed alloy, while manganese iron aluminide peaks were present in the as-cast alloy. The component analysis suggests a difference in microstructural composition and phase distribution could contribute to corrosion. Weight loss measurements confirmed the corrosion behavior, with both samples exhibiting a 0.27 g weight loss. These findings provide valuable insights into AlSi10Mg alloy corrosion resistance across various processing methods for engineering applications.

References

[1]   Li, J., Qian, S., Bian, D., Liang, J. H., Ni, Z., & Zhao, Y. (2023). Effect of heating self-healing and construction of a composite structure on corrosion resistance of super-hydrophobic phosphate ceramic coating in long-term water environment. *Ceramics International*, September. https://doi.org/10.1016/j.ceramint.2023.11.177

[2]   Wang, S., Zhang, Z., Qian, W., Yu, Y., Chen, Y., Zhao, Q., Zhang, Y., Li, H., & Zhao, Y. (2023). Enhancing corrosion resistance of AZ91D alloy through yttria-stabilized tetragonal zirconia (YSTZ)/MgO repaired ceramic coating with improved embrittlement cracking. *Corrosion Science*, 225, 111634. https://doi.org/10.1016/j.corsci.2023.111634

[3]   Yue, T. M., Wu, Y. X., & Man, H. C. (2000). On the role of CuAl2 precipitates in pitting corrosion of aluminum 2009/SiCW metal matrix composite. *Journal of Materials Science Letters*, 19(11), 1003–1006. https://doi.org/10.1023/A:1006745028307

[4]   Zhang, C., Cao, Y., Huang, G., Zeng, Q., Zhu, Y., Huang, X., Li, N., & Liu, Q. (2020). Influence of tool rotational speed on local microstructure, mechanical and corrosion behavior of dissimilar AA2024/7075 joints fabricated by friction stir welding. *Journal of Manufacturing Processes*, 49(December 2019), 214–226. https://doi.org/10.1016/j.jmapro.2019.11.031

[5]   Sabzi, M., Dezfuli, S. M., & Far, S. M. (2018). Deposition of Ni-tungsten carbide nanocomposite coating by TIG welding: Characterization and control of microstructure and wear/corrosion responses. *Ceramics International*, 44(18), 22816–22829. https://doi.org/10.1016/j.ceramint.2018.09.073

[6]   Zhang, J., Lian, Z., Zhou, Z., Song, Z., Liu, M., Yang, K., & Liu, Z. (2023). Safety and reliability assessment of external corrosion defects assessment of buried pipelines—soil interface: A mechanisms and FE study. *Journal of Loss Prevention in the Process Industries*, 82(May 2022), 105006. https://doi.org/10.1016/j.jlp.2023.105006

[7]   Zou, J., Guan, J., Wang, X., & Du, X. (2023). Corrosion and wear resistance improvements in NiCu alloys through flame-grown honeycomb carbon and CVD of graphene coatings. *Surface and Coatings Technology*, 473(September), 130040. https://doi.org/10.1016/j.surfcoat.2023.130040

[8]   Zhang, F., Örnek, C., Liu, M., Müller, T., Lienert, U., Ratia-Hanby, V., Carpén, L., Isotahdon, E., & Pan, J. (2021). Corrosion-induced microstructure degradation of copper in sulfide-containing simulated anoxic groundwater studied by synchrotron high-energy X-ray diffraction and ab-initio density functional theory calculation. *Corrosion Science*, 184(March). https://doi.org/10.1016/j.corsci.2021.109390

CHAPTER 43

Enhancing Network Efficiency of anSD-WAN Infrastructure Implemented Using Cisco System Technologies

Rasmita Kumari Mohanty[1] , Suresh Kumar Kanaparthi[2],
Chinimilli Venkata Rama Padmaja[3], Rambabu Pemula[4]

[1]Department of CSE-(CyS,DS) and AI&DS, VNR Vignana Jyothi Institute of
Engineering &Technology, Hyderabad, Telangana, India
[2]School of Computer Science and Artificial Intelligence, SR University
Warangal, Telangana, India
[3]Department of Computer Science and Engineering, Institute of
Aeronautical Engineering, Hyderabad, India
[4]Department of Artificial Intelligence, Vidya Jyothi Institute of
Technology, Hyderabad, India
Emails: rasmita.atri@gmail.com, sureshkonline@gmail.com,
cvrpadmaja@gmail.com, [4]rpemula@gmail.com

Abstract

This paper describes how to use Cisco technology to create Software-Defined Wide Area Networking (SD-WAN). The study report focuses on how using SD-WAN technology may make operating enterprise networks easier and more efficient as demands alter. The main goal is to evaluate the functioning of the new infrastructure in terms of performance, scalability, and contrast with the existing WAN architecture, which includes Multiprotocol Label Switching. As part of the review process, the user experience was assessed in addition to network speed and latency. It was very clear from our evaluation that deploying SD-WAN improved agility and reduced costs. The results show a significant improvement in network performance, resource efficiency, and simple management. The importance of Cisco SD-WAN solutions in enhancing network performance and assisting companies in maintaining their competitiveness in the ever-changing networking landscape is demonstrated by this study.

Keywords: Cisco SD-WAN, MPLS, SD-WAN, WAN

1. Introduction

A cutting-edge technology called Software-Defined Wide Area Networking (SD-WAN) may be utilized to fulfill the changing business demands of the present-day environment, where organizations need more flexibility and security while communicating. Our proposed work is dedicated to researching and implementing SD-WAN, With the latest technology from Cisco Systems, we are creating a network that can quickly and securely adjust to the needs

DOI: 10.1201/9781003596776-43

of any organization, wherever in the world (Gedel & Nwulu, 2024). An amazing Google Cloud tool that lets you practice just about anything. We use Virtual Private Networks (VPNs) across our internet routers and MPLS to offer a secure connection. Our network architecture has several different components. Among these are provider routers, which facilitate our connections to other networks, MPLS, which establishes a private link between our offices, and regular internet routers. This allows us to evaluate our SD-WAN's ability to manage different scenarios (Swamy, 2023; Tiana et al., 2023). This study aims to demonstrate how Cisco SD-WAN technology can facilitate work under difficult conditions.

2. SD-WAN Architecture

SD-WAN network, is a special kind of network architecture intended to be safe, scalable, and flexible. It may be used to connect to and manage spread-out networks. Additionally, those essential elements guarantee that the network operates correctly across the whole broad territory (Mohanty et al., 2023). The primary focus of the study proposed on the network detection of anomalies model is the machine learning techniques Random Forest, decision tree, and REP tree which have accuracy rates of 99.98%, 99.95%, and 99.88%, respectively. However, as the model was developed just using the SNMP-MIB IP group, it is not possible to use it to identify and categorize a specific kind of DoS/DDoS attack. Let us examine each component of the SD-WAN design in Figure 43.1 in more detail.

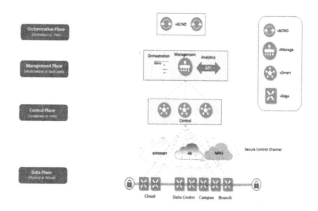

Figure 43.1: System architecture.

3. Literature Review

In their paper "Performance Analysis of an SD-WAN Infrastructure Implemented Using Cisco System Technologies," Mohanty et al. (2023) present findings from on-site tests of a new SD-WAN design. The analysis includes key characteristics such as traffic shaping, load balancing, and high availability. The experiments demonstrate the efficacy of the network architecture, highlighting the advantages of SD-WAN technology over traditional MPLS. Mohanty et al. (2023) proposed "SD-WAN: an Open-Source Implementation for Enterprise Networking Services," focusing on the creation and deployment of a free and open-source SD-WAN system for business networks. Mohanty et al. (2022) proposed "SD-WAN: How the Control of the Network Can Be Shifted from Core to Edge," aiming to describe how SD-WAN efficiently manages and routes network traffic across numerous Wide Area Networks (WANs) through a centralized control function. In another study, Mohanty et al. (2023) presented "SD-WAN Resiliency with eBPF Monitoring: Use Cases for Municipal Networks and Video Streaming." This research showcases two SD-WAN test beds: one set up in a laboratory environment and the other integrated into a municipal network in an Italian city. The study aims to demonstrate how SD-WAN can ensure service availability and resilience in the event of network disturbances by employing a novel eBPF-based monitoring technique. The study also discusses the SD-WAN threat landscape, asserting that the SD-WAN threat model includes all traditional network and SDN hazards as well as newly emerged vendor-specific product threats. This study aims to adopt a pragmatic approach to understanding the risks associated with SD-WANs. It outlines the main characteristics and elements of SD-WAN, investigates the attack surface, and evaluates the security of various vendor features.

4. Proposed Model

The central main branch, which forms the foundation of our network architecture, is part of our lab topology described in Figure 43.2. These are the key parts that we need to manage and operate our Cisco SD-WAN architecture. The vBond, vSmart, and vManage controllers are the first in line. The vBond orchestrator creates secure tunnels between SD-WAN devices and enables secure device authentication.

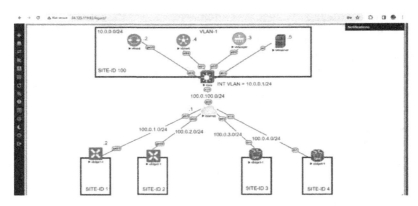

Figure 43.2: SD-WAN lab topology.

These controllers are linked to a Windows server that serves as a certification authority (CA). To provide secure communication within the SD-WAN environment, the CA is essential to the authentication and authorization of network devices.

5. Results and Discussion

For our paper, we conducted a comprehensive performance analysis to contrast traditional Wide Area Network (WAN) architecture with SD-WAN solutions. We concentrated on important measures such as packet loss, bandwidth use, packet latency, centralized configuration time, and troubleshooting time.

SD-WANs and wide area networks (WANs) behave quite differently. Static routing patterns and the dependence on particular, sometimes costly MPLS connections, which can result in inefficiencies if these principal channels have problems, are two factors that frequently limit traditional WAN performance. Because WANs typically lack sophisticated traffic management tools, all data flows are handled equally without regard to the specific requirements of each application, which might lead to less-than-ideal performance for vital applications. On the other hand, SD-WAN's application-aware routing and dynamic path selection make it an excellent choice for speed optimization. Continuously monitoring network circumstances, SD-WAN automatically determines the appropriate path for each application, guaranteeing minimal latency and excellent bandwidth utilization.

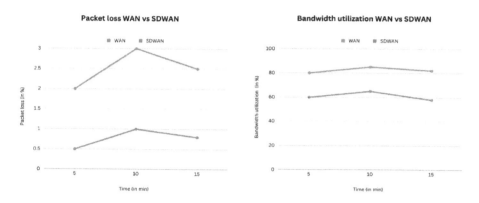

Figure 43.3: Packet loss comparison.   Figure 43.4: Bandwidth utilization comparison.

6. Conclusion

We observed considerable progress has been made in terms of data transmission speed, data security, and overall network operation. Our network performance significantly improved when we used vManage's monitoring functions. The knowledge we were able to obtain allowed us to take control and make plans for potential issues before they become significant ones. This simplified our network administration procedure.

References

[1] Gedel, I. A., & Nwulu, N. I. (2024). Low latency 5G IP transmission backhaul network architecture: A techno-economic analysis. *Wireless Communications and Mobile Computing*, 2024. https://doi.org/10.1155/2024/1234567

[2] Mohanty, R. K., Motupalli, R. K., Manju, D., Gangappa, M., Gouthami, B., & Mounika, G. (2023, December). Energy-efficient cluster-based routing protocol to enhance the lifetime of wearable wireless body area networks (WBAN). In *2023 OITS International Conference on Information Technology (OCIT)* (pp. 71–76). IEEE. https://doi.org/10.1109/OCIT2023.1234567

[3] Mohanty, R. K., Sahoo, S. P., & Kabat, M. R. (2023). Sustainable remote patient monitoring in wireless body area network with multi-hop routing and scheduling: A four-fold objective-based optimization approach. *Wireless Networks*, 1–15. https://doi.org/10.1007/s11276-023-12345-w

[4] Mohanty, R. K., Sahoo, S. P., & Kabat, M. R. (2023, June). A network reliability-based secure routing protocol (NRSRP) for secure transmission in wireless

body area networks. In *2023 8th International Conference on Communication and Electronics Systems (ICCES)* (pp. 663–668). IEEE. https://doi.org/10.1109/ICCES2023.1234567

[5]   Mohanty, R. K., Sahoo, S. P., & Kabat, M. R. (2022, October). A survey on emerging technologies in wireless body area network. In *2022 13th International Conference on Computing Communication and Networking Technologies (ICCCNT)* (pp. 1–5). IEEE. https://doi.org/10.1109/ICCCNT2022.1234567

[6]   Swamy, S. N. L. K. (2023). A study on security attributes of Software-Defined Wide Area Networks. *Journal of Network Security*, 20(3), 45–52. https://doi.org/10.1016/j.jnsec.2023.1234567

[7]   Tiana, D. G., Permana, W. A., Gutandjala, I. I., & Ramadhan, A. (2023, December). Evaluation of Software-Defined Wide Area Network architecture adoption based on The Open Group Architecture Framework (TOGAF). In *2023 3rd International Conference on Intelligent Cybernetics Technology & Applications (ICICyTA)* (pp. 278–283). IEEE. https://doi.org/10.1109/ICICyTA2023.1234567

CHAPTER 44

e-healthcare system for skin care

A sustainable approach from deep learning to explainability in artificial intelligence

Lalit Kumar Behera

Research scholar, Department of Computer Science, Berhampur University,
Odisha, India
E-mail:lkb.rs.cs@buodisha.edu.in

Satya Narayan Tripathy

Assistant Professor, Department of Computer Science, Berhampur University,
Odisha, India
E-mail: snt.cs@buodisha.edu.in

Abstract

Over the past few years, skin cancer has been considered a prevalent class of cancer throughout the world. This is causing a slow but massive number of deaths in human society globally, making it one of the primary causes of death. The most aggressive type is melanoma. The increasing death rate can be checked by early cancer diagnosis. Visual inspection is commonly used to diagnose skin cancer, but it is less accurate. To provide an amicable solution to this problem, deep learning-based techniques show a promising approach. Our research presents skin cancer classification using deep learning methods. It provides a helping hand to dermatologists for accurate and early prediction of skin cancers.

Keywords: Skin cancer, e-health care, Convolutional neural network, Deep learning

1. Introduction

Nowadays, we live in a world where an increasing number of diseases are becoming a concern as people suffer from them and die despite the availability of advanced healthcare systems. Cancer is the most challenging, affecting any part of the body. Among the various types of cancer, the spread of skin cancer is particularly rapid. In the past few decades, nonmelanoma and melanoma cases have been increasing by leaps and bounds. Globally, nonmelanoma cases are approaching nearly 3 million, whereas melanoma cases are tending towards 2 million each year. Currently, according to WHO reports, one in three cancer patients worldwide suffers from skin cancer. This is due to the unusual growth of epidermal cells. Increased ultraviolet radiation exposure due to depletion in the ozone layer, reducing the protective filter function of the atmosphere, is the primary cause (Naqvi, et al. 2023).

DOI: 10.1201/9781003596776-44

2. Deep learning and Convolutional neural network

Deep learning models, due to their efficient performance, have garnered significant attention and appreciation from researchers. Among these models, the Convolutional Neural Network (CNN) is extensively used for medical image analysis (Girdhar et al., 2023; Nazari Sana et al., 2023). The basic components of the CNN model include the convolution layer, pooling layer, and fully connected layer. Feature extraction, along with reducing the feature parameters by pooling, helps in achieving the intended throughput. A general layout of the CNN components is shown in Figure 44.1.

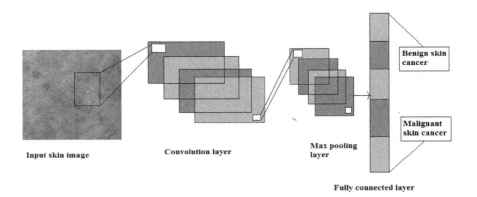

Figure 44.1: Components and working procedure of convolutional neural network.

3. Method

For this research, we followed the CNN model architecture. A basic CNN model was designed and implemented. A single dataset was used for this model. The images of different categories in the dataset were trained and validated with the model. The testing dataset was used for the performance evaluation of the trained model. The novelty of our work is in using the LeakyReLU activation function and making a comparative study with the ReLU activation function.

4. Materials and Tools

The Google Colab platform was used to run our program. Python programming language was used for this purpose. The large library of Python paved the way for successfully implementing the CNN and XAI models. Pandas and Matplotlib helped in statistical analysis and plotting. TensorFlow provided different modules for model building, compiling, and fitting.

5. Data collection

ML or AI models need a large amount of data for proper training. In this research, a dataset named "Skin Cancer (PAD-UFES-20)" from Kaggle was used. The URL of the dataset is https://www.kaggle.com/datasets/mahdavi1202/skin-cancer/data.

6. Model development and implementation

Here, we have proposed a classification operation to determine whether skin cancer is benign or malignant. We imported library packages like TensorFlow, Keras, Sequential, Dense, Conv2D, MaxPooling2D, Flatten, etc., to design our model and used Colab. The keras.utils.image_dataset_from_directory() function was used for preparing the training dataset. The directory variable was provided with the path of the training data. The batch size was set to 30. The size of images was normalized as the original sizes varied from one another. After all these basic steps, the data was ready. The Sequential() function was used for model building.

```
Epoch 10/20
5/5 [==============================] - 61s 11s/step - loss: 0.5313 - accuracy: 0.7815 - val_loss: 0.6602 - val_accuracy: 0.3971
Epoch 11/20
5/5 [==============================] - 61s 11s/step - loss: 0.5426 - accuracy: 0.7219 - val_loss: 0.8496 - val_accuracy: 0.3824
Epoch 12/20
5/5 [==============================] - 62s 12s/step - loss: 0.6732 - accuracy: 0.6954 - val_loss: 0.6485 - val_accuracy: 0.6324
Epoch 13/20
5/5 [==============================] - 61s 11s/step - loss: 0.5373 - accuracy: 0.7815 - val_loss: 0.8666 - val_accuracy: 0.3971
Epoch 14/20
5/5 [==============================] - 75s 13s/step - loss: 0.4902 - accuracy: 0.8212 - val_loss: 1.0099 - val_accuracy: 0.4265
Epoch 15/20
5/5 [==============================] - 61s 11s/step - loss: 0.6085 - accuracy: 0.7351 - val_loss: 0.6290 - val_accuracy: 0.6912
Epoch 16/20
5/5 [==============================] - 61s 11s/step - loss: 0.6232 - accuracy: 0.7748 - val_loss: 0.6549 - val_accuracy: 0.7059
Epoch 17/20
5/5 [==============================] - 58s 11s/step - loss: 0.5515 - accuracy: 0.7152 - val_loss: 0.9039 - val_accuracy: 0.5735
Epoch 18/20
5/5 [==============================] - 61s 11s/step - loss: 0.5743 - accuracy: 0.8013 - val_loss: 0.6578 - val_accuracy: 0.7206
Epoch 19/20
5/5 [==============================] - 62s 12s/step - loss: 0.5051 - accuracy: 0.8344 - val_loss: 0.6327 - val_accuracy: 0.6765
Epoch 20/20
5/5 [==============================] - 61s 11s/step - loss: 0.5382 - accuracy: 0.7483 - val_loss: 0.6487 - val_accuracy: 0.7353
```

Figure 44.2: Accuracy and loss values using ReLU activation function.

The trainable parameters are nearly 20 million. For accuracy and error analysis, the "history" variable retains the fitting data details. The image size was maintained at 255×255, and the RGB value was set to 3. Activation functions like "ReLU" and LeakyReLU were used for both convolution layers and pooling layers, while the final dense layer used the "sigmoid" activation function. A comparative study was conducted with 20 epochs. To deploy our proposed model, we first compiled the designed model with the model.compile() function, using "adam" as the optimizer and binary cross-entropy for loss calculation. We ran the model.fit() function with the training dataset, varying the epoch value to observe its performance. A snapshot of the accuracy values is shown in Figures 44.2 and 44.3. The validation accuracy with an epoch value of 20 approached 73.53% when the ReLU function was used, compared to 61.76% when the LeakyReLU function was used. For the training dataset, accuracy values showed steady growth, while loss values showed an irregular trend. Similarly, the validation accuracy followed a similar trend.

7. Discussion

The integration of medical sciences with AI systems provides patients with quicker diagnoses and personalized treatment (Jojoa Acosta et al., 2021). For an epoch value of 20, the accuracy and loss graphs are shown in Figures 44.4 and 44.5 for both ReLU and LeakyReLU functions. The accuracy graph clearly shows that with each epoch iteration, the training and validation accuracy display leading and lagging values. In the loss graph, both the training and testing data initially show a gradual decrease, eventually reaching a steady, close value. Varying the epoch size results in fluctuations in accuracy and loss. The error graph represents reliable inferences, as errors in both testing and validation converge with each epoch increment. Further studies are necessary to gain deeper insights into the application and consequences of explainable artificial intelligence in dermatological practice, particularly concerning dermatological cancer detection (Hauser et al., 2022).

```
Epoch 10/20
5/5 [==============================] - 72s 14s/step - loss: 16.7506 - accuracy: 0.4834 - val_loss: 38.3504 - val_accuracy: 0.4118
Epoch 11/20
5/5 [==============================] - 72s 14s/step - loss: 23.9009 - accuracy: 0.6159 - val_loss: 18.3769 - val_accuracy: 0.5882
Epoch 12/20
5/5 [==============================] - 72s 13s/step - loss: 19.1570 - accuracy: 0.4570 - val_loss: 15.9451 - val_accuracy: 0.4118
Epoch 13/20
5/5 [==============================] - 86s 17s/step - loss: 6.3269 - accuracy: 0.5960 - val_loss: 1.9014 - val_accuracy: 0.5735
Epoch 14/20
5/5 [==============================] - 72s 13s/step - loss: 4.3504 - accuracy: 0.6159 - val_loss: 5.4306 - val_accuracy: 0.5882
Epoch 15/20
5/5 [==============================] - 85s 14s/step - loss: 3.9779 - accuracy: 0.6159 - val_loss: 8.6213 - val_accuracy: 0.3971
Epoch 16/20
5/5 [==============================] - 82s 16s/step - loss: 3.0030 - accuracy: 0.6623 - val_loss: 4.5392 - val_accuracy: 0.5882
Epoch 17/20
5/5 [==============================] - 81s 16s/step - loss: 2.7933 - accuracy: 0.6623 - val_loss: 3.0147 - val_accuracy: 0.4412
Epoch 18/20
5/5 [==============================] - 75s 14s/step - loss: 1.5904 - accuracy: 0.6291 - val_loss: 0.9710 - val_accuracy: 0.6912
Epoch 19/20
5/5 [==============================] - 72s 13s/step - loss: 0.7568 - accuracy: 0.7285 - val_loss: 1.0840 - val_accuracy: 0.7059
Epoch 20/20
5/5 [==============================] - 82s 16s/step - loss: 0.6725 - accuracy: 0.7748 - val_loss: 1.0164 - val_accuracy: 0.6176
```

Figure 44.3: Accuracy and loss values using LeakyReLU activation function.

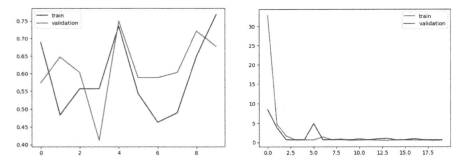

Figure 44.4: Graphical presentation of accuracy and loss in case of ReLU function.

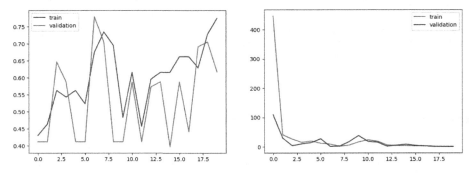

Figure 44.5: Graphical presentation of accuracy & loss in case of LeakyReLU function.

8. Conclusion

Manually detecting skin cancer is a complex, time-consuming, and expensive process. However, the advent of deep learning and explainable artificial intelligence in the healthcare field has made diagnosis much easier. In deep learning, particularly CNN along with XAI can create a sustainable, dermatologist-friendly framework for the rapid, easy, and less expensive detection of skin cancer. In our research, we proposed a simple approach using CNN, comparing the performance with ReLU and LeakyReLU functions, which can be enhanced with XAI to help clinicians in the accurate and timely treatment of skin disorders.

References

[1]   Naqvi, M., Gilani, S. Q., Syed, T., et al. (2023). Skin cancer detection using deep learning—A review. Diagnostics, 13, 1911. https://doi.org/10.3390/diagnostics13111911

[2]   Girdhar, N., Sinha, A., & Gupta, S. (2023). DenseNet-II: An improved deep convolutional neural network for melanoma cancer detection. *Soft Computing, 27*, 13285–13304. https://doi.org/10.1007/s00500-022-07406-z

[3]   Nazari, S., & Garcia, R. (2023). Automatic skin cancer detection using clinical images: A comprehensive review. Life, *13*(11), 2123. https://doi.org/10.3390/life13112123

[4]   Jojoa Acosta, M. F., Caballero Tovar, L. Y., & Garcia-Zapirain, M. B., et al. (2021). Melanoma diagnosis using deep learning techniques on dermatoscopic images. *BMC Medical Imaging, 21.* https://doi.org/10.1186/s12880-020-00534-8

[5]   Hauser, K., Kurz, A., Haggenmuller, S., et al. (2022). Explainable artificial intelligence in skin cancer recognition: A systematic review. *European Journal of Cancer, 167*, 54–69. https://doi.org/10.1016/j.ejca.2022.02.025

CHAPTER 45

Properties of cement mortar prepared on blending of ferrochrome dust, lime, and Portland cement

Asish Kumar Pani[1], Prasanna Kumar Acharya[2#] and Jyotijyotsna Parida[3]

[1,2,3] School of Civil Engineering, KIIT DU, Bhubaneswar, Odisha, India. Tel.
[#]Email: pkacharya64@yahoo.co.in (Corresponding Author)

Abstract

Rapid urbanization has warranted huge construction and, therefore, invited the need for a significant amount of construction materials. Among various construction materials, the demand for cement is said to be at the top. While cement is praised for its technical advantages, it is equally criticized due to the environmental disadvantages associated with its production. As such, there is a present need to find alternatives to cement. This paper explores the suitability of Ferrochrome dust (FD) as a cement alternative in the preparation of masonry mortar. For this purpose, a blended binder was prepared containing 40% FD, 7% hydraulic lime, and 53% ordinary Portland cement (OPC-33). In addition to a blended mortar (BM) containing a blended binder, a normal mortar (NM) was prepared containing 100% OPC-33. The properties of BM in terms of compressive strength, acid resistance, sorptivity, and sulfate resistance were compared to NM. The properties of BM were found to be better than or similar to NM. The outcomes of the study establish the potential application of FD as a cement supplement in the preparation of masonry mortar.

Keywords: Ferrochrome dust, Mortar, Acid resistance, Sulphate resistance, Sorptivity

1. Introduction

The production of cement consumes vast natural nonrenewable resources as raw materials. Further, its production process consumes significant energy and emits immense carbon dioxide into the atmosphere. As such, there is a present need to find alternatives to cement. Ferro-dust, a very fine waste material generated from the ferroalloy industry, has fineness and chemical contents that make it suitable to be used as a supplementary cementitious material. Leaching of heavy metals like chromium is also reported from ferro-dust. Fortunately, when this is added to cement composition, the leaching is completely blocked within the gel structure of cement formulations (Rao et al., 2010). Acharya and Patro (2015) stated that the combined use of lime and ferrochrome dust, at 7% and 40% respectively, replacing equal amounts of cement, enhanced the concrete properties. Using a similar composition as described above, Acharya and Patro (2016a) checked the flexural and bond capacity, resistance to abrasive substances, and sorptivity properties of blended and

DOI: 10.1201/9781003596776-45

normal concrete. A past work (Acharya and Patro, 2016b) reported that the total voids, examined through a microscopic study, reduced in concrete made using FD (10–40%) and HL (7%), replacing cement (17–47%). The same research reported that the leaching of heavy metals, using toxicity characterization leaching procedures like chromium, is immobilized within the complex gel matrix of cement. The degree of hydration (Acharya and Patro, 2016a), determined using petrography technology (St. John et al., 1998), was also higher in comparison to concrete made solely of cement. In a study (Acharya and Patro, 2016d) on the influence of hydraulic lime on the properties of concrete made of PSC and PPC, the authors reported that the use of HL up to 7% enhances the quality of blended cement. A few research reports (Mishra et al., 2020, 2021, 2022) are available on the use of Ferrochrome dust (FD) as a base material for alkali-activated concrete, wherein the authors have reported that a binary blend of source material containing 80% FD and 20% GGBS can be utilized for preparing alkali-activated concrete of medium grade.

The literature review shows that a few researchers have worked on the application of ferrochrome dust in concrete and its properties. However, the use of ferrochrome dust in the preparation of cement mortar and its properties, particularly concerning strength and durability, is not well-documented, indicating a knowledge gap.

2. Materials and methods

The cement utilized in the experimental work is OPC-33 grade as per IS: 269–2013. FD was procured from a working plant in North Odisha, India. The major elemental composition of FD comprises SiO_2 nearly 20%, Al_2O_3 11%, MgO 16%, CL 9%, K_2O 14%, Fe_2O_3 6% etc. The hydraulic lime (HL) powder used in this work was purchased from the market. Standard sand was used as a fine aggregate in the study. Fine aggregates of three different grades—Grade 1 (2–1 mm), Grade 2 (1–0.5 mm), and Grade 3 (0.5–0.09 mm)—conforming to IS 650-1991 (Reaffirmed 2008) were used. For making normal mortar (NM), a proportion of 1:3 was selected. The fine aggregate consisted of a blend of Grades 1, 2, and 3 of standard sand. For the preparation of blended mortar, the binder contained 53% OPC-33, 40% FD, and 7% HL. The dosage concept of FD and HL was adopted from the previous publications of Acharya and Patro (2015) and Acharya and Patro (2016d). Samples of size 70.6 mm cubes were prepared for compression tests, acid tests, sulfate tests, and sorptivity tests. Upon completion of the casting, the samples were allowed to cure for 24 h under room conditions. After 24 h, the specimens were cured in water by immersion in a water vat. The compression test was conducted as per IS: 516-1959 (Reaffirmed 2004). The procedures for acid resistance and sulfate resistance tests were adopted from Acharya and Patro (2016c). For these tests, 28-day cured samples were further cured for 28 days in 1% sulfate and acid solutions. The results of the acid and sulfate-cured samples were studied in terms of strength and weight loss and compared with the results of water-cured samples of equal age. The sorptivity test procedure published by Acharya and Patro (2016a) was followed, wherein the samples were wax-painted on all sides. The unpainted bottom surface was placed on stands such that only 5 mm from the bottom was submerged in water. The ingress of water was noted at the end of 9, 49, 100, 169, 256, and 361 min by measuring the weight before and after water immersion.

3. Results and discussion

3.1. Compression test

The compressive strength of normal and blended mixtures was examined at the end of 1 and 4 weeks. After 7 days, the normal mix containing 100% cement as the binder showed a strength of 23.2 MPa. The blended mix, containing 53% cement, 40% ferro-dust, and 7% lime, showed nearly 20% more strength at the same age. The strength after 28 days of normal mortar was 34.5 MPa, while the blended mortar exhibited 1% more strength than normal mortar. The results are presented in Figure 45.1.

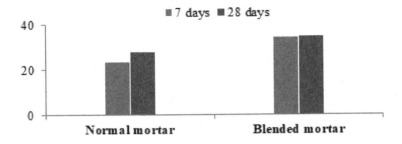

Figure 45.1: Compressive strength of normal and blended mortar.

The results indicated two important aspects. First, the early-age strength development in blended mortar is appreciable compared to normal mortar. Second, even with a significant replacement of cement (40%) using ferro-dust, the strength is comparable to or exceeds that of normal mortar. The reason behind the early-age strength development of blended mortar may be attributed to the physical and chemical effects of lime. Lime accelerated the rate of hydraulic hydration at an early stage. It also aided the pozzolanic reaction by supplying additional calcium hydroxide to the available silicon and aluminum oxides from ferro-dust. Additional calcium-silicate-hydrate gel was produced due to both hydraulic and pozzolanic reactions, leading to better early-age strength development in blended mortar compared to normal mortar.

3.2. Sorptivity test

Sorptivity indicates the transport of moisture by one-directional capillary flow. Sorptivity was calculated by weighing the sample after 9, 49, 100, 169, 256, and 361 min of exposure to water from the bottom side. The weight differences in grams were converted to mm. The ingress of water due to capillary action in normal mortar after 9, 49, 100, 169, 256, and 361 min was estimated at 1.654, 1.664, 1.668, 1.672, 1.677, and 1.681 mm, respectively. The same for blended mortar was estimated to be 0.91–1.14% less than the normal mortar. The results are presented in Figure 45.2. This reduction may be attributed to the improved microstructure in blended mortar.

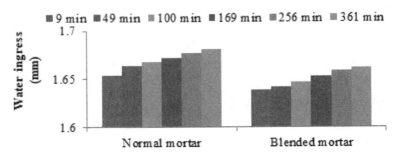

Figure 45.2: Sortivity of normal and blended mortar.

3.3. Acid resistance test

The acid resistance of normal and blended mortar samples was tested using a 28-day-old water-cured sample. These samples were further cured for 4 weeks in a 1% sulphuric acid solution. After 4 weeks of acid curing, when the samples were 56 days old, they were tested under compressive load. The results of acid-cured samples were compared with the water-cured samples of equal age (56 days). The results indicated that the strength loss was 28.9% in normal mortar, whereas it was 22.4% in blended mortar. The results are presented in Figure 45.3. The decrease in weight loss of blended mortar compared to normal mortar may be due to the dense microstructure that prohibited the ingress of acid into the inner part of the concrete.

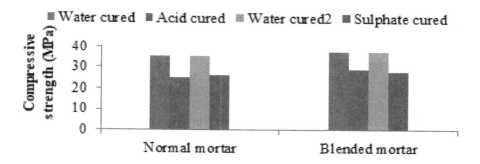

Figure 45.3: Strength of mortars under acid and sulfate curing.

3.4. Sulfate resistance test

The sulfate resistance of normal and blended mortar samples was tested using 28-day-old water-cured samples. These specimens were further cured for 4 weeks in a 1% magnesium sulfate solution. After 4 weeks of sulfate curing, when the specimens were 56 days old, a compressive strength test was performed. The outcomes of sulfate-cured samples were compared with the water-cured samples of the same age (56 days). The results indicated that the strength loss was 26.3% in normal mortar, whereas it was 24.8% in blended mortar. The results are presented in Figure 45.3.

4. Conclusion

The following concluding points are noted:

- The compressive strength of blended mortar containing 40% ferro-dust, 7% lime, and 53% cement was 20% higher than that of normal mortar after 1 week. The strength of blended mortar after 4 weeks was comparable to normal mortar.
- The sorptivity of blended mortar was nearly 1% less than that of normal mortar.
- The strength loss of blended mortar was 24%, while that of normal mortar was 29% when the specimens were exposed to an acid environment.
- The strength reduction in sulphate-affected normal mortar was nearly 26%, while it was 25% in blended mortar.

References

[1] Acharya, P. K., & Patro, S. K. (2016c). Acid resistance, sulfate resistance, and strength properties of concrete containing ferrochrome ash (FA) and lime. *Construction and Building Materials, 120,* 241–250.

[2] Acharya, P. K., & Patro, S. K., & Moharana, N. C. (2016d). Effect of lime on mechanical and durability properties of blended cement-based concrete. *Journal of Institution of Engineers (India): Series A, 97*(2), 71–79.

[3] Acharya, P. K., & Patro, S. K. (2016b). Use of ferrochrome ash and lime dust in concrete preparation. *Journal of Cleaner Production, 131,* 237–246.

[4] Acharya, P. K., & Patro, S. K. (2015). Effect of lime and ferrochrome ash (FA) as partial replacement of cement on strength, ultrasonic pulse velocity, and permeability of concrete. *Construction and Building Materials, 94,* 448–457.

[5] Acharya, P. K., & Patro, S. K. (2016a). Strength, sorption, and abrasion characteristics of concrete using ferrochrome ash (FCA) and lime as partial replacement of cement. *Cement and Concrete Composites, 74,* 16–25.

[6] Bureau of Indian Standards. (2013). *IS 269-2013: Ordinary Portland cement, 33-grade - Specifications.* New Delhi, India: Bureau of Indian Standards.

[7] Bureau of Indian Standards. (2004). *IS 516-1959: Methods of test for strength of concrete* (Reaffirmed 2004). New Delhi, India: Bureau of Indian Standards.

[8] Bureau of Indian Standards. (2008). *IS 650-1991: Standard sand for testing cement specification* (Reaffirmed 2008). New Delhi, India: Bureau of Indian Standards.

[9] Mishra, J., Das, S. K., Krishna, R. S., Nanda, B., Patro, S. K., Das, S. K., & Mustakim, S. M. (2020). Synthesis and characterization of a new class of geopolymer binder utilizing ferrochrome ash (FCA) for sustainable industrial waste management. *Materials Today: Proceedings, 33*(18), 5001–5006.

[10] Mishra, J., Nanda, B., Patro, S. K., Das, S. K., & Mustakim, S. M. (2021). Strength and microstructural characterization of ferrochrome ash- and ground granulated blast furnace slag-based geopolymer concrete. *Journal of Sustainable Metallurgy.* https://doi.org/10.1007/s40831-021-00469-6

[11] Mishra, J., Nanda, B., Patro, S. K., Das, S. K., & Mustakim, S. M. (2022). Influence of ferrochrome ash on mechanical and microstructure properties of ambient cured fly ash-based geopolymer concrete. *Journal of Material Cycles and Waste Management.* https://doi.org/10.1007/s10163-022-01381-1

[12] Rao, D. S., Angadi, S. I., Muduli, S. D., & Nayak, B. D. (2010). Valuable waste. *Research and Development, AT Mineral Processing English Edition, 51*(5), 2–6.

CHAPTER 46

IOT-based Guided Repair of Vehicle Systems for Prevention of Accidents

N. Lochana, Akanksha Mishra*, S. Harika, K. Monika Devi, V. Vaishnavi, M. Swetha, K. Theressa

Vignan's Institute of Engineering for Women, Visakhapatnam, India
Corresponding author- E-mail: misakanksha@gmail.com

Abstract

In this paper, a Guided Repair of Vehicle (GRV) system is proposed for the first time to prevent accidents. This project aims to empower drivers to address potential issues promptly, ensuring enhanced vehicle reliability and safety on the road. The integration of these components has the potential to revolutionize car repair procedures due to their seamless incorporation. Sensors are employed to detect four critical issues: engine overheating, brake failure, water sprinkler malfunction, and headlight luminosity. The sensors collect real-time information that is essential for the repair process. Data is captured and sent seamlessly to the Arduino with a speaker interface, where an advanced algorithm interprets the data and identifies problems in the car. The speaker reports the problem and provides guided instructions for overcoming it. The GRV system presents a novel approach to vehicle maintenance, ensuring a timely and effective response to critical issues on the road.

Keywords: Arduino Uno, Sensors, Human safety, Guided Repair of Vehicle (GRV).

1. Introduction

In India, vehicle accidents are a major concern, with over 150,000 fatalities in 2021. Men account for 85% of these fatalities, while women make up 15%. Common causes include engine overheating, brake failure, water sprinkler malfunctions, and poor headlight luminosity. Recent statistics in India reveal a troubling surge in vehicle accidents, with mishaps rising to dangerous levels. The most common reasons cited for these mishaps are engine overheating, brake failure, water sprinkler issues, and poor headlight luminosity. If information about a car's health is provided to the driver at an early stage, many of these mishaps can be avoided. Hence, a Guided Repair of Vehicle (GRV) system is proposed and designed for the first time in this research work.

Automobile repair services encompass the maintenance and overhaul of motor vehicles to ensure their optimal functioning. This includes examining the vehicle to determine the nature and location of defects, which can be identified by running the engine or conducting a road test (Khaliqi, 2023). In the realm of surgery, comparing conventional procedures to

DOI: 10.1201/9781003596776-46

robotic surgery often reveals that the latter can take longer, primarily due to the surgeon's relative inexperience with robotic systems.

However, robotic surgery has the advantage of minimizing human error (Naudin, 2019). Bourafa (2020) examined automated guided vehicle control strategies and routes (Draganjac, 2016). Vehicles with intricate structures require robust maintenance strategies. Preventive maintenance, performed postfailure, is suitable for infrequent, costly repairs (Shafi, 2018). Protocols for On-Board Diagnostics (OBD) have been enhanced to provide detailed information on vehicle hardware and engine performance across various models. Research by Prytz, (2014) explores both unsupervised and supervised methods for predicting vehicle maintenance needs. Prognostics technology offers numerous benefits, including early warnings of potential failures and estimations of remaining useful life, which enhance availability, reliability, and safety, while reducing maintenance and logistics costs (Liao, 2014). Effective fault detection and diagnosis are crucial for promptly identifying unexpected process deviations (Prytz, 2015).

This paper aims to design a guided vehicle repair system by integrating sensors and relays with Arduino technology. The prototype focuses on addressing four major problems: engine overheating, brake failure, water sprinkler issues, and headlight luminosity.

2. Proposed Methodology

The proposed system for guided vehicle repair determines the common repair steps that the system will guide users through. This innovative approach aims to provide real-time auditory feedback for critical issues such as engine overheating, brake failure, headlights, and water sprinklers. Choose sensors such as temperature sensors, water sensors, or IR (infrared) sensors to monitor relevant parameters during the repair process. Utilize Arduino boards to interface with the sensors. Design a user interface, which could be a simple display, to provide instructions and feedback to users throughout the repair process. Develop the logic for guiding users through each repair step based on sensor data and user inputs. Thoroughly test the system to ensure sensors are accurately detecting conditions.

3. Proposed Algorithm

The algorithm describing the main functionality of the Arduino code, including reading sensor values, controlling LEDs based on sensor readings, and implementing delays for periodic execution, is mentioned below:
During Installation

1. Define pin numbers for the IR sensor, water level sensor, temperature sensor, and luminosity sensor (TSL2561).
2. Initialize the temperature sensor object with the specified pin and type.
3. Set up pin modes for the IR sensor, temperature sensor, water level sensor, and luminosity sensor (TSL2561).
4. Begin serial communication at a baud rate of 9600.
5. Initialize temperature sensor communication.

During vehicle start-up

6. Take the IR sensor, water level sensor, and temperature sensor values and save them.

7. Verify if the value of the IR sensor is HIGH:
   • If yes, turn on the IR alert message for brake failure.
   • If no, turn off the alert message.

8. Check the level of water:

If the level of water is
   • above or equal to 330, turn on the alert message for the water sprinkler and turn off the others.
   • between 300 and 330, turn on the alert message and turn off the others.
   • below 300, turn on the alert message and turn off the others.

9. Check the temperature level:

   • If the temperature is above 35°C, turn on the temperature alert message.
   • If not, turn off the temperature alert message. Add a delay of 1000 ms (1 s) before the next iteration of the loop.

4. Modeling and Results

The research focuses on four frequent problems occurring in cars, namely: engine over-heating, brake failure, headlights, and water sprinklers. The GRV system has been implemented as shown in Figure 46.1.

Figure 46.1: Hardware implementation of guided repairing of vehicles.

5. Overheating of Engine

When the Peltier cell produces heat and the temperature exceeds 35°, the temperature sensor (W1209 module) detects the fault. The signal is sent to the Arduino UNO R3 with a speaker interface, indicating engine overheating. The APR33A3 Voice Recorder turns on and sends a signal to the speaker. The speaker reports that the engine is overheating and provides directions on how to cool it down.

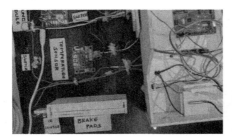
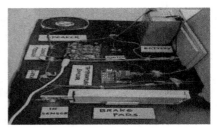

Figure 46.2: Engine overheating of guided repairing of vehicles (a). Brake failure of guided repairing of vehicles.

6. Brake failure

When an object enters the detection area of an IR sensor, it disturbs or reflects the infrared beam it emits, causing a change in the sensor's output. IR sensors are widely employed in various sectors for proximity sensing, object detection, motion detection, and obstacle avoidance. These applications encompass fields such as robotics, automation, security systems, and consumer electronics, as illustrated in Figure 46.2b.

7. Water Sprinklers

A water level depth detection sensor is an electronic device designed to detect the water level or depth in a container or reservoir. When submerged in water, the sensor provides an output signal that corresponds to the water level, which can be used for monitoring, control, or alarm purposes, as shown in Figure 46.3a.

Figure 46.3: Water sprinklers of guided repairing of vehicles (a). Headlights guided repairing of vehicles (b).

If the water level in the tank is below the requirement, the sensor ESC034 will detect the problem, and the APR33A3 Voice Recorder will turn on and send a signal to the speaker. The speaker reports that the water level is running low.

8. Head Lights

It is very important to check and maintain headlights to ensure they are functioning properly for safety while driving, especially at night. Here, we are creating a problem through a switch. A DIP LED is a compact LED variant with a rectangular block-like shape. It features two rows of pins, enabling convenient integration into circuit boards, as illustrated in Figure 46.3b.

The modeling and results section demonstrates the functionality of the GRV system in addressing four major vehicle issues: engine overheating, brake failure, water sprinkler malfunction, and headlight failures. Hardware implementation and sensor integration enable real-time detection and feedback, empowering drivers to handle emergencies promptly and effectively.

9. Conclusion

The GRV system shows considerable progress in automotive safety technology, offering a proactive approach to vehicle maintenance and repair. By leveraging sensor integration and real-time feedback mechanisms, the GRV system enables timely detection and resolution of critical issues, thereby enhancing road safety and vehicle reliability. It was observed that:

- The GRV system accurately detects problems and provides specific instructions for methodically locating and fixing issues.
- It provides technicians with the necessary knowledge and tools to quickly fix problems and get the cars back in top operating condition.
- It offers detailed guidance for effectively addressing issues, including step-by-step instructions and relevant technical information to assist technicians during the repair process.
- Furthermore, by guaranteeing precise and effective repairs, this method creates a satisfying service experience.
- Real-time feedback mechanisms are incorporated to ensure technicians are following the correct procedures and making progress in resolving the fault.

References

[1] Khaliqi, R., & Iulian, C. (2023). A deep learning approach to vehicle fault detection based on vehicle behavior.

[2] Naudin, I., Pham, M. T., Moreau, R., & Leleve, A. (2019). A robotic platform for endovascular aneurysm repair. In *2019 IEEE/ASME International Conference on Advanced Intelligent Mechatronics (AIM)* (pp. 996–1001). IEEE.

[3] Bourafa, R., & Siegfried, P. (2020). A review of the automated guided vehicle systems: Dispatching systems and navigation. Institute of Technology, MATE Hungarian University of Agriculture and Life Science, Hungary University of Applied Sciences Trier, Germany.

[4] Draganjac, I., Miklic, D., Kovacic, Z., Vasiljevic, G., & Bogdan, S. (2016). Decentralized control of multi-AGV systems in autonomous warehousing applications. *IEEE Transactions on Automation Science and Engineering, 13*(4), 1433–1447.

[5] Shafi, U., Safi, A., Shahid, A. R., Ziauddin, S., & Saleem, M. Q. (2018). Vehicle remote health monitoring and prognostic maintenance system. *Journal of Advanced Transportation*, 2018, 1–10.

[6] Liao, L., & Kottig, F. (2014). Review of hybrid prognostics approaches for remaining useful life prediction of engineered systems, and an application to battery life prediction. *IEEE Transactions on Reliability, 63*(1), 191–207.

[7] Prytz, R., Nowaczyk, S., Rögnvaldsson, T., & Byttner, S. (2015). Predicting the need for vehicle compressor repairs using maintenance records and logged vehicle data. *Engineering Applications of Artificial Intelligence, 41*, 139–150.

CHAPTER 47

Application of recycled concrete aggregate and fly ash for low-volume road construction

Deepak Kumar Mohanty[1,*], Madhusmita Ekka[2], Sudhanshu Sekhar Das[3]

[1,2,3]Department of Civil Engineering, VSSUT, Burla, Odisha, India
Corresponding author E-mail: deepakmohanty185@gmail.com

Abstract

The global construction boom has significantly increased construction waste, posing a pressing sustainability challenge worldwide. Construction and demolition waste (C&DW) adversely affects the environment, economy, productivity, and societal well-being. To address this issue, the utilization of recycled concrete aggregates (RCA) has become increasingly vital to conserve natural resources. In this study, different proportions of RCA (10%, 20%, 30%, 40%, and 50%) replaced natural crushed stone aggregate, while fly ash replaced cement in M30 grade pavement quality concrete (PQC) preparation. Compression and split tensile tests were compared with the controlled design mix. The findings revealed that a 20% replacement of RCA derived from C&DW, combined with fly ash (FA) as a cement substitute, effectively met the required PQC strength. Furthermore, substituting 10% of cement with FA notably increased PQC's compressive and split tensile strengths, attributable to the pozzolanic properties of FA.

Keywords: Waste, Compressive Strength, Split Tensile Strength, RCA, Fly Ash

1. Introduction

Rural roads support 69% of India's population, facilitating agricultural growth, education, healthcare, and socio-economic development. Conventional highway construction heavily relies on natural resources like aggregates and bitumen. However, this leads to significant construction waste globally, especially in India, where 99% of the 150 million tons of Construction and Demolition Waste (C&DW) annually ends up in landfills, impacting the environment, economy, and society. In India, the usual composition of C&DW includes 36% soil, sand, and gravel; 31% bricks and masonry; 23% concrete; 5% metals; 2% bitumen; 2% wood; and 2% other materials, totaling 85% of the overall waste (Tam et al., 2018). Enhancing sustainability through recycling aggregates includes using crushed highway concrete as recycled concrete aggregates (RCA) in the construction of new concrete or asphalt pavements (Albayati et al., 2018; Zhang et al., 2016). Despite successes in structural applications, RCA's use in pavements is hindered by characterization uncertainties. Nonetheless, it supplements natural aggregates (NA) in fresh concrete, easing shortages and landfill costs (Hwang et al., 2006). Properly treated RCA finds applications in various construction elements such as barriers, pavements, embankments, and bridge frames, subject to meeting specific suitability standards for each use (Grdic et al., 2010; Kou et al., 2009). This study aims to examine the efficacy of varying percentages of RCA when combined with fly ash (FA) at varying percentages, replacing the cement.

DOI: 10.1201/9781003596776-47

2. Experimental Methodology

The study comprised four phases, as shown in Figure 47.1. Phase one involved material collection and characterization according to standard procedures. Phase two prepared the control mix of PQC for the M30 grade as per IRC 44-2017. Phase three replaced RCA with natural coarse aggregates (NCA) at 10–50% increments alongside OPC to assess strength properties. Phase four replaced FA with cement at 10–30% increments, based on the optimal RCA percentage from phase three, and re-evaluated strength properties.

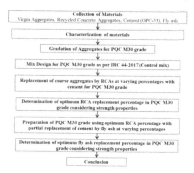

Figure 47.1: Research outline.

2.1. Laboratory Investigations

2.1.1. Materials Collection

Coarse aggregates were collected from a nearby crusher, and fine aggregates were collected from a nearby construction site of VSSUT, Burla. Cement was collected from Sambalpur. Fly ash was collected from Hindalco, Hirakud, and RCA was collected from Durgapali Flyover, Sambalpur, India.

2.1.2. Materials Characterization

All the materials were characterized according to the codal provisions of IRC: 44-2017 and IRC SP62:2014, meeting the criteria as depicted in Table 47.1.

Table 47.1: Physical property of NA, RCA, cement, and fly ash.

Sl. No.	Tests conducted	Results for aggregates		Results for OPC (53) cement	Results for fly ash	Limits (According to MORD specification 2004)
		NA	RCA			
1.	Specific gravity	2.81	2.5	3.15	2.10	2.5–3.0
2.	Impact value	6.21%	16.51%	-	-	30%
3.	Abrasion resistance	18.76%	27.33%	-	-	35%
4.	Water absorption	0.66%	2.66%	-	-	5%
5.	Initial setting time	-	-	58 min		-
6.	Final setting time	-	-	268 min	-	-
7.	Standard consistency	-	-	32%	-	-

Table 47.2: Gradation of Aggregate for PQC

19 mm Nominal size (as per IRC: 44-2017) and IRC: SP 62-2014					
Sieve size (mm)	Cumulative percentage by weight passing	Mean cumulative passing (%) (adapted)	Cumulative retained (%)	Individual retained (%)	Material required per 1 m³ of concrete (kg)
26.5	100	100	0	0	0
19.0	90–100	95	5	5	112.14
9.50	48–78	63	37	32	717.71
4.75	30–58	44	56	19	426.14
0.600	8–35	22	78	22	354
0.150	0–12	6	94	16	257.45
0.075	0–2	0	100	6	96.54

Grading of aggregate was done following IRC: SP 62-2014, as shown in Table 47.2.

2.1.3. Mix Design for M30 Grade PQC (control mix)

The mix design for PQC M30 grade was done using natural crushed stone aggregate according to IRC: 44-2017, and its properties were further studied. This mix was considered the control mix.

2.1.4. Preparation of M30 Grade PQC Specimens and Testing

Concrete was mixed with local sand, OPC-53 cement, and aggregates in a rotating mixer at 55 rpm. Water and superplasticizer were added for 3 min. RCA replaced NA at varying percentages. Specimens with optimal RCA and varying FA percentages to replace cement were tested for mechanical properties.

2.1.5. Workability Test on M30 Grade PQC

A workability test was done on cement mixes with 0–50% RCA. Slump test results indicated a slight decrease as RCA content increased, with values ranging from 32 mm (0% RCA) to 23 mm (50% RCA), as shown in Table 47.3. The mix stayed within acceptable workability limits across this RCA range. Further tests may be needed to determine the optimal RCA content for specific uses.

Table 47.3: Slump values for different RCA mix.

Mix of RCA	0%	10%	20%	30%	40%	50%
Slump value (mm)	32	32	29	26	24	23

2.1.6. Compressive Strength Test of PQC Specimens

The compressive strength test assesses a concrete specimen's axial load capacity. Cylinders or cubes are cured and then loaded in a compression machine until failure. Compressive strength is calculated by dividing the maximum load by the cross-sectional area, ensuring the concrete meets structural requirements.

Figure 47.2 illustrates that the control mix, comprising 100% natural crushed stone aggregate (0% RCA), satisfies the 7-day and 28-day compressive strength criteria. Substituting natural aggregates with 10%–50% RCA decreases strength. However, 10% and 20% of RCA replacements exhibit similar strengths, while 30%–50% fail to meet requirements, suggesting RCA's inferiority in PQC. To enhance RAC, fly ash replaces OPC cement at 10%–30%, with mechanical properties evaluated in Figure 47.3. PQC with 10% FA and 90% cement shows greater compressive strength than the control mix and PQC with 10% and 20% RCA without FA. PQC with 20% FA also surpasses the control mix, identifying 10% FA and 90% cement as the optimal mix.

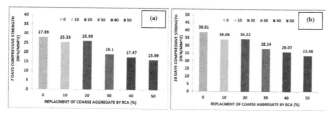

Figure 47.2: Compressive strength test results of RAC-PQC specimens (a) 7 days (b) 28 days.

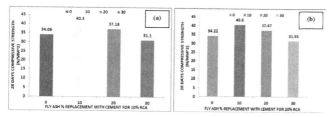

Figure 47.3: Compressive strength test results RAC-PQC specimens at varying fly ash % for (a) 10% RCA (b) 20% RCA.

2.1.7. Split Tensile Strength Test for PQC Specimens

The tension resistance of concrete is assessed using the tensile strength test. A common method is the split cylinder test, loading a cylinder until failure to induce perpendicular tensile stress. The maximum load determines tensile strength, crucial for designing structures against cracking and deformation.

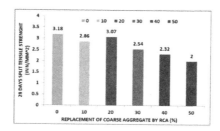

Figure 47.4: Split tensile strength test results of RAC-PQC specimens.

The split tensile strength test for RAC with RCA replacement (0–50%) yields the following results: 3.18 N/mm² at 0% RCA, 2.86 N/mm² at 10%, 3.07 N/mm² at 20%, 2.54 N/mm² at 30%, 2.32 N/mm² at 40%, and 2.00 N/mm² at 50%. Generally, as the RCA percentage increases, tensile strength decreases. Notably, 20% RCA shows a slight improvement over 10%, but higher RCA content reduces strength significantly. Results from Figure 47.5 show that split tensile strength improves with up to 10% fly ash replacement for both 10% and 20% RCA. However, beyond 10% fly ash, tensile strength decreases for both RCA percentages. The optimal mix for the highest strength is 10% fly ash replacement.

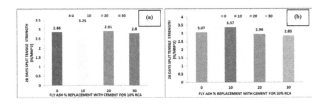

Figure 47.5: Split tensile strength test results in RAC samples at varying fly ash % for (a) 10% RCA, (b) 20% RCA.

3. Conclusion

RCA demonstrates inferior behavior compared to natural crushed stone aggregate; nevertheless, up to 20% RCA can be effectively integrated into conventional PQC for rural road construction. Fly ash, a pozzolanic material enhances concrete strength when utilized in optimal quantities. Specifically, incorporating up to 10% fly ash as a cement replacement, along with 20% RCA, proves advantageous in M30 grade PQC preparation, promoting sustainable construction practices.

References

[1]   Albayati, A., Wang, Y.u., Wang, Y., & Haynes, J. (2018). A sustainable pavement concrete using warm mix asphalt and hydrated lime-treated recycled concrete aggregates. *Sustainable Materials Technologies 18*, e00081.

[2]   Grdic, Z. J., Toplicic-Curcic, G. A., Despotovic, I. M., & Ristic, N. S. (2010). Properties of self-compacting concrete prepared with coarse recycled concrete aggregate. *Construction and Building Materials, 24*(7), 1129–1133.

[3]   Hwang, S. D., Khayat, K. H., & Bonneau, O. (2006). Performance-based specifications of self-consolidating concrete used in structural applications. *ACI Materials Journal, 103*(2), 121.

[4]   Guidelines for Cement Concrete Mix Design for Pavements (Third Revision) IRC: 44-2017.

[5]   Kou, S. C., & Poon, C. S. (2009). Properties of self-compacting concrete prepared with coarse and fine recycled concrete aggregates. *Cement and Concrete Composites, 31*(9), 622–627.

[6]   Ministry of Rural Development Specification 2004 for Plain Cement Concrete Pavement.

[7]   Tam, V. W. Y., Soomro, M., & Evangelista, A. C. J. (2018). A review of recycled aggregate in concrete applications (2000–2017). *Construction and Building Materials, 172*, 272–292.

[8]   Zhang, Z., Wang, K., Liu, H., & Deng, Z. (2016). Key performance properties of asphalt mixtures with recycled concrete aggregate from low-strength concrete. *Construction and Building Materials, 126*, 711–719.

CHAPTER 48

Evaluating 3D Concrete Printing Effectiveness through a Comparative Study using SVM, Decision Trees, and Random Forest Techniques

P. Latha[1], V. A. Shanmugavelu[2], A. Tamilmani[3], D. Thayalnayaki[4], J. Santhosh[5], S. J. Princess Rosaline[6]

Department of Civil Engineering, Periyar Maniammai Institute of Science and Technology, Thanjavur, Tamil Nadu, India
Corresponding author: lathap@pmu.edu

Abstract

This study explores the intricate dynamics of 3D concrete printing within the construction sector, emphasizing the key factors that influence its effective application. By simulating 3D concrete printing processes for both a 2-floor and a 12-floor building, a comprehensive dataset was generated. Machine learning models, including Support Vector Machines, Decision Trees, and Random Forests, were employed to analyze this dataset. Among these, the Random Forest model proved to be the most effective, achieving an accuracy of 94.2%, precision of 93.4%, recall of 94.0%, and an F1-score of 93.7%. Variables such as extrusion speed and robot location were identified as critical to the printing process. The findings highlight the potential of machine learning in optimizing 3D concrete printing, offering valuable insights for the construction industry. This research paves the way for a more efficient, data-driven approach to 3D concrete printing, underscoring the synergy between construction and technology.

Keywords: 3D concrete printing, machine learning, construction optimization, extrusion speed, robot location

1. Introduction

Three-dimensional (3D) technology, once the domain of futuristic imaginations, has increasingly found sophisticated applications across various fields, with the construction sector emerging as a notable beneficiary. 3D concrete printing, an innovative development within this technological revolution, is at the forefront of modern construction practices, redefining how buildings and structures are conceived, designed, and realized (Khan et al., 2020).

The rise of 3D concrete printing in construction has been transformative. Traditional construction methods have typically involved long durations, significant manual labor, substantial material wastage, and various environmental challenges (Rizzieri et al., 2023). In contrast, 3D concrete printing offers numerous benefits, including reduced completion times, and minimized

DOI: 10.1201/9781003596776-48

environmental impact due to its precision, automation, and adaptability. By eliminating the need for formworks, this technology accelerates construction while reducing resource waste. Furthermore, the design freedom it provides enables architects and builders to create structures that would be challenging to achieve using conventional methods. However, the full potential of 3D concrete printing can only be realized when the factors influencing its success are thoroughly understood and optimized. Key parameters, such as optimal extrusion speed, material extrusion paths, and precise robot movements, are still being studied and refined. Addressing these factors not only enhances the process's efficiency but also impacts the durability, safety, and longevity of the resulting structures (Liu et al., 2023).

Given this context, it is essential to assess these influencing factors. A thorough investigation can bridge the gap between the theoretical benefits of 3D concrete printing and its practical implementation, ensuring that the resulting structures are not only innovative but also reliable and long-lasting. Accordingly, this study aims to carefully examine the factors that affect the efficacy of 3D concrete printing in the construction industry. We intend to use machine learning to identify these factors, understand their interactions, determine their hierarchy, and assess their significance in various construction scenarios. By doing so, we seek to provide the construction industry with the knowledge, confidence, and clarity needed to advance with 3D concrete printing.

The historical expansion of 3D concrete printing demonstrates that technological excellence and innovation can coexist. The concept emerged in the latter half of the 20th century as additive manufacturing, or layer-by-layer construction, gained popularity across various industries. However, it was not until the turn of the millennium that researchers began to seriously consider applying this approach to concrete, a critical building material (Mollah et al., 2023). The early stages of 3D concrete printing were characterized by simple prototypes and small-scale applications. Most of these initial projects were exploratory, driven by curiosity and the potential for a transformative construction process. Over time, these experimental applications gave rise to larger, more complex structures. Collaborative efforts among academic institutions, tech startups, and construction firms led to numerous pilot projects worldwide, each pushing the limits of what could be achieved with printed concrete (Yin et al., 2023).

The appeal of 3D concrete printing lies in its multiple benefits. Automation significantly reduces labor costs, while the precision of automated printing minimizes material waste, leading to substantial cost savings and a reduced environmental impact. Moreover, the flexibility in design allows architects to create structures that blend aesthetics with functionality in ways previously unattainable (Wang et al., 2023). The technology also holds promise for rapid construction, especially in areas affected by natural disasters, offering the potential for quick infrastructure recovery However, despite the numerous advantages, 3D concrete printing faces several challenges. These include the need for specialized equipment, high initial setup costs, and the requirement for skilled personnel to operate and maintain the machinery. Material science also presents challenges; concrete mixtures must be precisely calibrated for printing to ensure structural integrity. Furthermore, building codes and regulations, which have not yet fully adapted to this new construction method, pose obstacles to widespread adoption (Chen et al., 2023).

Traditionally, the assessment of 3D concrete printing processes has relied on empirical and experimental methods. Prototype structures have been built and tested for durability, strength, and other structural parameters. Some pioneering research has also attempted to digitally simulate construction scenarios to predict potential challenges and assess design feasibility. However, these assessments were often isolated and lacked a systematic approach that considered the numerous variables involved. A significant gap in the literature exists regarding a comprehensive, data-driven analysis of the factors influencing 3D concrete printing. While empirical assessments offer valuable insights, they often fail to capture the complexities and interactions of multiple variables simultaneously. There is also a noted lack of modern data science and machine learning techniques, which have the potential to model complex processes, make predictions, and uncover hidden patterns (Ayyagari et al., 2023).

This study aims to address these existing gaps. Recognizing the limitations of isolated empirical assessments, we pursue a comprehensive, machine-learning-driven analysis of 3D concrete printing. The aim is to harness the predictive power of algorithms to offer a nuanced understanding of the factors that influence the effective application of this technology in the construction sector, thus bridging the gap between innovation and practical reliability.

2. Methodology

2.1. Description of the buildings

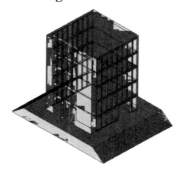

Figure 48.1: 3D printed building model.

In this study, two distinct building structures were selected for simulation to represent both modest and large-scale construction projects. This approach enabled us to assess the scalability and versatility of 3D concrete printing, as depicted in Figure 48.1. The first building, a relatively compact 2-floor structure, covers an area of 405.45 m^2. It was designed as a prototype of a medium-scale residential building, simulating the challenges and scenarios associated with constructing such a unit using 3D concrete printing. This 2-floor structure was selected to explore the nuances of 3D concrete printing in urban and suburban settings.

The second building, a 12-floor structure, was designed to emulate modern multistory commercial or residential complexes, chosen for its complexity and the challenges it presents. This 12-floor structure was intended to test the upper limits of 3D concrete printing in larger, more ambitious construction projects. The study sought to explore the breadth and depth of 3D concrete printing's potential and challenges across diverse architectural paradigms. The selection of the 12-floor structure was deliberate, as it introduced vertical stacking, load-bearing considerations, and logistical complications related to robot movement and material extrusion at height.

3. Explanation of the chosen machine learning models

This research utilized a combination of machine learning models, including Support Vector Machines (SVM), Decision Trees, and Random Forests, to provide a comprehensive analysis of 3D concrete printing processes. SVM is a supervised learning algorithm known for its ability to handle high-dimensional data and classification problems, particularly in nonlinear scenarios. Its precision and adaptability make it a promising tool for exploring 3D concrete printing processes. Decision Trees, which are flowchart-like structures representing features, decision rules, and outcomes, are valuable for understanding and visualizing the decision-making

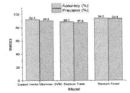
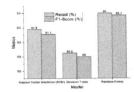
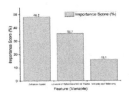

Figure 48.2: Accuracy and prediction.

Figure 48.3: Recall and F1 score.

Figure 48.4: Feature importance.

process. They help identify specific factors that play a critical role in the 3D concrete printing process. Random Forest, an ensemble learning method, builds upon Decision Trees by creating a forest of decision trees during training and then outputting the class or mean prediction of individual trees for regression tasks. Its enhanced accuracy and ability to manage overfitting make it an ideal choice for 3D concrete printing processes, as it considers diverse decision trees and derives a consensus for a more balanced, accurate prediction.

The analysis focused on variables such as extrusion speed and robot location, which impact structural integrity, surface finish, and overall quality. The robot's position determines the reach, angle, and efficiency of the printing process, especially in multistory buildings. The choice of machine learning models was based on their individual strengths and collective ability to provide a comprehensive analysis of the chosen variables. The precision of SVM, the transparency of Decision Trees, and the robustness of Random Forests ensure deep and wide-ranging insights.

4. Dataset collection

Table 48.1: Dataset collection.

Building type	Extrusion speed (mm/s)	Robot location (m)	Structural integrity score	Surface finish score
2-floor	50	2.5	88	92
12-floor	45	5.0	82	87
2-floor	55	3.0	91	90
12-floor	40	6.0	79	85

From the tabulated data in Table 48.1, several patterns emerge. There appears to be a correlation between extrusion speed and structural integrity. For instance, in the 2-floor building simulations, an increase in extrusion speed from 50 to 55 mm/s led to a 3-point improvement in structural integrity. However, it is important to note that there may be an optimal range for extrusion speed, beyond which returns may diminish or even negatively impact the building's properties. The location of the robot, particularly in the 12-floor simulations, shows varying outcomes. As the robot's location is adjusted to greater heights (likely corresponding to higher floors), there is a slight decline in both structural integrity and surface finish scores. This observation suggests the complexities and challenges introduced when 3D printing is at greater vertical extents. When comparing the data from the 2-floor and 12-floor buildings, the latter consistently shows slightly lower scores in both structural integrity and surface finish. This underscores the inherent challenges of large-scale constructions and highlights the importance of refining printing parameters for larger projects.

The dataset provides a snapshot, a microcosm of the broader results. The actual dataset, likely more expansive, would further illuminate the intricate relationships between variables and outcomes. While this provides certain insights, a deeper statistical analysis, aided by our chosen machine learning models, would yield more concrete conclusions and actionable insights for the construction industry.

5. Results and Discussion

The performance of the three machine learning models was evaluated using standard metrics such as accuracy, precision, recall, and F1-score, as shown in Figures 48.2 and 48.3. These metrics provide a comprehensive understanding of how well each model performed, considering various aspects of prediction quality.

Among the three models, Random Forest emerged as the top performer in terms of accuracy. This can be attributed to its ensemble nature, leveraging multiple decision trees to arrive at predictions, which generally increases accuracy and reduces the likelihood of overfitting. SVM and Random Forest models showed closely competitive precision and recall values, indicating that both models not only predicted outcomes with high accuracy but also maintained a balanced rate of true positive predictions.

Decision Trees, while still delivering commendable results, showed a slight dip in these metrics. The F1-score, which is the harmonic mean of precision and recall, offers insight into a model's balance between precision and recall. Random Forest again clinched the top spot, closely followed by SVM, indicating their strong predictive capabilities without heavily compromising either precision or recall.

The performance metrics presented in tabulated form provide a quantitative assessment of the chosen models in a 3D concrete printing project. While Random Forest showed an edge in the simulation, the choice of model would depend on project requirements such as interpretability, processing time, and data distributions. Feature importance is crucial in machine learning, particularly with algorithms like Decision Trees and Random Forest. "Extrusion Speed" emerged as the most significant feature, as it plays a critical role in determining the success and efficiency of the 3D concrete printing process. The "Location of Robot Mounted on Tracks" is the next significant feature, as it influences the ease and precision of the print and the quality of the final construct. However, in real-world applications, the interplay of multiple factors determines success. Even if "Extrusion Speed" is optimized, a poor robot location might compromise the print quality. This tabulated feature importance serves as a guide for practical applications, helping practitioners prioritize focus, optimize parameters, and make informed decisions during the 3D concrete printing process. It also provides direction for future research, highlighting areas that might benefit from more in-depth exploration.

The integration of 3D concrete printing with construction technology has revolutionized building processes. This research explores the nuances of 3D concrete printing using machine learning models. The results indicate strong performance metrics, particularly from Random Forest, which identifies discernible patterns and relationships in data related to the 3D concrete printing process. This suggests that despite the complexities of real-world construction scenarios, there is an inherent structure and predictability to how factors influence the printing outcome. 3D concrete printing has significant implications for the construction sector, offering rapid, sustainable solutions, environmental benefits, reduced waste, and cost savings. However, the research also has limitations, including the virtual representation of construction sites and the reliance on data-driven models. The reliability of machine learning models depends on the quality of data input, and the sector's deep-rootedness in traditional methods presents challenges in adapting to new technologies. The transition to 3D concrete printing will require workforce reskilling, infrastructure investments, and a paradigm shift in construction planning and execution. Despite these challenges, the prospect of a future where structures are built using printers is becoming increasingly plausible, as the construction sector stands on the cusp of this transformative era.

6. Conclusion

This study examined 3D concrete printing and used machine learning to understand the complexities of this novel construction process. SVM, Decision Trees, and Random Forest models were applied to data from simulations of a 2-floor and a 12-floor building. The Random Forest model stood out with remarkable performance metrics, including an accuracy of 94.2%, precision of 93.4%, recall of 94.0%, and an $F1$-score of 93.7%. These statistics, though quantitative, suggest that 3D concrete printing may possess structure and predictability. The positioning of the robot and its extrusion speed were identified as critical aspects, indicating areas of particular interest for practitioners. When considering the study as a whole, it presents an optimistic view of the construction industry's future, where data-driven insights can guide rapid, cost-effective, and environmentally responsible construction practices. This research leads to the conclusion that the future of construction, enabled by 3D printing and supported by machine learning insights, is not only promising but also imminent.

References

[1] Khan, M. S., Sanchez, F., & Zhou, H. (2020). 3-D printing of concrete: Beyond horizons. *Cement and Concrete Research*, *133*(March). https://doi.org/10.1016/j.cemconres.2020.106070

[2] Rizzieri, G., Cremonesi, M., & Ferrara, L. (2023). A 2D numerical model of 3D concrete printing including thixotropy. *Materials Today: Proceedings*, July. https://doi.org/10.1016/j.matpr.2023.08.082

[3] Liu, Z., Li, M., Quah, T. K. N., Wong, T. N., & Tan, M. J. (2023). Comprehensive investigations on the relationship between the 3D concrete printing failure criterion and properties of fresh-state cementitious materials. *Additive Manufacturing*, *76*(September), 103787. https://doi.org/10.1016/j.addma.2023.103787

[4] Mollah, M. T., Comminal, R., Leal da Silva, W. R., Šeta, B., & Spangenberg, J. (2023). Computational fluid dynamics modeling and experimental analysis of reinforcement bar integration in 3D concrete printing. *Cement and Concrete Research*, *173*(July). https://doi.org/10.1016/j.cemconres.2023.107263

[5] Yin, Y., Huang, J., Wang, T., Yang, R., Hu, H., Manuka, M., Zhou, F., Min, J., Wan, H., Yuan, D., & Ma, B. (2023). Effect of Hydroxypropyl methyl cellulose (HPMC) on rheology and printability of the first printed layer of cement activated slag-based 3D printing concrete. *Construction and Building Materials*, *405*(July), 133347. https://doi.org/10.1016/j.conbuildmat.2023.133347

[6] Wang, L., Ye, K., Wan, Q., Li, Z., & Ma, G. (2023). Inclined 3D concrete printing: Build-up prediction and early-age performance optimization. *Additive Manufacturing*, *71*(January), 103595. https://doi.org/10.1016/j.addma.2023.103595

[7] Chen, Y., Zhang, Y., Zhang, Y., Pang, B., Zhang, W., Liu, C., Liu, Z., Wang, D., & Sun, G. (2023). Influence of gradation on extrusion-based 3D printing concrete with coarse aggregate. *Construction and Building Materials*, *403*(August), 133135. https://doi.org/10.1016/j.conbuildmat.2023.133135

[8] Ayyagari, R., Chen, Q., & García de Soto, B. (2023). Quantifying the impact of concrete 3D printing on the construction supply chain. *Automation in Construction*, *155*(March 2022), 105032. https://doi.org/10.1016/j.autcon.2023.105032

CHAPTER 49

Predictive Analysis of Fiber Reinforced Concrete Using Construction and Demolition Waste with Insights from SVR, Decision Trees, and Random Forests

V. A. Shanmugavelu[1], A. Tamilmani[2], D. Thayalnayaki[3], J. Santhosh[4], S. J. Princess Rosaline[5], P. Latha[6]

Department of Civil Engineering Periyar Maniammai Institute of Science and Technology, Thanjavur, Tamil Nadu, India
Corresponding author: shanmugavelu@pmu.edu

Abstract

This study embarked on an exploratory investigation into fiber-reinforced concrete (FRC) integrated with construction and demolition (CD) waste as a partial replacement for fine aggregate. A structured experimentation with mixes A1, A2, B1, and B2 revealed several significant findings. For instance, Mix A1, as predicted by Support Vector Regression (SVR), exhibited a promising compressive strength of 34.8 MPa, while Mix B2, according to Random Forest predictions, indicated a porosity of 8.7%, suggesting its potential vulnerability to certain environmental conditions. The results, which span mechanical attributes such as compressive strength (32.7–36.3 MPa) and durability metrics like porosity (7.7%–8.8%), underscore the feasibility of incorporating CD waste into FRC mixes. These findings were further validated by the close alignment of predictions across three advanced machine-learning methodologies. Overall, the research presents a compelling case for sustainable construction, demonstrating that waste materials can be effectively integrated into concrete, balancing environmental sustainability with concrete performance.

Keywords: Fiber Reinforced Concrete, Construction and Demolition Waste, Sustainable Construction, Machine Learning Predictions, Compressive Strength

1. Introduction

The construction industry has long sought to optimize the properties of concrete, leading to the development and widespread adoption of fiber-reinforced concrete (FRC). This unique form of concrete incorporates fibers—whether steel, polypropylene, or other synthetic materials—into the mix, significantly enhancing its tensile strength and ductility (Chu et al., 2023). While concrete is traditionally known for its excellent compressive strength, it can be brittle and prone to cracking under tensile stresses. The inclusion of fibers helps bridge these microcracks, preventing them from widening into larger, more problematic cracks. Consequently, FRC has been used in a wide array of structures that require increased resilience against dynamic loads, such as pavements, tunnels, and earthquake-resistant structures (Tran et al., 2023).

DOI: 10.1201/9781003596776-49

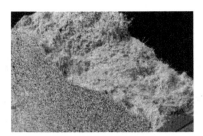

Figure 49.1: Fiber-reinforced concrete (FRC).

The construction sector faces significant sustainability concerns, particularly due to the carbon emissions associated with Portland cement production. As infrastructure demand grows, especially in developing economies, there is a pressing need to mitigate this impact. Utilizing recycled materials in construction can help reduce the carbon footprint and address waste disposal challenges (Zhou et al., 2023). By repurposing waste products or by-products, the industry can reduce its reliance on virgin raw materials and tackle the growing problem of waste management.

In this context, construction and demolition (CD) waste presents a viable alternative to traditional aggregates. Every year, millions of tons of CD waste, including broken bricks, concrete, tiles, and other construction materials, are generated. If left untreated, this waste can pose environmental challenges. However, when properly processed, CD waste can replace fine aggregates in concrete production. Preliminary studies have shown that this not only diverts waste from landfills but can also produce concrete with satisfactory mechanical properties (Roshani et al., 2023). The idea of integrating CD waste into FRC is particularly intriguing. Fibers enhance the concrete's tensile strength, while the use of CD waste can potentially reduce its environmental impact. The combined benefits of these innovations could lead to a new era of sustainable and resilient infrastructure. This research aims to explore the mechanics and viability of FRC made by partially replacing fine aggregates with CD waste, paving the way for eco-friendly construction solutions (Dai et al., 2023).

FRC has been extensively studied for its unique structural properties, including enhanced tensile strength, reduced shrinkage, and increased crack resistance. Its performance under various conditions, such as elevated temperatures and aggressive environments, depends on the alignment, distribution, and volume fraction of fibers within the matrix. FRC's versatility is evident in its applications, ranging from pavements to precast elements (Jin et al., 2023). On the other hand, the use of CD waste as a substitute for natural aggregates is being explored to reduce environmental impact and conserve natural resources. Studies have shown that CD waste can improve workability, strength, and durability, achieving compressive strength values comparable to conventional concrete. However, concerns about potential impurities in CD waste have been raised, as these could affect the consistency and integrity of the mix (Qiao et al., 2023).

The durability of recycled aggregates, such as those derived from CD waste, includes resistance to chemical attack and environmental aggressors. However, the increased permeability of recycled aggregates may heighten susceptibility to chemical ingress. Factors like water absorption, porosity, and chloride ion penetration can influence durability. With the right mix design and best practices, recycled aggregate concrete can achieve durability metrics comparable to those of conventional concrete. Artificial intelligence (AI) has revolutionized concrete research, enabling more accurate predictions of its behavior by analyzing the complex relationships between mix proportions, curing conditions, and properties (Al-Rousan et al., 2023). AI can predict FRC durability by considering variables such as fiber type, dosage, and

CD waste proportion. This research represents a promising frontier in concrete studies, combining FRC, CD waste utilization, durability assessments, and AI for sustainable construction.

1.1. Methodology

1.1.1. Materials and Mix Proportions

This study employed a diverse range of materials to achieve the desired mix characteristics and investigate the effects of CD waste in FRC, as shown in Figure 49.1. The core constituents of the mixes included cement, water, natural fine aggregates, fibers, and CD waste, with each material selected for its specific quality and adherence to standards. Ordinary Portland Cement (OPC) of grade 53, conforming to international standards, was used due to its consistent quality and wide acceptance in the construction industry. River sand, chosen for its well-graded nature and minimal impurities, served as the natural fine aggregate. The sand was sieved to remove debris and had a fineness modulus of approximately 2.6, ensuring optimal workability and strength in the concrete mix.

Two distinct types of fibers were incorporated to provide a comprehensive understanding of their effects on FRC. Steel fibers, known for their high tensile strength and excellent bond with concrete, were used in some mixes. These fibers were straight, with a length of 30 mm and a diameter of 0.5 mm. Polypropylene fibers, recognized for their resistance to alkali and improved impact strength, were also included. These synthetic fibers had a length of 12 mm and were fibrillated to promote better dispersion within the mix. The CD waste primarily comprised crushed concrete, broken bricks, tiles, and mortar. Before use, the CD waste underwent meticulous processing, including sieving to separate particles below 4.75 mm, in line with the definition of fine aggregates. The garbage was rinsed thoroughly after any apparent pollutants or foreign objects were carefully removed. For tabulation purposes, the following mix proportions for some of the experimental groups are shown in Table 49.1.

Table 49.1: FRC composition.

Mix ID	Cement (kg/m3)	Water (kg/m3)	River Sand (kg/m3)	CD Waste (kg/m3)	Steel Fiber (%)
A1	300	150	600	0	1
A2	300	150	450	150	1
B1	300	150	600	0	0
B2	300	150	450	150	0

2. Experimental Groups and Testing Methods

The primary objective of this study was to gain a comprehensive understanding of how FRC behaves when mixed with varying proportions of recycled fine aggregate derived from CD waste. This necessitated the development of 15 experimental groups, each comprising 225 specimens. This rigorous methodology enabled a thorough investigation of concrete behavior under different conditions.

The significant difference between the groups was the amount of recycled fine aggregate derived from CD waste used in place of natural fine aggregate. These percentages were predetermined at intervals of 0%, 25%, 50%, 75%, and 100%, ranging from standard concrete to a mix with fine aggregate entirely composed of CD waste. For each percentage level, three experimental groups were formed to account for variables such as fiber type and dosage. The specimens within each subgroup varied based on the type and composition of the fibers. Steel fiber reinforcement was used as a control in the first group, polypropylene fiber reinforcement in the second, and no reinforcement in the third. This diversity was crucial for understanding the interaction between the recycled fine aggregate and the fibers, as well as their combined effect on the mechanical properties of the concrete.

For example, in the specimens with 50% recycled fine aggregate:

One group contained steel fibers. The next group had polypropylene fibers. The third group had no fibers, providing a direct comparison to understand the sole influence of the recycled aggregate at that percentage.

3. Experimental Results and its Prediction

Table 49.2: Data collected—Set 1.

Mix ID	Compressive Strength (MPa)	Splitting Tensile Strength (MPa)	Flexural Strength (MPa)
A1	35.0	3.1	5.2
A2	33.2	3.0	5.0
B1	36.5	3.3	5.4
B2	32.8	2.9	4.9

Table 49.3: Data collected—Set 2

Mix ID	UPV (m/s)	Modulus of Elasticity (GPa)	Water Absorption (%)	Porosity (%)
A1	3650	29.0	2.5	8.0
A2	3600	28.4	3.0	8.5
B1	3665	29.5	2.3	7.8
B2	3590	28.1	3.2	8.7

The results are presented in a tabulated format in Tables 49.2 and 49.3, providing an organized view of the data and making it easier to discern patterns and draw conclusions. Each Mix ID represents a unique combination of variables (such as fiber type, percentage of CD waste, etc.), offering a comprehensive view of the performance metrics of each mix. This systematic representation aids in understanding the effects of different mix proportions and compositions on the overall behavior of FCC.

Delving deeper into concrete research, Table 49.4 and Figures 49.2–4 serve as a testament to the integration of traditional civil engineering practices with modern computational techniques. The table presents the predicted mechanical and durability properties of modified FCC mixes (A1, A2, B1, and B2) using three machine-learning methodologies: Support Vector Regression, Decision Trees, and Random Forests. The properties range from compressive strength to porosity, focusing on the material's ability to resist compressive loads. The table also includes predictions for splitting tensile strength and flexural strength, which are crucial for understanding the effectiveness of fiber incorporation in mitigating concrete's natural tensile weaknesses.

Table 49.4: Machine learning predicted results for 4 mix ids.

Mix ID	Property	SVR Prediction	Decision Tree Prediction	Random Forest Prediction
A1	Compressive Strength (MPa)	34.8	35.1	35.0
A1	Splitting Tensile Strength (MPa)	3.0	3.1	3.1
A1	Flexural Strength (MPa)	5.1	5.2	5.2
A1	UPV (m/s)	3645	3648	3647
A1	Modulus of Elasticity (GPa)	28.9	29.0	29.0

A1	Water Absorption (%)	2.4	2.5	2.5
A1	Porosity (%)	7.9	8.0	8.0
A2	Compressive Strength (MPa)	33.0	33.2	33.1
A2	Splitting Tensile Strength (MPa)	2.9	3.0	3.0
A2	Flexural Strength (MPa)	4.9	5.0	5.0
A2	UPV (m/s)	3595	3600	3598
A2	Modulus of Elasticity (GPa)	28.2	28.3	28.3
A2	Water Absorption (%)	3.1	3.2	3.2
A2	Porosity (%)	8.5	8.6	8.6
B1	Compressive Strength (MPa)	36.3	36.5	36.4
B1	Splitting Tensile Strength (MPa)	3.2	3.3	3.3
B1	Flexural Strength (MPa)	5.3	5.4	5.4
B1	UPV (m/s)	3663	3665	3664
B1	Modulus of Elasticity (GPa)	29.4	29.5	29.5
B1	Water Absorption (%)	2.2	2.3	2.3
B1	Porosity (%)	7.7	7.8	7.8
B2	Compressive Strength (MPa)	32.7	32.8	32.8
B2	Splitting Tensile Strength (MPa)	2.8	2.9	2.9
B2	Flexural Strength (MPa)	4.8	4.9	4.9
B2	UPV (m/s)	3588	3590	3590
B2	Modulus of Elasticity (GPa)	28.0	28.1	28.1
B2	Water Absorption (%)	3.3	3.4	3.4
B2	Porosity (%)	8.7	8.8	8.8

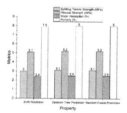 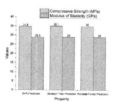 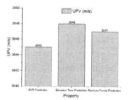

Figure 49.2: Set 1.   **Figure 49.3**: Set 2.   **Figure 49.4**: Set 3.

Ultrasonic Pulse Velocity (UPV) is a non-destructive testing metric that provides insights into concrete quality. Higher UPV values often indicate denser, more homogeneous, and higher-quality concrete. UPV predictions can help engineers forecast structural deflections and serviceability conditions for structures built with specific mixes. The Modulus of Elasticity measures concrete's stiffness, which is crucial for structural designers. Water absorption and porosity are interrelated properties that affect concrete's permeability, impacting the long-term performance of structures exposed to water and chemical ingress.

Three machine learning methods—SVR, Decision Trees, and Random Forests—were employed to predict these properties. SVR aims to find the best hyperplane to fit the data, while Decision Trees offer a visual and hierarchical approach. Random Forests combine the predictions of multiple trees for a more stable and generalized result. The variations in predictions across these methods highlight the complexities of predicting concrete behavior, and the fact that no single method consistently provides the highest or lowest predictions for all properties or mixes underscores the challenge. In summary, the table serves as a valuable repository of knowledge, offering a comparative framework for understanding the interplay of variables in modified FRC mixes. It exemplifies the advancement in construction research, where the synthesis of traditional experimental methodologies with modern computational techniques paves the way for more informed, efficient, and sustainable construction practices.

4. Conclusion

In the quest to understand the viability of using CD waste in FRC, a comprehensive analysis was undertaken, blending traditional concrete testing with modern machine learning algorithms. The study's findings, captured in the detailed table, highlight the nuanced behavior of modified FRC mixes. For instance, Mix A1 exhibited an SVR-predicted compressive strength of 34.8 MPa, showcasing the potential strength of FRC even with waste incorporation. Mix B2, on the other hand, was predicted to have a porosity of 8.7% according to Random Forests, suggesting considerations for its use in environments prone to water or chemical ingress. These predictions, ranging from compressive strength of 32.7–36.3 MPa to porosity values between 7.7%–8.8%, underscore the delicate balance between mix composition, mechanical properties, and durability. The results suggest that, with careful design, CD waste can be effectively integrated into FRC without significantly compromising its performance. Furthermore, the close agreement between the predictions of different machine learning models reaffirms the reliability of the analysis. Ultimately, this research illuminates a promising path towards sustainable construction practices, harnessing waste materials without sacrificing structural integrity and performance.

References

[1]   Chu, S. H., Unluer, C., Yoo, D. Y., Sneed, L., & Kwan, A. K. H. (2023). Bond of steel reinforcing bars in self-prestressed hybrid steel fiber reinforced concrete. *Engineering Structures*, *291*(June), 116390. https://doi.org/10.1016/j.engstruct.2023.116390

[2]   Tran, D. T., Pham, T. M., Hao, H., San Ha, N., Vo, N. H., & Chen, W. (2023). Precast segmental beams made of fibre-reinforced geopolymer concrete and FRP tendons against impact loads. *Engineering Structures*, *295*(September), 116862. https://doi.org/10.1016/j.engstruct.2023.116862

[3]   Zhou, M., He, X., Wang, H., Wu, C., & He, J. (2023). Mesoscale modeling of polypropylene fiber reinforced concrete under split tension using discrete element method. *Construction and Building Materials*, *404*(September), 133274. https://doi.org/10.1016/j.conbuildmat.2023.133274

[4]   Roshani, H., Yousefi, M., Gharaei-Moghaddam, N., & Khatibi, S. H. (2023). Flexural performance of steel-concrete-steel sandwich beams with lightweight fiber-reinforced concrete and corrugated-strip connectors: Experimental tests and numerical modeling. *Case Studies in Construction Materials*, *18*(January), e02138. https://doi.org/10.1016/j.cscm.2023.e02138

[5]   Dai, L., Zhu, Z., Zhang, C., & Zhu, D. (2023). Experimental study on the influence of glass fiber reinforced concrete isolation layer on the seismic dynamic response

of tunnels. *Case Studies in Construction Materials, 19*(July), e02303. https://doi.org/10.1016/j.cscm.2023.e02303

[6] Jin, Z., Mao, S., Zheng, Y., & Liang, K. (2023). Pre-treated corn straw fiber for fiber-reinforced concrete preparation with high resistance to chloride ions corrosion. *Case Studies in Construction Materials, 19*(July), e02368. https://doi.org/10.1016/j.cscm.2023.e02368

[7] Qiao, L., Miao, P., Xing, G., Luo, X., Ma, J., & Farooq, M. A. (2023). Interpretable machine learning model for predicting freeze-thaw damage of dune sand and fiber reinforced concrete. *Case Studies in Construction Materials, 19*(September), e02453. https://doi.org/10.1016/j.cscm.2023.e02453

[8] Al-Rousan, E. T., Khalid, H. R., & Rahman, M. K. (2023). Fresh, mechanical, and durability properties of basalt fiber-reinforced concrete (BFRC): A review. *Developments in the Built Environment, 14*(December 2022). https://doi.org/10.1016/j.dibe.2023.100155

CHAPTER 50

Voice-Driven ChatGpt Interface for Divyang Individuals

Chapala Jhansi Durga Sree[1], K. Therissa[2], Akanksha Mishra[3], Boga Mary Pushpalatha[4]

[1]Scholar Electrical and Electronics Engineering, Vignan's Institute of Engineering for Women, Visakhapatnam, India E-mail: Eswaridurga2001@gmail.com
[2]Electrical and Electronics Engineering, Vignan's Institute of Engineering for Women, Visakhapatnam, India E-mail: k.therissa@gmail.com
[3]Electrical and Electronics Engineering, Vignan's Institute of Engineering for Women, Visakhapatnam, India E-mail: misakanksha@gmail.com
[4]Electrical and Electronics Engineering, Vignan's Institute of Information and Technology, Visakhapatnam, India E-mail: pushpalathaboga1@gmail.com

Abstract

Nowadays, ChatGPT is one of the most trending AI tools that utilizes natural language processing to enhance the understanding of queries and generate responses according to the user in real time, ensuring a smooth conversational flow within a text-based interface. To augment user experience with voice interactions, we developed a voice-driven ChatGPT interface. Voice interactions are becoming increasingly important in user experience design as they allow users to communicate using natural language and speech, eliminating the need for hands-free communication. The voice input for this interface is converted into text using Application Programming Interface keys. Subsequently, this textual input undergoes processing via the ChatGPT interface, ultimately delivering responses in the form of voice output, catering to the need for speaking and listening instead of text. The goal is to enhance the user experience while maintaining efficiency through this interface.

Keywords: Natural Language Processing, User Experience Design, Text-Based Interfaces, ChatGPT Interface, Application Programming Interface Keys

1. Introduction

This topic has gained significant attention since November 2022, when ChatGPT, an AI-driven language model, was released by the US company OpenAI. The manufacturer claims that ChatGPT-4 outperforms humans in various professional and academic benchmarks.

Voice assistants are AI-powered digital tools that understand and respond to spoken commands. Voice-mediated service experiences are made possible by these voice-based interfaces, which integrate natural language processing and automated speech recognition. Although many text-based interfaces exist, ChatGPT's interface and features are designed to provide a conversational and engaging experience for users, contributing to its rising popularity among students. Its ability to understand context, generate coherent responses, and offer an extensive spectrum of information makes it a versatile tool for various

DOI: 10.1201/9781003596776-50

purposes, from learning and exploring topics to obtaining personalized assistance and recommendations.

The voice-driven ChatGPT interface includes the features provided by the standard ChatGPT interface, enhancing the capabilities of voice assistance according to the user needs and make studying and learning more accessible, interactive, and efficient by harnessing the power of voice technology and artificial intelligence.

2. Literature Review

The study by Andreas et al. (2022) summarizes that AI systems exhibit various capabilities, ranging from human-like behavior in communication to providing autonomy and contextual understanding. Additionally, AI systems can collaborate with humans as teammates, augment human intelligence, and often operate as opaque "black box" systems, making their inner workings challenging to understand or verify.

Further, the study by Gopal Vishwas Patil and Vidya Dhamdhere (2022) highlights the transition from text-based to voice-based interfaces in chatbots, incorporating hot word detection like Alexa for activation, which initiates the TensorFlow JS model for speech recognition. To achieve this, Python-based API keys convert the audio into text through speech recognition algorithms. An NLP-based interpreter recognizes text with the actual data that was predefined.

The survey by Lawal et al. (2024) indicates that regardless of the product's outcome, the final decision-maker is the user; therefore, the priority is to consider user needs, which helps in advancing user requirements.

The survey by Wu et al. (2023) provides an outline of ChatGPT, discussing its prototype, features, limitations, social influence, and potential future development. It explores fundamental methods such as large-scale language models and reinforcement learning, highlighting the emergence phenomenon in AI.

Additionally, the study by Vikas Hassija et al. (2023) signifies that large language models (LLMs) are overtaking NLP but face challenges in addressing problems. It provides the capabilities, and limitations of LLMs in voice companions and underlying techniques such as deep learning and reinforcement learning.

3. Correlation Analysis

India boasts a diverse population with over 22 official languages. Despite this linguistic diversity, ChatGPT has gained significant traction in the country. India ranks second after the United States in terms of ChatGPT users, accounting for 8.37% of all users. ChatGPT is currently utilized in India primarily for text-based applications, including customer support and content generation.

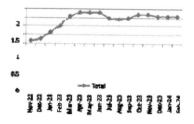

Figure 50.1: ChatGPT monthly visits progressed over time.

By mimicking the format of familiar messaging apps, users feel comfortable engaging with it. This approach helps bridge the gap between humans and technology, fostering trust and acceptance of AI-powered chat interfaces. One key feature of ChatGPT is its ability to adapt its responses to match the tone and sentiment of user input. This ensures that interactions feel personalized and relevant, enhancing the overall user experience. Moreover, ChatGPT is equipped with content filters to uphold ethical guidelines and community standards. These filters prevent the generation of responses that could be inappropriate, offensive, or harmful. By incorporating these safeguards, ChatGPT promotes responsible use of AI technology and maintains a safe environment for users.

The voice input is captured by the microphone and processed by the ESP32, where it is converted into text and subsequently used by the cloud speech client. The text is processed, and the output from the cloud speech client is in the form of text. This text is then converted into voice. At the final stage, the voice output is amplified, and users can hear it through the speaker.

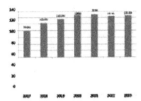

Figure 50.2: Voice assistant users progressed over time.

Voice-based interfaces have gained renown despite the rise of text-based interfaces due to their advanced abilities and capabilities. They help make studying and learning more accessible, interactive, and efficient by leveraging the power of voice technology.

4. Methodology

The process is dichotomized into two core phases. The initial phase covers the conversion of speech to text, and the second phase covers the conversion of text to speech. For converting speech to text and text to speech, API keys are leveraged. These APIs encompass a wide array of domains, spanning computation, storage, machine learning, data analytics, and beyond. Google Cloud API offers abundant documentation, tutorials, and support infrastructure, empowering developers to harness the platform's full potential.

Some of the available API keys for converting speech to text include Google Cloud Speech-to-Text API. Google Cloud's Text-to-Speech API offers extensive language support, robust customization options, seamless integration, accuracy, scalability, and competitive cost—all of which make Google Cloud API key the optimal choice for our project.

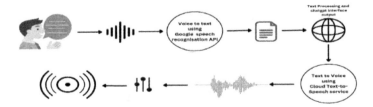

Figure 50.3: Working principle of voice-driven Chatgpt.

5. Results and Discussion

The voice-driven ChatGPT is designed as shown in Figure 50.1a and b. The voice input is captured by the microphone and processed by the ESP32, where it is converted into text and subsequently used by the cloud speech client. The text is processed, and the output from the cloud speech client is in the form of text. This text is then converted into voice. At the final stage, the voice output is amplified, and users can hear it through the speaker (Figure 50.2).

Figure 50.4: (a) Design of voice-driven Chatgpt using a breadboard. (b) Design of voice-driven Chatgpt using PCB.

6. Conclusion

Voice assistants have transcended to become indispensable assets woven into the fabric of our everyday existence, offering unparalleled convenience and efficiency through their ability to understand and respond to spoken commands. The rise of conversational ChatGPT interfaces stands out for their engaging experiences, making them increasingly popular among users seeking interactive and personalized assistance through text-based interfaces.

This voice-based ChatGPT interface leverages ChatGPT's remarkable ability to understand context, generate coherent responses, and offer a vast array of information while enabling seamless voice output. It will be a valuable companion for learning, exploring, and accessing personalized recommendations. As technology continues to evolve, the integration of conversational AI into voice assistants promises to further enhance their functionality and usability.

References

[1]   Gopal Vishwas, P., & Vidya. D. (2022). Research and Analysis on Voice-Based Systems with Machine Learning. *10th IEEE International Conference on Emerging Trends in Engineering & Technology-Signal and Information Processing (ICET-ET-SIP-22)*. https://doi.org/10.1109/ICETET-SIP-2254415.2022.9791498

[2]   Holzinger, A., Kargl, M., Kipperer, B., Regitnig, P., Plass, M., & Heimomuller, M. (2022). Personas for Artificial Intelligence (AI): An Open Source Toolbox. *IEEE Access*. https://doi.org/10.1109/ACCESS.2022.3154776

[3]   Hassija, V., Chakrabarti, A., Singh, A., Chamola, V., & Sikdar, B. (2023). Unleashing the Potential of Conversational AI: Amplifying ChatGPT's Capabilities and Tackling Technical Hurdles. *IEEE Access*. https://doi.org/10.1109/ACCESS.2023.3339553

[4]   Lawalibrahim Dutsinma, F., Babakerkhell, M. D., Mongkolnam, P., Chongsuphajaisiddhi, V., Funilkul, S., & Pal, D. (2024). A review of subjective scales measuring the user experience of voice assistants. *IEEE Access*. https://doi.org/10.1109/ACCESS.2024.3358423

[5]   Wu, T., He, S., Liu, J., Sun, S., Liu, K., Han, Q.-L., & Tang, Y. (2023). A brief overview of ChatGPT: The history, status quo, and potential future development. *IEEE Journal of Automation and Systems*. https://doi.org/10.1109/JAS.2023.123618

CHAPTER 51

Optimizing Geopolymer Concrete Properties through Red Mud and Fly Ash with Recycled Water for Sustainable Construction

Manoranjan Rupa[1], V. Madhava Rao[2], Kaliprasanna Sethy[3]

[1]Ph.D. Scholar, GIET University, Gunpur, India
[2]Professor, GIET University, Gunpur, India
[3]Assistant Professor, GCEK, Bhawanipatna, India
E-mail: manoranjan.rupa@giet.edu, profvmrao@giet.edu, drkps@gcekbpatna.ac.in

Abstract

The rapid development of urban areas and industries has resulted in the accumulation of significant quantities of industrial waste, including red mud (RM) and fly ash (FA), which are often disposed of indiscriminately, causing environmental pollution. Concurrently, heightened construction activities exacerbate the environmental impact through the production and use of traditional building materials like cement. To address these challenges and ensure a sustainable future, this study explores the potential of geopolymerization as an eco-friendly substitute for traditional concrete. Through rigorous testing of geopolymer concrete incorporating RM and FA with normal water (NW) and recycled water (RW), various mechanical and physical characteristics, such as compressive strength, flexural strength, and split tensile strength, were evaluated. The findings suggest that the geopolymerization of industrial wastes presents a promising and workable solution to mitigate the environmental consequences of traditional concrete production.

Keywords: Red mud, fly ash, geopolymer concrete, sustainable

1. Introduction

As a construction material, concrete is globally valued for its durability and adaptability. Its composition includes aggregate, cement, and water, allowing it to solidify into desired shapes (Viyasun et al. 2021). The construction industry's impact on waste and greenhouse gas emissions is notable, primarily due to cement production's heavy resource usage and emission of CO_2 and NO_2 gases, exacerbating global warming (Ahmad et al. 2021). Geopolymer binders, a promising substitute for Portland cement, produce geopolymer concrete by reacting silicon and aluminum-rich minerals or industrial wastes, offering an eco-friendly and faster-setting alternative to conventional cement (Johilingam et al. 2021). Geopolymer, a distinct synthetic aluminosilicate binding phase, is formed through the interaction of alkaline solutions with mineral sources like metakaolin and fly ash. Its notable benefits, including a reduced carbon footprint and the repurposing of industrial waste, have propelled it into the spotlight as an encouraging alternative in construction materials

DOI: 10.1201/9781003596776-51

(Tuan et al. 2022). Red mud, characterized by trace elements and six primary oxides, including calcium oxide, silicon dioxide, iron oxide, aluminum oxide, titanium dioxide, and sodium oxide, is considered a cementitious material due to its composition similar to cement (Chavan et al. 2021). Red mud remains a feasible substitute for raw cement, with red mud concrete demonstrating enhanced durability over time compared to conventional concrete, despite decreased workability, as evidenced by increased compressive and flexural strength. Red mud (RM) can be rendered environmentally sustainable through two methods: neutralization or inertness. While neutralization addresses alkalinity, it may not ensure complete safety. The optimal approach involves using the residue in applications that maintain its inert nature, thus preventing environmental leaching.

2. Literature Review

Cement's vital role in concrete is marred by its significant carbon dioxide emissions, necessitating the exploration of eco-friendly substitutes. The environmental impact of widespread development, coupled with substantial waste generation, highlights the urgent need for long-term fixes (Kanyal et al., 2021). Utilizing recycled materials in construction diminishes reliance on virgin resources. Quarry dust, foundry sand, and other waste materials, when integrated into concrete, serve as sustainable alternatives to Portland cement, leading to the conservation of resources and substantial economic gains. To mitigate the greenhouse gas emissions from Portland cement production, industrial waste such as RM is utilized to produce geopolymer materials and alkali-activated binders (AAB). These geopolymers, derived from residual and oxidizable heavy metals, exhibit enhanced stability and contribute to environmental preservation (Kumar et al. 2021). Ongoing research focuses on utilizing RM in diverse applications due to its abundant metallic oxides, large surface area, and stable dispersion, offering a wide array of mechanical, physical, and thermal qualities, rendering it appropriate for various uses. A geopolymer binder, synthesized using Bayer red mud, exhibited enhanced long-term strength with the inclusion of 20–30 wt. percent SF and lower solution/solid ratios added to strength improvement, reaching a compressive strength of 31.5 MPa after 28 days. Liquefied aluminosilicate and silica geopolymerization yielded condensed matrices in the geopolymer (Nan et al. 2014). Geopolymer concrete reduces carbon emissions by utilizing industrial waste like FA and RM. Adding recycled rainwater to geopolymer concrete enhances sustainability, conserving freshwater resources.

3. Materials Used

The investigation made use of the following resources:
- OPC—43 Grade (by IS: 8112-2013)
- Natural sand (by IS: 383-2016)
- Coarse aggregates (20 mm down) (by IS: 383-2016)
- RM and fly ash (FA)
- Alkali activator mixture, with a ratio of 2:0 for sodium hydroxide to sodium silicate (NaOH/ Na_2SiO_3). A solution was created by mixing one part Na_2SiO_3 with two parts NaOH with NW and RW.

4. Methodology and Model Specifications

- A binder mix ratio of 60:40 (RM: FA), the mix percentage was prepared by the standard concrete mix design for M20 grade, which complies with IS 10262:2009.
- The liquid-to-binder ratio was kept at 0.65%.

- A 2:1 combination of sodium silicate solution and sodium hydroxide with a molarity of 6M was added to the alkali activator solution.
- Specimens were cured using ambient temperature.

5. Test Results of Mechanical Properties

The compressive strengths of conventional and geopolymer concrete were compared at 7 and 28 days, by IS 516: 1959.

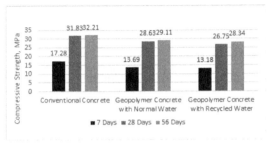

Figure 51.1: Compressive strength values for conventional, geopolymer concrete with NW and RW.

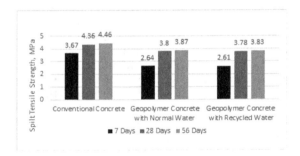

Figure 51.2: Spilt tensile strength values for conventional, geopolymer concrete with NW and RW.

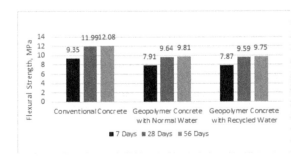

Figure 51.3: Flexural strength values for conventional, geopolymer concrete with NW and RW.

6. Conclusion

1. In the current mix proportion, due to the implementation of RM and FA, it is observed that economical concrete can be designed compared to conventional concrete.
2. The workability of the developed concretes showed reduced values compared to conventional concrete.
3. The mechanical (compressive) strength results of RM and FA-based geopolymer concrete with NW and RW showed a marginal difference from the corresponding strength observed in conventional concrete.
4. The target compressive strength of 29.11 N/mm^2 was easily achieved at 56 days with the RM and FA ratio used for the mix design.
5. The liquid-to-binder ratio plays a vital role in achieving the splitting tensile strength, which was observed as 3.87 N/mm^2.
6. The flexural strength test carried out using FTM was obtained as 9.81 N/mm^2 for geopolymer concrete, whereas 12.08 N/mm^2 was obtained for conventional concrete at 28 days.

References

[1]   Viyasun, K., Anuradha, R., Thangapandi, K., Santhosh Kumar, D., Sivakrishna, A., & Gobinath, R. (2021). Investigation on the performance of red mud-based concrete. *Journal of Materials in Civil Engineering, 39*(1), 796–799.

[2]   Almutairi, A. L., Tayeh, B. A., Adesina, A., Isleem, H. F., & Zeyad, A. M. (2021). Potential applications of geopolymer concrete in construction: A review. *Journal of Building Engineering, 15*, e00733.

[3]   Jothilingam, M., & Preethi, V. (2021). Feasibility, compressive strength, and utilization of red mud in geopolymer concrete for sustainable constructions. *Journal of Cleaner Production, 45*(7), 7016–7022.

[4]   Nguyen, T. A. H., Guo, X., You, F., Saha, N., Wu, S., Scheuermann, A., Ren, C., & Huang, L. (2022). Co-solidification of bauxite residue and coal ash into indurated monolith via ambient geopolymerisation for in situ environmental application. *Journal of Environmental Management, 422*, 126925.

[5]   Chavan, S. P., Salokhe, S. A., Nadagauda, P. A., Patil, S. T., & Mane, K. M. (2021). An investigational study on properties of concrete produced with industrial waste red mud. *Journal of Cleaner Production, 42*(2), 733–738.

[6]   Kanyal, K. S., Agrawal, Y., & Gupta, T. (2021). Properties of sustainable concrete containing red mud: A review. *Journal of Scientific Research and Reports, 27*(9), 15–26.

[7]   Kumar, A., Saravanan, T. J., Bisht, K., & Kabeer, K. I. S. A. (2021). A review on the utilization of red mud for the production of geopolymer and alkali-activated concrete. *Construction and Building Materials, 302*, 124170. Elsevier BV.

[8]   Ye, N., Yang, J., Ke, X., Zhu, J., Li, Y., Xiang, C., Wang, H., Li, L., & Xiao, B. (2014). Synthesis and characterization of geopolymer from Bayer red mud with thermal pretreatment. *Journal of the American Ceramic Society, 97*(5), 1652–1660.

CHAPTER 52

Reclaimed Asphalt Pavement as a Sustainable Material for Pavement Application

Debakinandan Naik[1,*], Sujit Kumar Pradhan[2], Bikash Chandra Panda[2]

[1]Research Scholar, BPUT, Rourkela and Department of Civil Engineering, IGIT Sarang, Odisha, India Email debakinandannaik@gmail.com
[2]Deptt. of Civil Engineering, IGIT Sarang, Odisha, India

Abstract

The global use of recycled asphalt pavement (RAP) has significantly increased, reducing the need for raw materials. The use of RAP as an alternative material in road construction has grown due to the scarcity of eco-friendly materials and natural aggregates. This research provides a comprehensive assessment of RAP aggregates, focusing on characterization, strength, durability, and permanent deformation. A progressive characterization process is used to analyze the physical and mechanical properties of RAP materials. It has been demonstrated that RAP aggregates possess fresh and hardened qualities, particularly in terms of durability, making them a viable partial replacement for natural aggregates. The primary objectives of this critical review are to outline the essential attributes required for sustainable pavement construction and to provide a comprehensive understanding of the current status of RAP utilization. The review highlights various research gaps related to material characteristics, strengths, and the environmental impacts of the circular economy. Moreover, optimal ratios of these alternative materials are suggested for sustainable flexible pavement systems. This thorough analysis will assist national and state highway authorities, pavement engineers, and researchers in efficiently utilizing RAP aggregates in flexible pavement construction.

Keywords RAP, Bituminous mixture, Mechanical properties, Environment, Life cycle cost assessment

1. Introduction

Since the beginning of the human revolution, people have continually sought to improve their lives through innovation. Every innovation stems from a need. New ideas, research, and techniques bridge these gaps. This issue presents two potential responses: the scarcity and high value of materials and the high cost of production. The problem can be addressed either by discovering new alternative materials to replace virgin materials or by recycling materials after their service life. Over the past decade or two, a growing number of countries have begun to recognize the importance of sustainability programs. Initially, these programs were limited to a few developed nations, but they have recently expanded to less developed countries as well.

This paper discusses RAP, a product of damaged pavement containing both aged asphalt and aggregate. RAP is obtained during road maintenance or by removing old pavement and

DOI: 10.1201/9781003596776-52

replacing it with new pavement (Al-Shujairi et al. 2021; Cosentino et al. 2003). Eighty percent of roads in the European road network are built with bitumen, consuming 950 million tons of asphalt. In 1970, it was established that RAP maintains its functionality when used in fresh bituminous mixtures without degradation (Antunes et al. 2019). The first use of RAP as a replacement for virgin materials occurred in 1973 at a rate of 3%; today, that rate has increased to 20%, 30%, and sometimes even more than 50% (Abdo 2016). RAP not only helps conserve natural resources but also benefits ecological environments (Zaremotekhases et al. 2022).

2. Benefits and Challenges

The use of RAP in new bituminous mixtures offers both economic and environmental advantages (Zaumanis et al. 2014). When a product is part of the circular economy, its value is preserved for as long as possible. From 2008 to 2014, approximately 300 million tons of asphalt mix were produced annually in Europe, including both hot and warm mixes. Due to various unfavorable perceptions and practical concerns, the RAP content is typically limited to 15–30%. These challenges are summarized based on the quality of RAP aggregates used, the technology employed in the plant, the mix design technique, and the outcomes of the recycled mixes. The presence of fine particles poses a risk to the required gradation, potentially limiting the maximum RAP concentration (Copeland 2011). Fine particles have a large surface area, which aids in retaining a significant amount of aged binder (Newcomb et al. 2007).

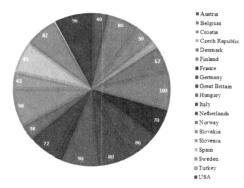

Figure 52.1: RAP application in various countries.

3. RAP Application and Limitations

RAP application in the production of hot and warm asphalt mixes in various countries is illustrated in Figure 52.1. The limitation of RAP quantities in different countries is evident as RAP recycling rapidly expands for various applications, such as producing new bitumen mixtures, unbound layers, and landfill material. The majority of the USA and many European countries utilize RAP in constructing both hot and warm mix asphalt. The recycling rate of RAP in new bituminous mixtures varies across countries, with the maximum incorporation rates differing by state, layer, and plant production capabilities in the USA. Some states specify maximum incorporation percentages for specific layers rather than a uniform maximum across all layers (Chesner et al. 2018). Certain state agencies in the USA require evaluation of RAP aggregate polishing and mineralogical composition when applying recycled mixtures to high-speed traffic surfaces (Daniel et al. 2005).

3.1. Performances Tests on Recycled Mixture

Many researchers have studied various performance properties, including moisture sensitivity, rutting, resilient modulus, fatigue, cracking resistance, and indirect tensile strength (Figure 52.2). This paper discusses rutting and fatigue properties in subsequent sections.

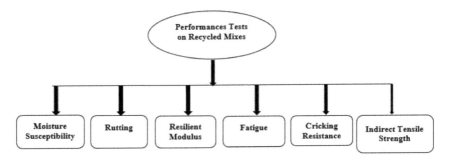

Figure 52.2: Various performance tests of recycled mixes.

3.1.1. Rutting

It has been observed that RAP increases the dynamic stability value in traditional approaches. This leads to enhanced resistance to rutting deformation, as the mixture's overall stiffness increases. Figures 52.3 and 52.4 illustrate the relationship between dynamic stability (DS) and rut depth concerning RAP content. According to the literature, deflection decreases with increasing RAP content (Shao et al. 2017; Wu et al. 2020). Consequently, resistance to rutting improves with higher RAP content (Zhang et al. 2021; Pradhan & Sahoo 2020). This improvement may result from increased stiffness due to RAP incorporation.

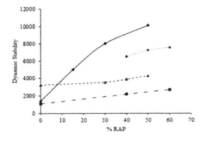
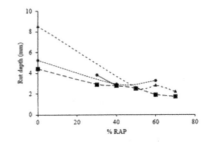

Figure 52.3: Dynamic stability vs % RAP.   **Figure 52.4:** Rut depth vs % RAP.

3.1.2. Fatigue

As the stress ratio rises, the fatigue life of high-modulus asphalt concrete mixtures decreases (Figure 52.5). A similar trend is observed in high-modulus asphalt concrete, though this type of concrete exhibits a slightly longer fatigue life than recycled mixtures. The mixture's modulus increases with the incorporation of more RAP and FRAP, making it more sensitive to fatigue cracking (Han et al. 2019). In both conditioned and unconditioned mixtures, RAP (40%) contributes to increased fatigue life due to the high stiffness of the mixed binder, adequate interaction between the virgin and aged binder, and the adhesive properties of silicon aggregates. However, as RAP content increases, fatigue life performance decreases.

Adding a rejuvenator to the recycled mixture improves fatigue performance by restoring adhesion between RAP and the new binder (Zhang et al. 2021). RAP diffusion affects the fatigue life of virgin mixes, primarily due to increased mixture stiffness with higher RAP content (Dong et al. 2017). Fatigue cracking test results indicate that the control mix has lower fatigue cracking resistance than the 15% and 30% RAP content mixes. Many researchers report that fatigue life improves up to a certain limit (Mullapudi et al. 2020; Pradhan & Sahoo 2020). However, Ziari et al. (2021) found contradictory results, possibly due to improper adhesion between the aged binder and virgin aggregates.

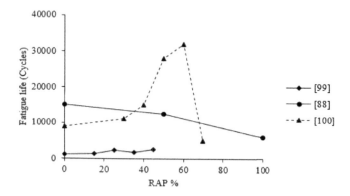

Figure 52.5: Fatigue life of RAP mixtures.

4. Conclusion

Increasing the quantity of RAP in asphalt mixtures enhances strength; however, the content percentage is generally limited to 30%. At 30% and above, the aged binder content becomes too high, leading to improper bonding between the asphalt binder and the aggregate dosage. This gap increases the likelihood of moisture damage in recycled mixtures. To address these issues, different rejuvenators or admixtures can be mixed with high RAP content mixtures, improving both mechanical performance and the rheological properties of the binder. Thus, while the growth in RAP content is not a significant concern, it is essential to monitor the mixture's ductility properties as RAP content increases.

References

[1]   Al-Shujairi, A. O., Al-Taie, A. J., & Al-Mosawe, H. M. (2021). Review on applications of RAP in civil engineering. *IOP Conference Series: Materials Science and Engineering.* https://doi.org/10.1088/1757-899X/1105/1/012092

[2]   Cosentino, P., Kalajian, E., Shieh, C., Mathurin, W., Gomez, F., Cleary, E., & Treeratrakoon, A. (2003). Developing specifications for using recycled asphalt pavement as base, subbase, or general fill materials. Phase II. *Report No. FL/DOT/RMC/06650-7754*, Florida Institute of Technology. https://rosap.ntl.bts.gov/view/dot/37263

[3]   Antunes, V., Freire, A. C., & Neves, J. (2019). A review on the effect of RAP recycling on bituminous mixtures properties and the viability of multi-recycling. *Construction and Building Materials.*

[4]   Abdo, A. M. A. (2016). Utilizing reclaimed asphalt pavement (RAP) materials in new pavements: A review. *International Journal of Thermal and Environmental Engineering.* https://doi.org/10.5383/ijtee.12.01.008

[5]   Zaremotekhases, F., Sadek, H., & Hassan, M. (2022). Impact of warm-mix asphalt technologies and high reclaimed asphalt pavement content on the performance of alternative asphalt mixtures. *Construction and Building Materials*. https://doi.org/10.1016/j.conbuildmat.2021.126035

[6]   Zaumanis, M., Mallick, R. B., & Frank, R. (2014). 100% recycled hot mix asphalt: A review and analysis. *Resources, Conservation and Recycling*. https://doi.org/10.1016/j.resconrec.2014.07.007

[7]   Copeland, A. (2011). Reclaimed asphalt pavement in asphalt mixtures: State of the practice. *Federal Highway Administration* (Report No. FHWA-HRT-11-021). https://rosap.ntl.bts.gov/view/dot/40918

[8]   Newcomb, D. E., Brown, E. R., & Epps, J. A. (2007). Designing HMA mixtures with high RAP content: A practical guide. *National Asphalt Pavement Association* (Quality Improvement Series 124).

[9]   Chesner, W., Collins, R., MacKay, M., & Emery, J. (2018). User guidelines for waste and byproduct materials in pavement construction. *Federal Highway Administration* (Report No. FHWA-RD-97-148https://www.fhwa.dot.gov/publications/research/infrastructure/structures/97148/rap132.cfm.

[10]   Shao, H., Sun, L., Liu, L., You, Z., & Yang, X. (2017). A novel double-drum mixing technique for plant hot mix asphalt recycling with high reclaimed asphalt pavement content and rejuvenator. *Construction and Building Materials*. https://doi.org/10.1016/j.conbuildmat.2016.12.077

[11]   Cheng, Y., Wang, W., Gong, Y., Wang, S., Yang, S., & Sun, X. (2018). Comparative study on the damage characteristics of asphalt mixtures reinforced with an eco-friendly basalt fiber under freeze-thaw cycles. *Materials*. https://doi.org/10.3390/ma11122488

[12]   Wu, Z., Zhang, C., Xiao, P., Li, B., & Kang, A. (2020). Performance characterization of hot mix asphalt with high rap content and basalt fiber. *Materials*. https://doi.org/10.3390/ma13143145

[13]   Zhang, J., Guo, C., Chen, T., Zhang, W., Yao, K., Fan, C., Liang, M., Guo, C., & Yao, Z. (2021). Evaluation of the mechanical performance of recycled asphalt mixtures incorporated with a high percentage of RAP and self-developed rejuvenators. *Construction and Building Materials*.

[14]   Pradhan, S. K., & Sahoo, U. C. (2022). Influence of softer binder and rejuvenator on bituminous mixtures containing reclaimed asphalt pavement (RAP) material. *International Journal of Transportation Science and Technology*. https://doi.org/10.1016/j.ijtst.2020.12.001

[15]   Ziari, H., Orouei, M., Divandari, H., & Yousefi, A. (2021). Mechanical characterization of warm mix asphalt mixtures made with RAP and para-fiber additive. *Construction and Building Materials*. https://doi.org/10.1016/j.conbuildmat.2021.122456

[16]   Dong, F., Yu, X., Xu, B., & Wang, T. (2017). Comparison of high temperature performance and microstructure for foamed WMA and HMA with RAP binder. *Construction and Building Materials*. https://doi.org/10.1016/j.conbuildmat.2016.12.106

[17]   Mullapudi, R. S., Noojilla, S. L. A., & Kusam, S. R. (2020). Effect of initial damage on healing characteristics of bituminous mixtures containing reclaimed asphalt material (RAP). *Construction and Building Materials*. https://doi.org/10.1016/j.conbuildmat.2020.120808

[18]   Pradhan, S. K., & Sahoo, U. C. (2020). Impacts of recycling agent on superpave mixture containing RAP. *5th International Symposium on Asphalt Pavement and Environment. Lecture Notes in Civil Engineering, Springer Nature, Switzerland*. https://doi.org/10.1007/978-3-030-29779-24

CHAPTER 53

State of the Art

Image Classification by Hybrid Tuning Deep Ensemble with XAI

Chapala Maharana[1], Ch. Sanjeev Dash[2], Bijan Bihari Mishra[3]

Faculty, CSE, Parala Maharaja Engineering College, Berhampur, Odisha, India
chapala.maharana@pmec.ac.in
Associate Professor Silicon Institute of Tech, Bhubaneswar, Odisha, India
sanjeevc@silicon.ac.in, misrabijan@gmail.com

Abstract

Many papers have been published on machine learning methods for healthcare issues, with a few focusing on image datasets. Efficient disease diagnosis models for image datasets using deep learning methods with higher accuracy are crucial. Therefore, we propose an ensemble deep learning model with the hybrid tuning of hyperparameters, combining feature extraction, classification by ensemble deep learning, and XGBoost as the ensemble classifier. This model exhibits higher generalizability and accuracy. The output is presented with explanations by explainable AI, making it more powerful and acceptable to experts.

Keywords: Blending, Deep Learning, Deep Ensemble Learning, Explainable AI, Feature Importance, Image Data, LIME, Shapley, Stacking, Quantization

1. Introduction

Machine learning, deep learning, ensemble learning, deep ensemble learning, and artificial intelligence are emerging technologies that address today's challenging medical issues in real-time. These methods are applied to image classification, object detection, image segmentation, image processing, natural language processing, time series medical data processing, and audio & video data processing. Ensemble learning is particularly significant for predicting disease diagnosis. The art of ensemble learning involves combining classifiers for optimal solutions. The deep ensemble method, especially for image datasets, combines deep neural networks and ensemble methods. In this proposed deep ensemble method, we used a convolutional neural network (CNN) for efficient feature extraction after hyperparameter tuning.

2. Literature Review

Challenges in histopathological data are often marked by incompleteness and subjectivity. Fatima et al. noted that such issues are unavoidable due to the difficulty in identifying all cells with certainty.

Proper identification of cell features is facilitated by powerful computational models that can accurately capture low-level, mid-level, and high-level features. Lilhore et al. found that deep learning is highly powerful, flexible, and simple, with applications across

DOI: 10.1201/9781003596776-53

many disciplines. DL technology is used for feature identification from MRI, X-ray, and ultrasound images, as well as for tasks such as tumor genetic mutation detection, cell classification, protein/protein interaction analysis, and diagnosing diabetic retinopathy. Ensemble deep learning can be constructed with more than one deep learning model and the dynamic selection of different machine learning models in various layers to achieve optimal solutions.

3. Related Work

Malebary et al. proposed a new RNN-LSTM CNN model for feature extraction. Misra et al. introduced a deep ensemble model that outperforms the weaknesses of individual models. Wang et al. proposed a multinetwork feature and dual-network orthogonal learning approach for image classification. Albert et al. developed novel segmentation and ensemble algorithms using deep learning with limited data. Tang et al. proposed a transform modal ensemble learning approach for breast tumor segmentation in ultrasound images. Monkam et al. presented an ensemble learning 3D-CNNs model for micro nodule identification in CT images, achieving accuracy, AUC, F-score, and sensitivity of 97.35%, 0.98, 96.42%, and 96.57%, respectively. Wang et al. proposed image classification by deep ensemble using multi-network features and dual-network orthogonal low-rank learning. Wentao proposed an efficient diversity-driven deep ensemble using a novel knowledge transfer method with selective transfer of previous generic knowledge. Radha et al. used ensemble DeepLabV3+ for lung image data segmentation. Sebastian et al. proposed generalized negative correlation learning for deep ensembles. Patrice et al. developed a multiple-view 3D-CNNs model using ensemble learning.

4. Proposed Work

The proposed work has been carried out as per the following Figure 53.1 and Figure 53.2.

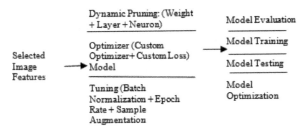

Figure 53.1: Deep learning

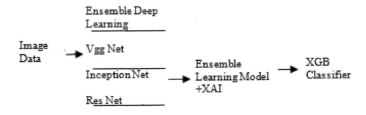

Figure 53.2: Proposed work (EDL)

The performance evaluation has been provided as per following table 53.1. The feature extraction has been placed for Lung Cancer with Train Data, Test Data, and Valid Data. The performance results in terms of VGG Net Accuracy, Inception Net Accuracy, Res Net Accuracy, Proposed Ensemble have been placed in the table 1 with respect to time.

Table 53.1: Performance result

Data set	Features Extracted	Vgg Net Accuracy	Inception Net Accuracy	Res Net Accuracy	Proposed Ensemble
Lung Cancer	Train Data: (123,32768) / (490,256,256,3) Test Data: (315,32768)/ (315, 256,256,3) Valid Date: (72,32768)/ (72,256,256,3)	Train Acc: 33.34% Test Acc: 24% Time: 5403	Train Acc: 91.56% Test Acc 31.11% Time:7003	Train Acc: 64.56% Test Acc:21.2% Time: 5153	Train Acc: 99% Test Acc: 64% Time: 2.67

The Prediction rate with respect to numbers of rows has been displayed in Figure 53.3. The data prepared after feature extraction by VGG Net value transform to two-dimensional train data and test data by the next phase of model.

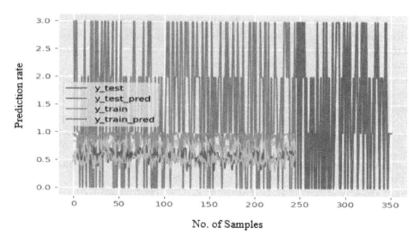

Figure 53.3: Train Test Data set

5. Result Evaluation

The proposed model is implemented with four class labels. The pretrained VGG Net model extracts features of 32768 from 1,96,608. The training accuracy of VggNet, InceptionNet, and ResNet are 33.34%, 91.56%, and 64.56%, respectively. The proposed model achieves 100% training accuracy and 64% test accuracy, with a time of 2.67 units.

6. Conclusion and Future Work

We found that deep ensemble learning is one of the most suitable methods for image classification. The deep ensemble model can be hyper-tuned by modifying the CNN classification model as the final classifier, making the deep ensemble faster and more flexible. We used the XGBoost ensemble classifier for the final classification.

References

[1]   Albert, B. A. (2020). Deep learning for limited training data: Novel segmentation and ensemble algorithms applied to automatic melanoma diagnosis. *IEEE*, *8*, 31254–31269. Bartz, E., Beielstein, B., Zaefferer, M., & Mersmann, O. (2022). Hyperparameter tuning for machine and deep learning with R: A practical guide. Springer.

[2]   Foysal, Md., Aowlad, H., & A. B. N. (2021). Covid-19 detection from CT images using ensemble deep convolutional neural network. *IEEE*, doi: 10.1109/INCET51464.2021.9456387.

[3]   Haq, I. U., Ali, H., Wang, H. Y., Lei, C., & Ali, H. (2022). Feature fusion and ensemble learning-based CNN model for mammographic image classification. *Journal of King Saud University-Computer and Information Sciences*, *34*, 3310–3318. Elsevier.

[4]   Karthik, R., Menaka, R., & Siddharth, M. V. (2022). Classification of breast cancer from histopathology images, using an ensemble of dep multiscale networks. *Elsevier*, *42*, 963–976. Lilhore, U. K., Dalal, S., Neetu, M., Martin, C., Chakraborty, T., Simaiya, S., Kumar, P., Thangaraju, P., & Velmurugan, H. (2023). Hybrid CNN-LSTM model with efficient hyperparameter tuning for prediction of Parkinson's disease. *Scientific Reports*, *13*, 14 [Hybrid CNN LSTM].

[5]   Malebary, S. J., & Hashmi, A. (2021). Automated breast mass classification system using deep learning and ensemble learning in mammograms. *IEEE*, *9*, 55312–55328.

[6]   Monkam, P., Qi, S., Xu, M., Li, H., Han, F., Teng, Y., & Qian, W. (2019). Ensemble learning of multiple view 3D-CNNs model for micro nodule identification in CT images. *IEEE*, *7*, 5564–5576.

[7]   Musaev, J., Anorboev, A., Seo, Y. S., Nguyen, N. T., & Hwang, D. (2023). ICNN-Ensemble: An improved convolutional neural network ensemble model. *IEEE*, *11*, 86285–86296.

[8]   Muthuswamy, A., Dewan, M. A. A., Murshed, M., & Parmar, D. (2023). Distraction classification using deep convolutional autoencoder and ensemble learning. *IEEE*, *11*, 71435–71448.

[9]   Ning, G., Zhang, X., Tu, C., Feng, Q., & Zhang, Y. (2019). Application to breast histopathological image. *IEEE*, *7*, 150910–150923.

[10]   Pradhan, K. S., Chawla, P., & Tiwari, R. (2023). High ranking deep ensemble learning-based lung cancer diagnosis model. *Expert Systems with Applications*. https://doi.org/10.1016/j.eswa.2022.118956. Saha, P., Sadi, M. S., & Islam, M. M. (2021). EMCNet: Automated Covid-19 diagnosis from X-ray images using convolutional neural network and ensemble of machine learning classifiers. *Elsevier*, *22*. doi: 10.1016/j.imu.2020.100505.

[11]   Shen, Q., Mo, L., Liu, G., Zhou, J., Zhang, Y., & Ren, P. (2023). Short term load forecasting based on multiscale ensemble deep learning neural network. 1 *IEEE*, *11*, 111963–111975. doi: 10.1109/ACCESS.2023.3322167.

[12]   Sheeba, A., Kumar, P. S., Ramamoorthy, M., & Sasikala, S. (2023). Microscopic image analysis in breast cancer detection using ensemble deep learning architectures integrated with the web of things. *Elsevier*, *79*. doi: 10.1016/j.bspc.2022.104048.

CHAPTER 54

Single-atom photocatalysts

Innovations in water purification process focusing on organic pollutant degradation

Amita Somya[a*], Amit Prakash Varshney[b]

[a]Department of Chemistry, Amity School of Applied Sciences, Amity University, Bengaluru, Karnataka, India
[b]IT Company, Bengaluru, India

Abstract

Single-atom photocatalysts (SAPs) offer promising applications in environmental contexts, particularly in water purification. This review presents recent progress in their synthesis, characterization, and application for organic pollutant degradation. SAPs' notable properties, including their large surface area, light absorption, and adsorption capabilities, enhance their efficacy in pollutant removal. Various synthesis methods, such as wet-chemical, atomic layer deposition, and electrochemical methods, are discussed alongside their respective advantages and limitations. The catalytic mechanisms of SAPs, including reactive oxygen species generation and charge-transfer complexes, are explored, with a focus on their performance in degrading organic pollutants and heavy metals. Finally, future directions in SAP development for water purification are discussed.

Keywords: Single-atom photocatalysts, water purification, organic pollutants, degradation, synthesis, characterization, catalytic mechanisms, performance, challenges, future directions

1. Introduction

Water contamination is a global environmental concern, posing substantial threats to both human health and ecosystems (Somya, 2024). Organic pollutants, such as volatile organic compounds (VOCs), dyes, and pharmaceuticals, are major contributors to water pollution, making their removal from water (Somya, 2023) crucial for maintaining water quality. Traditional water purification methods, such as filtration, ion exchange (Somya, 2019), and chemical oxidation, are often ineffective in removing certain organic pollutants, particularly those resistant to degradation. Consequently, there is growing interest in developing advanced materials for water purification, with a focus on the degradation of organic pollutants.

Single-atom photocatalysts (SAPs) (Gao et al., 2020; Khandelwal et al., 2022) have emerged as a promising class of materials for water purification due to their unique properties and high catalytic activity. SAPs consist of individual metal atoms, such as titanium, platinum, or gold, anchored onto a support material, such as graphene or metal-organic frameworks. The large surface area of the support material provides numerous active sites for catalytic reactions, while the metal atoms serve as efficient catalytic centers for the degradation of organic pollutants. Additionally, SAPs have enhanced light absorption properties, enabling them to utilize solar energy for catalytic reactions, and they can be engineered to have specific catalytic properties for the selective degradation of certain pollutants.

DOI: 10.1201/9781003596776-54

This review aims to summarize recent progress in SAPs for water purification, covering synthesis methods, catalytic mechanisms, and their role in degrading organic pollutants. We will also examine their performance and the factors influencing catalytic activity, including metal atom choice, support material, and reaction conditions.

2. Synthesis of single-atom photocatalysts

The synthesis of SAPs involves creating a support material, such as graphene or metal-organic frameworks, and depositing individual metal atoms onto it (Liu et al., 2017; Li et al., 2023). Li et al. (2023) reviewed synthesis methods and characterization techniques aimed at advancing SAPs' industrialization, highlighting wet-chemical synthesis, atomic layer deposition (ALD), and electrochemical deposition. While wet-chemical synthesis allows control over metal atom distribution, it may involve toxic chemicals and result in SAPs with limited stability. ALD offers precise control over composition but is complex and costly, whereas electrochemical deposition, though scalable, may produce less durable SAPs.

3. Characterization of single-atom photocatalysts

Gao et al. (2020) discussed advanced characterization techniques and theoretical studies that enhance our understanding of SAPs for photocatalysis applications. Their review covered metal-support interactions and characterization methods such as TEM, XAFS, STM, and XPS, providing detailed insights into SAPs' composition and catalytic behavior, which are crucial for optimizing their performance in various applications.

4. Catalytic mechanisms of single-atom photocatalysts

The catalytic mechanisms of (SAPs) leverage their unique properties to enhance light absorption and charge separation, driving efficient pollutant degradation. Zhang et al. (2021) highlighted theoretical advancements in SAPs, emphasizing their role in producing reactive oxygen species and reducing electron-hole recombination, thus enhancing photocatalytic efficiency. The atomic dispersion of metal atoms facilitates targeted adsorption and activation of molecules, improving selectivity and efficiency in pollutant degradation, collectively enhancing SAPs' performance in environmental remediation.

4. Applications of single-atom photocatalysts in wastewater treatment

Degradation of organic pollutants

SAPs excel in degrading various organic pollutants in wastewater, including dyes, pharmaceuticals, pesticides, and industrial contaminants. Their unique catalytic properties, driven by isolated metal atoms, effectively facilitate the degradation process. Cheng et al. (2024) explored SAPs' effectiveness in degrading waterborne organic pollutants, examining key factors influencing their efficiency and discussing mechanisms and potential applications in pollutant remediation. Zeng et al. (2023) scrutinized iron-based single-atom catalysts (Fe-SAPs) in water treatment, assessing their efficacy across three reaction stages and suggesting avenues for improvement. Their study provides valuable insights into leveraging Fe-SAPs for water pollution mitigation. Gamelas et al. (2023) examined the efficiency and recyclability of phthalocyanines in wastewater treatment, highlighting their oxygen species generation and byproduct identification. Their study featured comparative tables of catalysts, with most degrading at least 5% of pollutants and over 50 exhibiting rates surpassing 50%, emphasizing their potential for effective pollutant degradation. Jasim et al. (2023) examined photocatalyst applications in environmental contexts, evaluating removal methods for pollutants and microbes. They discussed operational factors such as photocatalyst dosage, pollutant concentration, and light conditions, highlighting SAPs' role in initiating photocatalytic reactions and facilitating pollutant degradation (Liu et al., 2023).

4.1. Reduction of heavy metals

SAPs facilitate the reduction of heavy metals in wastewater via redox reactions, addressing environmental concerns. Liu et al. (2023) discussed the potential of SAPs in water treatment, highlighting their stability and catalytic activity. Sun et al. (2022) emphasized the need for innovative photocatalysts like SAPs, reviewing their development, mechanisms, performance, and future research directions, and noting their advantages for wastewater treatment and environmental remediation.

4.2. Disinfection

SAPs contribute to wastewater disinfection by generating reactive oxygen species under light, with Wang et al. (2022) highlighting their role in microorganism inactivation. They emphasized the potential of SAP photocatalysis in decomposing organic matter and improving water quality. Gusain et al. (2020) underscored the significance of photocatalysis in addressing water scarcity, exploring nanomaterials for pollutant degradation, and optimizing factors such as pH and temperature for enhanced efficacy.

5. Future directions in the development of single-atom photocatalysts for water purification

Future development of SAPs for water purification will focus on advanced synthesis methods for better metal atom dispersion, exploring new support materials, and enhancing light absorption. Efforts will target a deeper mechanistic understanding, improved stability, and the design of multifunctional catalysts. Transitioning from lab research to practical applications through pilot studies and industry collaborations, along with environmental and economic assessments, will be crucial. These advancements will enhance the efficiency, scalability, and practicality of SAPs in water purification.

6. Conclusion

SAPs represent a significant advancement in water purification, offering high efficiency in degrading organic pollutants, reducing heavy metals, and disinfecting water. Leveraging the unique properties of isolated metal atoms on support materials, their catalytic activity is influenced by factors such as support material, metal atom type and dispersion, synthesis methods, light absorption, reaction conditions, and stability. Ongoing research focuses on optimizing these factors through advanced synthesis methods, new support materials, and better light utilization. Practical application requires improved stability, multifunctional catalysts, and transitioning from lab to real-world use through pilot studies and industry collaborations. SAPs thus present a promising solution for scalable and environmentally beneficial water purification, with continued research poised to enhance their effectiveness and sustainability.

7. Acknowledgments

The authors are thankful to the Chancellor, Vice-Chancellor, Vice-President (RBEF), and HOI, Amity University, Bengaluru, for providing all facilities to complete this project.

References

[1] Cheng, Y., Ouyang, Z., Wang, Z., Zhang, Y., He, C., Zhang, S., & Yu, J. (2024). Harnessing the power of single-atom catalysts: A promising solution for organic pollutant remediation under light exposure. *Materials Today Sustainability, 27,* 100858. https://doi.org/10.1016/j.mtsust.2024.100858

[2] Gamelas, S. R. D., Tomé, J. P. C., Tomé, A. C., & Lourenço, L. M. O. (2023). Advances in photocatalytic degradation of organic pollutants in wastewaters: Harnessing the power of phthalocyanines and phthalocyanine-containing materials. *RSC Advances, 13*(48), 33957–33993. https://doi.org/10.1039/d3ra06598g

Gao, C., Low, J., Long, R., Kong, T., Zhu, J., & Xiong, Y. (2020). Heterogeneous single-atom photocatalysts: Fundamentals and applications. *Chemical Reviews, 120*(21), 12175–12216. https://doi.org/10.1021/acs.chemrev.9b00840

[3] Gusain, R., Kumar, N., & Ray, S. S. (2020). Factors influencing the photocatalytic activity of photocatalysts in wastewater treatment. In *Photocatalysts in Advanced Oxidation Processes for Wastewater Treatment* (pp. 229–270). Wiley. https://doi.org/10.1002/9781119631422.ch8

[4] Jasim, N. A., Ebrahim, S. E., & Ammar, S. H. (2023). A comprehensive review on photocatalytic degradation of organic pollutants and microbial inactivation using Ag/AgVO$_3$ with metal ferrites based on magnetic nanocomposites. *Cogent Engineering, 10*(1). https://doi.org/10.1080/23311916.2023.2228069

[5] Khandelwal, A., Maarisetty, D., & Baral, S. S. (2022). Fundamentals and application of single-atom photocatalyst in sustainable energy and environmental applications. *Renewable and Sustainable Energy Reviews, 167,* 112693. https://doi.org/10.1016/j.rser.2022.112693

[6] Li, S., Kan, Z., Wang, H., Bai, J., Liu, Y., Liu, S., & Wu, Y. (2023). Single-atom photocatalysts: Synthesis, characterization, and applications. *Nano Materials Science.* https://doi.org/10.1016/j.nanoms.2023.11.001

[7] Liu, J., Bunes, B. R., Zang, L., & Wang, C. (2017). Supported single-atom catalysts: Synthesis, characterization, properties, and applications. *Environmental Chemistry Letters, 16*(2), 477–505. https://doi.org/10.1007/s10311-017-0679-2

Liu, Q., Zhao, Y., Wang, J., Zhou, Y., Liu, X., Hao, M., Chen, Z., Wang, S., Yang, H., & Wang, X. (2023). Application of single-atom-based photocatalysts in environmental pollutant removal and renewable energy production. *Critical Reviews in Environmental Science and Technology, 54*(12), 909–930. https://doi.org/10.1080/10643389.2023.2284646

Somya, A. (2019). Surfactant-based hybrid ion exchangers. *Research Journal of Chemistry and Environment, 23*(3), 96.

Somya, A. (2023). Metal phosphates: Their role as ion exchangers in water purification. In *Metal Phosphates and Phosphonates* (pp. 341–356). Springer International Publishing. https://doi.org/10.1007/978-3-031-27062-8_19

[8] Somya, A. (2024). An investigation study on water quality of a few lakes of Bengaluru in Industry 4.0 with modern technology. In S. Sethi, M. Mahmud, S. K. Pradhan, & R. Sethi (Eds.), *Proceedings of the International Conference on Emerging Trends in Engineering and Technology, Industry 4.0 (ETETI-2023)* (1st ed.). CRC Press. http://dx.doi.org/10.1201/9781003450924-37

[9]   Sun, H., Tang, R., & Huang, J. (2022). Considering single-atom catalysts as photocatalysts from synthesis to application. *iScience, 25*(5), 104232. https://doi.org/10.1016/j.isci.2022.104232.

[10] Wang, J., Li, Y., & Shao, H. (2022). What insights can the development of single-atom photocatalysts provide for water and air disinfection? In *ACS ES&T Engineering, 2*(6), 1053–1067. https://doi.org/10.1021/acsestengg.1c00454.

[11] Zhang, W., Fu, Q., Luo, Q., Sheng, L., & Yang, J. (2021). Understanding single-atom catalysis given theory. *JACS Au, 1*(12), 2130–2145. https://doi.org/10.1021/jacsau.1c00384 Zeng, H., Chen, Y., Xu, J., Li, J., Li, D., & Zhang, J. (2023). Preparation, characterization of iron-based single-atom catalysts and their application for photocatalytic degradation of contaminants in water. *Journal of Environmental Chemical Engineering, 11*(5), 110681. https://doi.org/10.1016/j.jece.2023.110681

CHAPTER 55

Intelligent Transport Traffic System of Smart Cities using Machine Learning Techniques

Sidhartha Sankar Dora, Prasanta Kumar Swain

Department of Computer Application,
Maharaja Sriram Chandra BhanjaDeo University, Odisha, India
E-mails: lpurna@gmail.com, prasantanou@gmail.com

Abstract

The Intelligent Transport Traffic Prediction System (ITTAPS) is a sophisticated application of machine learning techniques designed to forecast traffic patterns and optimize traffic in transport systems. This system leverages historical traffic data, real-time information, and various machine-learning algorithms to predict traffic congestion, travel times, and optimal routes. Benefits of ITTAPS include improved traffic management, enhanced user experience, reduced environmental impact, and increased economic efficiency. Simulations show that traffic volume increases on working days and decreases on holidays. Among the classification algorithms used, gradient boosting provides the best accuracy results.

Keywords: Machine learning, Regression model, Intelligent transport, Dataset, Traffic flows

1. Introduction

The transportation network is an essential system that facilitates the movement of people, products, and services between locations. It includes various modes that serve distinct requirements and capacities locally or worldwide, including air, sea, rail, and road. Efficient transportation infrastructure is crucial for economic development, trade, and social connectivity. Traditional transport systems, while serving society for many years, have several drawbacks and limitations. Machine learning (ML) finds numerous applications in transport systems, revolutionizing the way transportation networks operate, optimize, and manage various aspects, as discussed by Alsheikh et al. (2014). Intelligent Traffic Transport Systems (ITTS) offer numerous advantages over traditional transport systems by leveraging advanced technologies such as artificial intelligence, data analytics, and communication networks, as mentioned by Quintero et al. (2018). We have designed a regression model to predict the number of vehicles at different times and under various weather conditions in smart cities using machine learning classification algorithms. The remainder of the paper is organized as follows: Section 2 discusses pertinent works, Section 3 explains the intended model and specifications, Section 4 provides details of the simulation work, Section 5 discusses the results, and Section 6 presents the conclusion and future scope.

DOI: 10.1201/9781003596776-55

2. Literature review

Technologies that guarantee road safety will be in high demand in future urban traffic environments. These technologies specifically need to monitor illicit activity, recognize traffic-related concerns, and avoid crashes. Quintero et al. (2018) contributed to the development and execution of a prototype for Intelligent Transportation Systems (ITS) using smart sensors. According to Larry et al. (2016), machine learning is a subfield of artificial intelligence impacted by biological learning algorithms, involving a multitude of intricate processes that comprise learning. The fundamental goal is to develop algorithms that can be trained with machine-readable data. We focus on embracing new machine learning-based statistical techniques, which will be essential for medical professionals in the twenty-first century, as narrated by Obermeyer and Emanuel (2016). According to Wang et al. (2017), Reynolds-averaged Navier-Stokes (RANS) equation-based numerical simulations of industrial flows heavily rely on turbulence modeling. Liu et al. (2017) state that interactive model analysis is essential for understanding, identifying, and improving machine learning models through interactive visualization, which helps users effectively address data mining and artificial intelligence problems in the real world. Mohsen et al. (2018) describe the term "Self-Organizing Network" (SON) as an autonomous cellular network designed to enhance operational tasks by configuring, optimizing, and healing itself. Machine learning techniques have demonstrated their utility in computer-aided diagnosis, particularly in the context of magnetic resonance imaging (MRI) glioma characterization, as proposed by Looze et al. (2018).

3. Methodology and model specifications

This system predicts future traffic volumes, allowing authorities to optimize traffic flow, plan infrastructure upgrades, and implement congestion mitigation strategies in smart cities effectively using machine learning.

Methodology

Regression models are more appropriate for predicting continuous variables, such as traffic flow volume, travel time, or congestion levels. Regression models can analyze historical traffic data along with various features such as time of day, weather conditions, road infrastructure, and special events to predict continuous outcomes. We have applied different classification algorithms for our proposed model.

Data set

This dataset comprises 48.1k (48,120) observations detailing the hourly count of vehicles across four distinct junctions, including Date, Time, Junction, Vehicles, and ID. The second weather dataset contains 31 features and 14593 instances. The data set is taken from data. open-power-system-data.org and Kagglewww.kaggle.com.

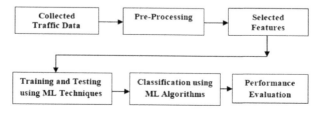

Figure 55.1: Proposed model.

Simulation environment

The simulation tasks were conducted using Python on a computer system equipped with an Intel Core i5 processor running at 2.00 GHz and 8 GB of RAM, operating on the Microsoft Windows 10 platform. For model training and testing, classification algorithms of ML are used. In machine learning, particularly in this context, the training and testing rules pertain to how data is partitioned and utilized during the model development and evaluation process. The performance of the trained regression model is assessed using appropriate evaluation metrics such as mean absolute error (MAE) and accuracy.

Result analysis

The results of the simulation conducted above are presented in terms of graphs and charts as follows.

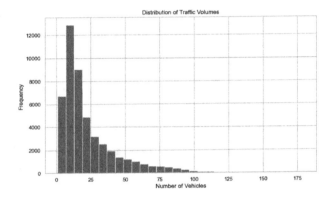

Figure 55.2: Traffic volume.

Figure 55.2 shows that the frequency of vehicles has increased over time. The maximum frequency is 13000, whereas the minimum frequency is 100. The trend of traffic increased in December and decreased in January, as shown in Figure 55.3. The minimum trend is 11, while the maximum trend is 31.

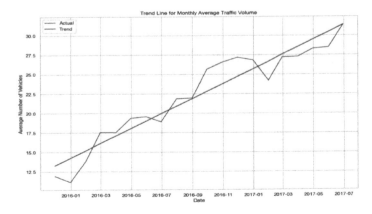

Figure 55.3: Trend of Traffic volume.

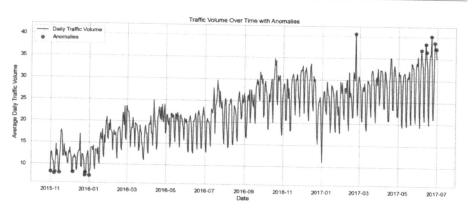

Figure 55.4: Traffic anomalies.

Table 55.1: Accuracy analysis of predicted traffic flow.

Classifier	Accuracy
Decision Tree	82.94
Random Forest	82.98
SVM	82.14
KNN	80.68
Logistic Regression	78.97
Gradient Boosting	83.08

Figure 55.4 illustrates how external variables affect traffic patterns by graphically showing trends and anomalies in traffic volume over time. In intelligent traffic flow prediction, accuracy analysis refers to the evaluation of how well a predictive model performs in forecasting traffic conditions compared to actual observed data. From Table 55.1, it is clear that gradient boosting provides the best result for the classification of traffic volume, while the rest of the algorithms provide different results. The simulation study of the model shows an MAE of 9.070056032057833.

4. Conclusion

The implementation of a regression model-based intelligent traffic management system demonstrates the effectiveness of data-driven approaches in optimizing urban mobility and alleviating traffic congestion in smart cities. Continuous monitoring, evaluation, and optimization of the regression model-based traffic management system are essential for adapting to changing traffic patterns, emerging technologies, and urban development trends. Integration with emerging technologies such as connected and autonomous vehicles holds promise for further enhancing urban mobility and safety.

References

[1] Abu Alsheikh, M., Lin, S., Niyato, D., & Tan, H. P. (2014). Machine learning in wireless sensor networks: algorithms, strategies, and applications. *IEEE Communications Surveys and Tutorials, 16*(4), 1996–2018.

[2] De Looze, C., Beausang, A., Cryan, J., Loftus, T., & Kearney, H. (2018). Machine learning: A useful radiological adjunct in the determination of a newly diagnosed glioma's grade and IDH status. *Journal of Neuro-Oncology, 139*(2), 1–9.

[3] Herrera-Quintero, L. F., Vega-Alfonso, J. C., Banes, K. B. A., & Zambrano, E. C. (2018). Smart ITS sensor for transportation planning based on IoT approaches using serverless and microservices architecture. *IEEE Intelligent Transportation Systems Magazine, 10*(2), 17–27. https://data.open-power-system-data.org/weather_data/ https://www.kaggle.com/datasets/fedesoriano/traffic-prediction-dataset

[4] Larry, D. J., Alevi, A. H., Gondomar, A. H., & Walker, A. L. (2016). Machine learning in geosciences and remote sensing. *Geosocial Front, 7*(1), 3–10.

[5] Liu, S., Wang, X., Liu, M., & Zhu, J. (2017). Towards better analysis of machine learning models: A visual analytics perspective. *Visual Informatics, 1*(1), 48–56.

[6] Mohsen, J., Garcia-Lozano, M., Geoponic, L., & Ruiz, S. (2018). Conflict resolution in mobile networks: A self-coordination framework based on non-dominated solutions and machine learning for data analytics. *IEEE Computational Intelligence Magazine, 13*(2), 52–64

[7] Obermeyer, Z., & Emanuel, E. J. (2016). Predicting the future—big data, machine learning, and clinical medicine. *New England Journal of Medicine, 375*(13), 1216–1219.

[8] Wang, J.-X., Wu, J.-L., & Xiao, H. (2017). Physics-informed machine learning approach for reconstructing Reynolds stress modeling discrepancies based on DNS data. *Physical Review Fluids, 2*(3), 034603

CHAPTER 56

Strength Performance of Lightweight Concrete Containing Steel Fibres and Silica Fume

Japthi Sravani[1*], P. Manoj Kumar[1], M. Srinivasula Reddy[2]

[1]Assistant Professor, G Pulla Reddy Engineering College (Autonomous),
Andhra Pradesh, India
[2]Principal Structural Engineer, Bangalore,
India Emails: jsravani85@gmail.com, putturumanojkumar@gmail.com,
srinivasreddy.iitr@gmail.com

Abstract

Lightweight aggregate concretes are as strong as regular-weight concrete and can be employed in structural applications. The utilization of this lightweight concrete principally decreases the structural dead loads or self-weight of a building. The current study's goal is to use lightweight expanded clay aggregate (LECA) to create lightweight concrete with medium strength. LECA is utilized in this study as a fine-weight, coarse-weight, and blended aggregate in concrete along with other aggregates of similar weight. Three mixes are made, one containing only lightweight aggregates and the other two combining lightweight aggregates with conventional weight aggregates. In addition to the above three mixes, further mixtures were made by substituting 10% of the silica fume with cement and 2% of the cement's weight with steel fibers. In-depth testing is done on concrete mixtures to assess their workability while still soft as well as their mechanical properties (compressive, split tensile, flexural strength, and water absorption).

Keywords: Lightweight aggregate, Lightweight expanded clay aggregate, Steel fibers, Silica fume.

1. Introduction

Concrete is among the most durable building materials. It is more fire-resistant and becomes stronger over time when compared to timber construction (Zhu et al., 2021). Different types of concrete are designed based on their purpose, such as compositions, finishes, and performance characteristics. These include various types of concrete such as self-compacting, high-strength, high-performance, pervious, glass, asphalt, rapid-strength, cellular, polymer, geopolymer or green, lightweight, and ultra-lightweight concrete. Lightweight concrete was first invented in 1923, primarily for use as an insulating material, and has a remarkably lengthy history. Concrete constructed with lightweight aggregates often weighs significantly less than gravel or crushed stone aggregates (Kim & Sadowski, 2022). The American Concrete Institute (ACI) specifies that structural

DOI: 10.1201/9781003596776-56

lightweight aggregate concretes must weigh no more than 1920 kg/m³ and have a 28-day compressive strength of at least 17 MPa (ACI 213R-03). Lightweight concrete is primarily used to lower the dead load of concrete structures, which enables structural designers to minimize the size of the column, footing, and other load-bearing components. The strength of a structural lightweight concrete mixture can be designed to be comparable to that of heavy concrete. For structural elements, lightweight structural concrete offers a better strength-to-weight ratio. When there are structural considerations, lightweight concrete is used. LECA or expanded clay and steel fibers are used as a replacement for coarse-weight aggregate, a fine-weight aggregate, and in combination with other aggregates of similar weight.

2. Literature Review

The study provides an overview of significant studies and findings regarding the application of LWA and St. F as extra substitutes in concrete. Nadesan and Dinakar (2017) found that by following the recommended procedure, SLWACs can retain air-dry densities below 2000 kg/m³ and attain compressive strengths of 28–70 MPa without the need for mineral additives. For sintered F.A. LWAC, a new strength-to-w/c connection was found. According to ASTM C 330, these concretes are suitable for structural applications and exhibit excellent ionic resistivity. Rumsys et al. (2017) found that specimens with silica fume had lower density due to the higher W/S ratio and increased viscosity, which prevents trapped air from escaping during mixing. Concrete samples with expanded clay aggregate have lower densities compared to those with LDPE crumbs. Yehia et al. (2016) investigated FRSCC's mechanical and durability qualities, finding that fibers had minimal impact on compressive strength and elasticity modulus. However, the flexural and splitting tensile strengths were 7%–26% and 12%–79% higher, respectively, than the control mix, indicating fibers' effectiveness in preventing macrocrack propagation and containing microcracks. According to Vakhshouri and Nejadi (2016), normal-weight coarse aggregate makes up about 30% of mixes, with a maximum weight ratio of 38.6%. The range of the cement and mineral powder ratios is 9.44% to 42.53% of the mix weight, and the water-to-cement ratio is 0.25 to 0.5. Using LWSCC in construction may pose additional challenges. However, using LWSCC in actual construction projects could lead to extra issues that need to be resolved.

3. Experimental Program

The following materials were used and tested according to the standards for the experimental program. OPC 53-grade cement was used in the present work with a specific gravity (S.G) of 3.14. Coarse aggregate used was from crushed aggregate along with lightweight coarse aggregate (4–8 mm size) and fine aggregate used was natural river sand along with lightweight fine aggregate (0–4.75 mm) from crushed expanded clay. Table 1 provides the various test results of aggregates employed for the mix. In addition to Portland cement, silica fumes with an S.G of 2.63 and having a 0.76 gm/cc pack density are used. Also, steel fibers with a double hooked type whose diameter is 0.65 mm, length = 30 mm, aspect ratio = 46, and density = 3900 kg/m³ are used. Chemical additives (SP and VMA) and clean water are used as the curing medium. Superplasticizers, specifically polycarboxylic ether-based Glenium B-233, are crucial for reducing water in high-strength concretes.

Table 56.1: Physical properties of aggregates.

Test on aggregates	C.A		F. A	
	Local	LECA	Sand	Crushed LECA
Specific gravity	2.808	1.36	2.63	1.29
Compacted bulk density (kg/m³)	1536	612.33	1674	668
Crushing test	13.83%	2.26%	-	-
Impact test	11.29%	-	-	-
Fineness modulus	-	-	2.86	2.6
Water absorption	-	-	1.2%	15.33%

4. Experimental Procedure

4.1. Mixture ratio and specimen preparation

The mix design adheres to ACI 221.2-98 standards for structural lightweight concrete, proposing different mixtures including all-lightweight and combined aggregates. Materials used are cement, expanded clay, crushed expanded clay, local aggregates, water, superplasticizer, silica fume, and steel fibers. Expanded clay is presoaked for 24 h to prevent water absorption during mixing, ensuring proper mix consistency. If not saturated, the aggregate absorbs mixing water, causing improper mix and delayed C–S–H gel formation, reducing strength. Superplasticizers aid in a 20% reduction in water content without compromising workability. Before being evaluated dry, specimens are water-cured for 28 days after being air-cured for 24 h.

Table 56.2: Proportions of compounds used in concrete.

Mix	Cement	LWCA	LWFA	NWFA	NWCA	S. P	Water	Silica Fume	Steel Fibers
Mix-1	766	249	406.71	0	0	7.7	113	0	0
Mix-2	689	249	406.71	0	0	7.7	113	76.53	15.30
Mix-3	718	329	0	736.52	0	7.2	137	0	0
Mix-4	646	329	0	736.52	0	7.2	137	71.78	14.35
Mix-5	718	0	443.82	0	549.49	7.2	115	0	0
Mix-6	646	0	443.82	0	549.49	7.2	115	71.78	14.35

4.2. Methodology

Tests for the mechanical properties of concrete using varied specimen sizes, all cured over 28 days. Each test uses specific dimensions: 100 × 100 × 100 mm for compressive, 100 × 150 mm for tensile, and 100 × 100 × 500 mm for flexural strength. Durability tests, including water absorption to measure the rate and percentage of voids in hardened concrete, and the Rapid Chloride Permeability Test (RCPT) to assess electrical conductance and chloride ion penetration resistance, were conducted.

5. Results and Discussions

5.1. Compressive strength

Following an appropriate 28-day curing time, the specimens' compressive strength was assessed. Several aggregate combinations, such as a blend of lightweight and regular aggregates and all-lightweight aggregates, were used to create these specimens. To improve the qualities, 2% steel fibers and 10% silica fume were also added. The values for each mix's compressive strength were then computed, revealing information on the efficiency of the additives and aggregate combinations. As seen in the graph, the blend of LWFA and NWCA with 2% steel fibers and 10% silica fume had the maximum compressive strength, measuring 68 MPa. On the other hand, the mix that contained just lightweight aggregates had the lowest strength, at 39 MPa.

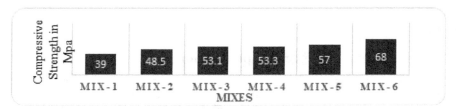

Figure 56.1: 28 days compressive strength for mix 1 to 6.

Split tensile strength

Comparing mixes M-1, M-3, and M-5 to M-2, M-4, and M-6, respectively, revealed lower split tensile values. The fact that mixes M-2, M-4, and M-6 have 2% more steel fibers in their cement weight can help to explain this disparity.

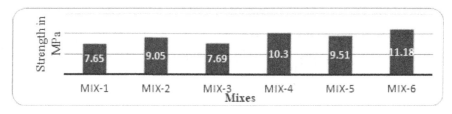

Figure 56.2: 28 days split tensile strength for mix 1 to 6.

Flexural strength

Mixes M-1 and M-3 displayed the lowest values in comparison to other blends, whereas mixes M-2, M-4, and M-6 displayed the highest values due to the presence of steel fibers.

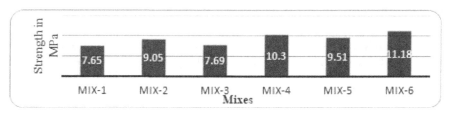

Figure 56.3: 28 days flexural strength for mix 1 to 6.

6. Conclusion

The current study achieves HSLW concrete. The investigation's findings lead to the following conclusions:

1. By adding 2% steel fibers, 10% silica fume, and LWFA and NWCA to the weight of cement, Mix-6 with the highest strength across all mechanical aspects is produced.
2. When steel fibers and silica fume admixture are added, compressive strength values between mixes M-1 and M-2 increase by the largest proportion (19.5%). The mixes M-3 and M-4 exhibit the greatest percentage differences in strength values (34% and 25%, respectively) for split tensile and flexural strengths. This is the result of mixing in silica fume additive with steel fibers.

References

[1] American Concrete Institute. (2003). *Guide for Structural Lightweight Aggregate Concrete (ACI 213R-03)*. ACI.

[2] Kim, J., & Sadowski, Ł. (2022). Properties of recycled aggregate concrete designed with equivalent mortar volume method. In P. O. Awoyera, C. Thomas, & M. S. Kirgiz (Eds.), *The Structural Integrity of Recycled Aggregate Concrete Produced with Fillers and Pozzolans* (pp. 365–381). Woodhead Publishing. https://doi.org/10.1016/B978-0-12-824105-9.00018-4

[3] Nadesan, M. S., & Dinakar, P. (2017). Mix design and properties of fly ash waste lightweight aggregates in structural lightweight concrete. *Case Studies in Construction Materials, 7,* 336–347. https://doi.org/10.1016/j.cscm.2017.09.005

[4] Rumšys, D., Bacinskas, D., Spudulis, E., & Meskenas, A. (2017). Comparison of material properties of lightweight concrete with recycled polyethylene and expanded clay aggregates. *Procedia Engineering, 172,* 937–944.

[5] Vakhshouri, B., & Nejadi, S. (2016). Self-compacting lightweight concrete: Mix design and proportions. *Structural Engineering and Mechanics, 58,* 143–161. https://doi.org/10.12989/sem.2016.58.1.143

[6] Yehia, S., Douba, A., Abdullahi, O., & Farrag, S. (2016). Mechanical and durability evaluation of fiber-reinforced self-compacting concrete. *Construction and Building Materials, 121,* 120–133. https://doi.org/10.1016/j.conbuildmat.2016.05.127

[7] Zhu, Y., Hussein, H., Kumar, A., & Chen, G. (2021). A review: Material and structural properties of UHPC at elevated temperatures or fire conditions. *Cement and Concrete Composites, 123,* 104212. https://doi.org/10.1016/j.cemconcomp.2021.104212

CHAPTER 57

M30 Grade Concrete's Strength Characteristics with a Partial Cement Replacement Using Dolomite and GGBS

[1]P. Manoj Kumar*, [1]Japthi Sravani, [3]M. Srinivasula Reddy,
[1]K. V. S. Gopala Krishna Sastry

[1]G Pulla Reddy Engineering College (Autonomous), Kurnool, Andhra Pradesh, India
[2]Principal Structural Engineer, Bangalore, India
E-mail: [1]putturumanojkumar@gmail.com, [2]jsravani85@gmail.com,
[3]srinivasreddy.iitr@gmail.com, [4]kodursastry@gmail.com

Abstract

Concrete's superior compressive strength and stability make it the most popular building material in civil engineering. To reduce the harmful carbon dioxide emissions associated with the production of cement, the concrete industry is searching for new cementitious materials or industrial waste. One such substance is dolomite powder, which is made by grinding the mineral dolostone. This study investigates the practical substitution of dolomite powder (DP) for some of the cement in the manufacturing of concrete. The mechanical characteristics of M30 grade concrete are examined in ratios of 5%, 10%, 15%, 20%, and 25% dolomite powder instead of cement. The optimum percentage of dolomite for maximum strength is identified. Ground granulated blast furnace slag (GGBS) substitution of extra cement in proportions of 5%, 10%, 15%, and 20% are investigated while maintaining this ideal percentage constant. According to test results, a mixture including 10% dolomite and 10% GGBS yields the highest strength. The behavior of M30 concrete with dolomite powder and granulated blast furnace slag in place of some of the cement is the main topic of this study.

Keywords: DP, GGBS, Strength, M30 Grade Concrete

1. Introduction

Concrete, used globally due to its strength, low cost, and versatility, contributes to 7% of global CO_2 emissions (Ali et al., 2011). Partial replacement of Portland cement with additives can reduce emissions, save energy, and enhance durability (Adesina, 2020). Supplementary cementitious materials improve concrete workability, durability, and strength in both fresh and hardened states (Bradley & Wilson, 2005). These materials, like fly ash, GGBS, dolomite, and silica fume, reduce cracking and heat generation, allow mixture customization, utilize industrial byproducts, and lower Portland cement consumption, thus conserving energy and reducing emissions. The study provides an overview of significant studies and findings from

DOI: 10.1201/9781003596776-57

the literature regarding using GGBS and dolomite as partial cement substitutes in concrete. Safan et al. (2012) evaluated self-compacting concrete's bond strength with dolomite powder, enhanced by silica fume or fly ash. Seven mixes were tested, showing increased bond strength with higher dolomite powder replacement of Portland cement, providing benefits for rapid construction. The study concluded that using dolomite powder improved the shear strength of reinforced concrete beams compared to conventional SCC. Barbhuiya (2011) investigated the impact of fly ash and dolomite powder on self-compacting concrete properties, aiming to substitute limestone powder with dolomite powder. Five different concrete mixes with different ratios of fly ash and dolomite powder were cast; the mix combining fly ash and dolomite powder in a 3:1 ratio satisfied the specifications set out by EFNARC for the construction of SCC. Mikhailova et al. (2013) studied how adding dolomite limestone powder affects concrete compressive strength. Their findings suggest that replacing 25% of cement with dolomite powder enhances compressive strength, indicating the potential for Portland dolomite limestone cement production.

2. Materials, Methods, and Testing

The materials employed to develop the medium-strength concrete are OPC cement, natural fine aggregate (FA), coarse aggregate (CA), dolomite powder (DP), GGBS, and water. Table 57.1 displays the physical characteristics of various materials. In this investigation, M30 grade concrete was utilized following the guidelines of IS: 10262-2009 and IS: 456-2000 for mix design. Cement replacement with dolomite powder ranged from 0% to 25% by weight, denoted by Mix A to Mix F. The optimum replacement percentage for dolomite powder was determined, and further cement replacement with GGBS ranged from 5% to 25%, denoted by Mix G to Mix K. Mix design proportions are detailed in Table 57.2.

Table 57.1: Material properties in their physical form.

Tests on materials	Cement	C.A.	F. A.	Dolomite	GGBS
Specific Gravity(G)	3.12	2.85	2.66	2.85	2.86
Compacted Bulk density(kg/m³)	-	1560	1602	-	-
Fineness Modulus	4%	4.6	3.22	-	-

Table 57.2: Mix proportions of M30 Grade (Mix A) Concrete (control mix in kg/m³).

Material	Cement	F. A.	C. A.	Water
Quantity	425.5	663.33	1210	191.5
Ratio of ingredients	1	1.55	2.84	0.45

3. Tests on Fresh Concrete

Test for Compaction Factor and Slump Cone: The slump test, commonly used to measure concrete consistency, is unsuitable for very wet or dry concrete and does not account for all workability factors. Despite its limitations, it conveniently indicates batch-to-batch uniformity. The compacting factor test measures the workability of fresh concrete by comparing the weights of partially and fully compacted concrete, as per IS: 1199–1959. Table 3 displays the results of the tests for the slump and compaction factors.

Table 57.3: Slump and compaction factor values for mixes A to K.

Designation	Mix Proportions	Slump (mm)	Compaction factor
Mix A	Control mix	70	0.89
Mix B	5% D.P.	72	0.91
Mix C	10% D.P.	73	0.90
Mix D	15% D.P.	75	0.90
Mix E	20% D.P.	77	0.92
Mix F	25% D.P.	80	0.91
Mix G	10% D.P. + 5% GGBS	72	0.89
Mix H	10% D.P. + 10% GGBS	74	0.90
Mix I	10% D.P. + 15% GGBS	75	0.92
Mix J	10% D.P. + 20% GGBS	74	0.91
Mix K	10% D.P. + 25% GGBS	76	0.91

4. Tests on hardened concrete

Compressive Strength: Compressive strength was determined by dividing the failure load by the specimen area using a 3000 kN CTM on 150 mm concrete cubes and applying 140 kg/sq. cm/min until failure, per IS 516 (Part 1/Sec1): 2021.

Split Tensile Strength: The split tensile test, per IS 516 (Part 1/Sec1): 2021, used a 200-ton UTM on 150 mm × 300 mm concrete cylinders, applying 140 kg/sq. cm/min until failure. Specimens were placed horizontally between plates with plywood strips, and the tensile strength was calculated from the failure load as per IS 5816-1999.

Flexural Tensile Strength: For the flexure test, specimens were simply supported on two rollers over a 50 cm span, with loading applied via a 15-ton precalibrated proving ring. Two 16 mm rods were placed 13.33 cm from each support, and as the load increased, cracks widened until collapse. The ultimate load was recorded, and flexural strength was calculated from these results.

5. Results and Discussion

The study evaluated the strengths of concrete mixes using dolomite and dolomite with GGBS as cement replacements, with results detailed in tables and graphs.

Dolomite as a Supplement to Cement in Concrete

Phase 1 involved the preparation of M30-grade concrete, with dolomite filling in for some of the cement. The strengths of the several dolomite mixtures and the reference Mix A are shown in Figures 57.1 – 57.3.

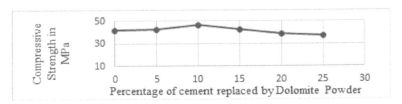

Figure 57.1: 28-day compressive strength for mixes A to F.

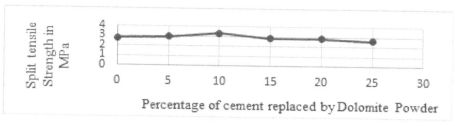

Figure 57.2: 28 days strength of split tensile for mixes A through F.

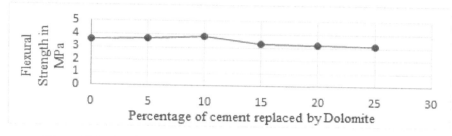

Figure 57.3: 28 days strength of flexural tensile for mixes A through F.

From the above figures, it is concluded that the mechanical properties of concrete show improvement with up to a 10% partial replacement of cement with dolomite. This enhancement could potentially be attributed to the finer particles present in dolomite powder compared to cement.

Mixture of Dolomite and GGBS for Concrete with Partial Replacement of Cement:

Phase 2 experimented with M30 grade concrete, maintaining a constant dolomite percentage while varying GGBS from 5% to 25% as a cement replacement, and the resulting strength values are presented in Table 57.4.

Table 57.4: Strength data for dolomite and GGBS combined as a cement substitute (MPa)

Mix Designation	Replacement Percentage of Materials	Compressive Strength	Split Tensile Strength	Flexural Strength
Mix G	10%DP+5%GGBS	47.23	3.37	3.90
Mix H	10%DP+10%GGBS	48.34	3.67	4.16
Mix I	10%DP+15%GGBS	44.45	3.92	4.34
Mix J	10%DP+20%GGBS	38.62	3.43	3.52
Mix K	10%DP+25%GGBS	35.48	3.14	3.26

The findings show that adding dolomite and GGBS in place of 10% and 15% of cement increases the concrete's compressive strength. This enhancement is likely due to the finer particles in dolomite powder and the higher SiO_2 content in GGBS compared to cement.

6. Conclusion

Conclusions from tests and experiments comparing normal concrete to blended mixes (with dolomite powder and GGBS) are outlined:

1. To partially replace cement in the preparation of concrete, dolomite powder, and GGBS can be utilized as pozzolanic materials up to 10%. Strengthening after 28 days could be achieved by combining GGBS with dolomite powder.
2. Dolomite and GGBS-containing concrete mixtures have workability qualities that are similar to those of regular, conventional concrete.
3. The addition of GGBS and dolomite to concrete can somewhat increase its strength properties, resulting in a combination that is less expensive than ordinary concrete.
4. Experimental studies show that adding dolomite powder and GGBS to concrete preparation significantly increases the material's strength properties.

References

[1]   Adesina, A. (2020). Recent advances in the concrete industry to reduce its carbon dioxide emissions. *Environmental Challenges, 1,* 100004. https://doi.org/10.1016/j.envc.2020.100004

[2]   Ali, M. B., Saidur, R., & Hossain, M. (2011). A review on emission analysis in cement industries. *Renewable and Sustainable Energy Reviews, 15,* 2252–2261.

[3]   Barbhuiya, S. (2011). Effects of fly ash and dolomite powder on the properties of self-compacting concrete. *Construction and Building Materials, 25,* 3301–3305. https://doi.org/10.1016/j.conbuildmat.2011.03.018

[4]   Bradley, B., & Wilson, M. L. (2005). Using supplementary cementitious materials. *58,* 34–41.

[5]   Mikhailova, O., Yakovlev, G., Maeva, I., & Senkov, S. (2013). Effect of dolomite limestone powder on the compressive strength of concrete. *Procedia Engineering, 57,* 775–780. https://doi.org/10.1016/j.proeng.2013.04.098

[6]   Safan, M. A., Kamal, M. M., & Al-Gazzar, M. A. (2012). Experimental evaluation of steel–concrete bond strength in low-cost self-compacting concrete. *Concrete Research Letters, 3,* 137438078.

[7]   IS 516 (Part 1/Sec1): 2021, "Hardened Concrete Methods of Test" Bureau of Indian Standards. New Delhi 2021.

[8]   IS 5816:1999, "Splitting Tensile Strength of Concrete - Method of Test" Bureau of Indian Standards. New Delhi 1999.

CHAPTER 58

Reclamation of Low-Lying Area with Legacy Ash

Jagma Priyadarshini[1], Aditya Kumar Bhoi[2], Kasmika Bag[3]

[1]Under Graduate Student, Civil Engineering Department, Indira Gandhi Institute of Technology Sarang, Odisha, India, E-mail: jagma2002@gmail.com [2]Assistant Professor, Civil Engineering Department, Indira Gandhi Institute of Technology Sarang, Odisha, India, E-mail: adityabhoi@igitsarang.ac.in [3]Post Graduate Student, Civil Engineering Department, Indira Gandhi Institute of Technology Sarang, Odisha, India, E-mail: kasmika1999@gmail.com

Abstract

Around 1734 million tonnes of legacy ash is available across India. It can serve as a filling material for the reclamation of low-lying areas in potential civil engineering projects. On the other hand, there is a major lack of comprehensive evaluation of legacy ash as a substitute filling material. The purpose of this study is to determine whether legacy ash can be used as a substitute in place of natural material. To facilitate this objective, several geotechnical examinations were conducted. Based on the findings of the geotechnical tests, it may be inferred that legacy ash is similar in quality to conventional fill material.

Keywords: legacy ash, filling material, geotechnical characterization

1. Introduction

India mainly depends on coal-based thermal power plants for electricity generation, which leads to the generation of 270 million tonnes of fly ash yearly. Recent policies target 100 percent utilization of fly ash; however, it is not yet possible to utilize the total ash generated at power plants. As a result, about 1734 million tonnes of legacy ash is available at ash ponds of 200 thermal power plants spread across India (CEA, 2022). The accumulation of excess ash in ash ponds sometimes leads to the breaching of embankments. Hence, some scholars advocate for the use of coal ash along with other combustion residues as a filler whenever possible, including the reclamation of low-lying land (Bhoi et al., 2020, 2021, 2023; Kumar & Mandal, 2019). The literature study reveals that many investigations were carried out on the application of fly ash to serve as filling material. Additionally, the FARC (2017) and Indian Roads Congress guidelines (SP: 58 and SP: 102) endorse the substitution of natural fill material with fly ash. Additionally, certain research indicates that legacy ash can be utilized as a filler material in embankments. However, the existing literature does not contain any studies conducted to study the appropriateness of using legacy ash as a fill material for reclaiming low-lying areas. Given the intended use of legacy ash as infill material for reclaiming low-lying areas, it is crucial to assess its particle size distribution (fraction of materials), plasticity, angle of internal friction, cohesion, hydraulic conductivity, along with compaction characteristics.

DOI: 10.1201/9781003596776-58

2. Methodology

Legacy ash was obtained from the National Thermal Power Corporation, Kaniha, Odisha, India. The material was in a dry state at the time of collection but contained some debris, which was removed before the assessment of geotechnical characteristics. The prospective material that will be utilized as fill material in low-lying areas is required to meet certain standards that are established in some internationally renowned standards (FARC, 2017, and Indian Roads Congress guidelines (SP: 58 and SP: 102)). According to the information provided by the Indian Roads Congress guidelines (SP: 102), the evaluation of particle size distribution and plasticity are the utmost essential factors, followed by hydraulic conductivity, the maximum density of probable filling materials, apparent cohesion, and internal friction angle.

ASTM D854 (2014) standard, specifically method A, was used to measure the specific gravity of legacy ash. The grain size distribution was determined in line with the ASTM D7928 (2017) and ASTM D6913 (2017) standards, respectively. The plastic limit tests were conducted per the ASTM D4318 (2017) standard. The cone penetration method was conducted to determine the liquid limit according to Indian Standard 2720 (Part 5). The optimum moisture content and maximum dry density were assessed using modified effort (ASTM D1557). IS 2720 (Part XL) was used to determine the free swelling index of legacy ash. The consolidation parameters were assessed using the ASTM D2435 (2011) standard. The coefficient of hydraulic conductivity was determined by ASTM D5856 (2015). The apparent cohesion and internal friction angle were assessed as specified by IS 2720 (13) (1986). To ascertain the acidic and alkaline characteristics, pH tests were conducted according to ASTM D4972 (2013) standard.

3. Results and discussion

The legacy ash was determined to have a specific gravity of 2.21. The fly ash should have a specific gravity between 1.6 and 2.6, according to FARC (2017), and the tested legacy ash satisfies this criterion.

Grain size analysis of legacy fly ash and particle size distributions recommended by FARC (2017) is presented as materials fraction in Table 58.1. The legacy fly ash was determined to have a coefficient of uniformity of 1.8, indicating that it had a uniform grade. Along the same lines, the coefficient of curvature was precisely 0.94. It was classified as sandy silt according to the Indian standard.

Table 58.1: Materials fraction of legacy ash compared with fly ash.

Particle size	Legacy ash (%)	Fly ash (IRC: SP: 58, 2001) (%)
Clay size	7.5	1–10
Silt size	47.01	8–85
Sand size	43.89	7–90
Gravel size	1.5	0–10
Coefficient of uniformity	1.8	3.1–10.7

The legacy ash started bleeding during the liquid limit test. At the same time, the thread required for the plastic limit test could not be made. As a result, it was classified as a nonplastic soil. FARC (2017) suggests the use of nonplastic soil to reclaim low-lying land; hence, the proposed material can be used for this purpose.

The compaction curve of legacy ash is shown in Figure 58.1. The optimum moisture content was determined to have a value of 23.5%. FARC (2017) recommends an optimum moisture content of fly ash of 18 to 40 to reclaim low-lying areas. The maximum dry density was determined to have a value of 1.1 g/cc. FARC (2017) recommends a maximum dry density of 0.9 to 1.6 g/cc to reclaim low-lying areas. The tested legacy ash satisfies this criterion; thus, it may be employed to reclaim low-lying areas.

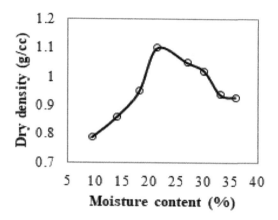

Figure 58.1: Compaction curve.

The legacy ash did not swell during the test. FARC (2017) recommends the usage of fly ash with a very low free swell index to reclaim low-lying land, and the legacy ash satisfies this criterion; hence, it can be used for this purpose.

The compression index and consolidation coefficient of legacy ash were found to be 0.026 and 3.1×10^{-4} cm²/s, respectively, which is in the suggested range for fly ash to reclaim low-lying areas (FARC, 2017); hence, legacy ash can be used for this purpose.

The coefficient of hydraulic conductivity was 3.1×10^{-5} cm/s, which falls in the range of 8×10^{-6} to 7×10^{-4} cm/s, as recommended by FARC (2017).Legacy ash had an internal friction angle of 33°. At the same time, the apparent cohesion of legacy ash was found to be 1.65 kN/m². According to FARC (2017), the suggested internal friction angle value is 28° to 42°, with minimal cohesion. Legacy ash may be considered for use as a filler material to reclaim low-lying regions based on shear strength norms.The legacy ash was alkaline, with a pH value of 8.1. Indian Roads Congress guidelines (SP: 58) recommend the usage of ash with a pH range from 5 to 10.

4. Conclusion

Laboratory experiments were undertaken on the legacy ash to determine its appropriateness as a filler material for low-lying land reclamation projects. The study focused on analyzing the key geotechnical features that influence the choice of infill material, including grain size distribution, plasticity, permeability, compaction behavior, and shear strength parameters. The findings of this study indicate that legacy ash meets the requirements set by FARC (2017) and Indian Roads Congress guidelines (Special Publication: 58 and 102) for using it as filling material to reclaim low-lying land.

References

[1] ASTM. (2011). Standard test methods for one-dimensional consolidation properties of soils using incremental loading. *ASTM D2435*, 1–15.

[2] ASTM. (2012). Standard test methods for laboratory compaction characteristics of soil using modified effort (56,000ft-lbf/ft3 (2,700 kN-m/m3)). *ASTM D1557*, 1–14.

[3] ASTM. (2013). Standard test method for pH of soils. *ASTM D4972*, 1–4.

[4] ASTM. (2014). Standard test methods for specific gravity of soil solids by water pycnometer. *ASTM D854*, 1–8.

[5] ASTM. (2015). Standard test method for measurement of hydraulic conductivity of porous material using a rigid-wall, compaction-mold permeameter. *ASTM D5856*, 1–9.

[6] ASTM. (2017). Standard test method for particle-size distribution (gradation) of fine-grained soils using the sedimentation (hydrometer) analysis. *ASTM D7928*, 1–25.

[7] ASTM. (2017). Standard test methods for particle-size distribution (gradation) of soils using sieve analysis. *ASTM D6913*, 1–34.

[8] ASTM. (2017). Standard test methods for liquid limit, plastic limit, and plasticity index of soils. *ASTM D4318*, 1–20.

[9] Bhoi, A. K., Mandal, J. N., & Juneja, A. (2020). Feasibility study of bagasse ash as a filling material. In L. Hoyos & H. Shehata (Eds.), *Advancements in Unsaturated Soil Mechanics. GeoMEast 2019. Sustainable Civil Infrastructures*. Springer, Cham. Bhoi, A. K., Mandal, J. N., & Juneja, A. (2021). Interface shear strengths between bagasse ash and geogrid. In S. K. Ghosh, S. K. Ghosh, B. G. Mohapatra, & R. L. Mersky (Eds.), *Circular Economy in the Construction Industry* (1st ed.). CRC Press. Bhoi, A. K., Juneja, A., & Mandal, J. N. (2023). Sugar factory ash as retaining wall backfill: A techno-economic trial. *Journal of Cleaner Production, 385*, 1–18.CEA. (2022). Report on fly ash generation at coal / lignite-based thermal power stations and its utilization in the country for the year 2021–22. *Central Electricity Authority, Ministry of Power, Government of India.*

[10] FARC. (2017). *Guidelines for reclamation of low-lying areas and abandoned quarries with ash*. Fly Ash Resource Centre. State Pollution Control Board, Odisha.

[11] IRC: SP: 58. (2001). *Guidelines for use of fly ash in road embankments*. New Delhi: Indian Roads Congress.

[12] IRC: SP: 102. (2014). *Guidelines for design and construction of reinforced soil walls*. New Delhi: Indian Roads Congress.

[13] IS: 2720 (Part 5) (1985). *Determination of liquid and plastic limit*. New Delhi: Indian Standards Institute.

[14] IS 2720 (Part XL) (1977). *Determination of free swell index of soils*. New Delhi: Bureau of Indian Standards.

[15] IS 2720 (13) (1986). *Direct shear test*. New Delhi: Bureau of Indian Standards.

[16] Kumar, A., & Mandal, J. N. (2019). Parametric studies on two-tiered model fly ash wall. *International Journal of Geotechnical Engineering, 16*(7), 815–825.

CHAPTER 59

Effectiveness of Nano-Titanium Dioxide on Abrasion Resistance, Fatigue Response, and Intrinsic Crack Repair Mechanisms in Cementitious Composites

A Review

*[a]Ashmita Mohanty, [b]Dr Dipti Ranjan Biswal

*[a]PhD scholar, Kalinga Institute of Industrial Technology, Bhubaneswar, India
E-mail: 2281065@kiit.ac.in
[b]Associate Professor, Kalinga Institute of Industrial Technology, Bhubaneswar, India
E-mail: dipti.biswalfce@kiit.ac.in

Abstract

This review examines the impacts of nano titanium dioxide (TiO_2) on cementitious composites, particularly focusing on abrasion resistance, self-healing, and fatigue behavior. Results show that nano TiO_2 significantly enhances abrasion resistance (up to 62.7% reduction in mass loss), improves fatigue life (up to 475.38% increase), and promotes self-healing through calcium carbonate crystal formation. These improvements stem from the ability of TiO_2 to accelerate cement hydration and create a more compact microstructure. The results highlight how nano TiO_2 may be used as an addition in transportation engineering to improve the sustainability and durability of concrete.

Keywords: Nano TiO_2, Abrasion Resistance, Crack-healing properties, Fatigue performance

1. Introduction

Cementitious composites, such as concrete, are essential for modern infrastructure but face challenges like mechanical wear, fatigue failure, and cracking, which reduce durability and increase maintenance costs. Nanotechnology offers innovative solutions, with nano titanium dioxide (TiO_2) emerging as a promising additive. Nano TiO_2 particles enhance mechanical strength, improving abrasion resistance against vehicular traffic and harsh environmental conditions (Florean et al., 2024). Its photocatalytic properties enable self-cleaning by breaking down organic pollutants on pavement surfaces, reducing maintenance and environmental impact. Additionally, nano TiO_2 promotes self-healing in concrete, mitigating cracks and prolonging pavement lifespan (Ghomi et al., 2023). This represents a significant advancement in enhancing the sustainability, resilience, and longevity of concrete pavements, leading to safer and more cost-effective transportation networks (Florean et al., 2024).

DOI: 10.1201/9781003596776-59

Properties of nano TiO₂ (adapted from Li et al., 2007; Jalal & Tahmasebi, 2015).

Item	Diameter (nm)	Specific surface area (m²/g)	Density (g/cm³)	Purity (%)	Phase
Nano TiO₂	1–25	240 ± 50	0.04–0.06	99.7	Anatase

2. Literature Review

Nano TiO₂ on abrasion resistance of cementitious composites

Partial replacement of cement with nano TiO₂ enhances abrasion resistance and durability due to its small size and high surface area, which improve the adhesion and cohesion of concrete. Figure 59.1 shows SEM images of control samples and samples with nano TiO₂. Figure 59.2 illustrates the enhanced abrasion resistance of concrete pavement with varying percentages of nano TiO₂. Florean et al. (2024) investigated the impact of nano TiO₂ on cementitious composites, finding substantial enhancements in abrasion resistance, with up to a 62.7% reduction in mass loss using 5% nano TiO₂. A more compact and refined microstructure is produced by adding 3% TiO₂ nanoparticles, which speed up cement hydration and cause the development of more calcium hydroxide (CH) crystals and calcium–silicate–hydrate (C–S–H) gels. Another research by Li et al. (2007) found the impact of nano TiO₂ particles on the abrasion resistance of concrete for pavements, finding notable enhancements. The surface index expanded by 180.7% with 1% nano TiO₂, 147.7% with 3%, and 90.4% with 5%. The side index showed improvements of 173.3%, 40.2%, and 86% for 1%, 3%, and 5% nano TiO₂, respectively. These improvements are attributed to the ability of nano TiO₂ to control crystallization during hydration, acting as nuclei to restrict the formation of crystals of calcium hydroxide (Ca(OH)₂). As a result, the cement matrix gets more compact and homogeneous, significantly boosting both abrasion resistance and overall strength.

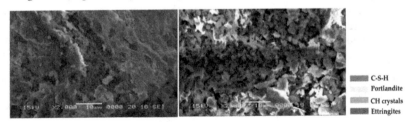

Figure 59.1: SEM image of control sample (a) and sample with 3% nano TiO₂ (Adapted from Florean et al., 2024).

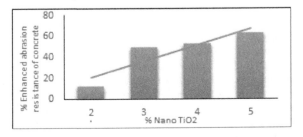

Figure 59.2: Abrasion resistance of concrete pavement w.r.t different percentages of nano TiO₂.

H2Nano TiO$_2$ on fatigue performance of cementitious composites

Fatigue failure in concrete occurs due to repeated loading, leading to eventual fracture through the accumulation of small cracks. Factors influencing fatigue life include concrete quality, load magnitude, frequency, and environmental conditions. Stress–strain fatigue models like the S–N curve predict concrete fatigue behavior. Adding 1% nano TiO$_2$ can enhance fatigue resistance, as analyzed using the S–N curve model (Figure 59.3). Jalal and Tahmasebi (2015) found that incorporating TiO$_2$ nanoparticles (1%–5%) into self-compacting concrete with nanomaterials (SCC-N) improved consistency and reduced bleeding and segregation. TiO$_2$ concentrations of up to 4% expedited the development of Ca–S–H gel by raising the percentage of crystalline calcium hydroxide (Ca(OH)$_2$) and improving the gel's resilience to flexural failure. Above 4% decreased flexural strength, however, since there is not enough crystalline Ca(OH)$_2$ for C–S–H gel formation. Li et al. (2007) investigated the impact of nanoparticles, such as nano TiO$_2$ and nano SiO$_2$, on the flexural fatigue performance of concrete. They found that concrete with 1% nano TiO$_2$ exhibited maximum flexural fatigue performance, with the greatest enhancement in regression parameter α and the smallest in β. Theoretical fatigue values increased at stress levels of 0.85 and 0.70 relative to plain concrete, by 475.38% and 267.22%, respectively, suggesting that increasing stress levels improved fatigue life. Concrete with 1% nano TiO$_2$ outperformed concrete with 1% nano SiO$_2$, 3% nano TiO$_2$, and combinations of nano TiO$_2$ with pp fiber, with conventional concrete showing the lowest fatigue life.

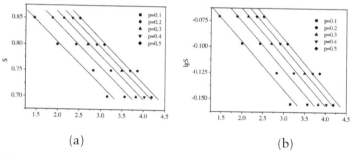

(a)   (b)

Figure 59.3: 1% Nano TiO$_2$ "on different failure probabilities (a) p–S–lgN curves; (b) p–lgS–lgN" curves (Adapted Li et al., 2007).

Nano TiO$_2$ on crack-healing properties of cementitious composites

Nano TiO$_2$) enhances the crack-healing properties of concrete by acting as nucleation sites for calcium carbonate crystals, facilitating the self-healing process of micro-cracks. This technology improves concrete durability and sustainability, reducing maintenance needs. Figure 59.4 shows a virgin fractured surface without nano TiO$_2$ and a healed refractured surface with nano TiO$_2$. Ravandi et al. (2024) investigated the impact of including 0.5 weight percent n-TiO$_2$ and chitosan on the crack-healing capabilities and mechanical properties of nanocomposites. The nanocomposite demonstrated enhanced fracture toughness and self-healing effectiveness (KIC-healed). The self-healing effectiveness varied from 58.46% to 62.31% during a period of 1–90 days following 3 months of water aging. Strong durability of the self-healing capabilities is suggested by the nanocomposite's self-healing efficiency, which remained steady at 59.84% even after 90 days of water aging.

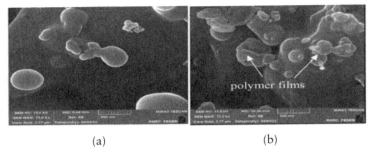

(a)   (b)

Figure 59.4: SEM image of fracture surface without nano TiO$_2$ (a) and healed and re-refractured surface with nano TiO$_2$ (b) (Adapted from Florean et al. 2024).

Choi et al. (2021) found that adding TiO$_2$ nanoparticles to epoxy nanocomposites (NCs) improved crack resistance and mechanical strength. The optimal mechanical properties were achieved with 1.0 wt.% TiO$_2$, while excessive nanoparticle content (>3.0 wt.%) decreased properties due to water interference during polymerization and nanoparticle agglomeration. TiO$_2$ nanotubes (TNT) were utilized by Vijayan and Al-Maadeed (2016) to encapsulate an amine curing agent and epoxy prepolymer for epoxy coatings on carbon steel. A 57% recovery of anticorrosive capabilities after 5 days was demonstrated by Electrochemical Impedance Spectroscopy (EIS), suggesting a self-healing capacity appropriate for long-term corrosion resistance. Xu et al. (2019) analyzed ECC (Engineered Cement Composites) with 0%–15% nano TiO$_2$ content. Increasing TiO$_2$ up to 10% enhanced peak strength and reduced crack opening. Important indicators such as complementary energy (J'_b) and crack tip toughness (J_{tip} = Km^2/ Em), where Em is the matrix Young's modulus and Km is a material constant, demonstrated robust tensile ductility. Regardless of the TiO2 content, ECC showed strong strain capacity (>3.5%); however, above 1% TiO$_2$, the initial cracking strength began to degrade. With increasing TiO$_2$ content, ultimate tensile strength and strain capacity first declined and subsequently increased.

3. Conclusion

In summary, this review underscores the transformative impact of nano TiO$_2$ on cementitious composites and nanocomposites. The integration of nano TiO$_2$ enhances abrasion resistance, extends fatigue life, and promotes self-healing properties through the facilitation of calcium carbonate crystal formation. These improvements are pivotal for advancing the durability and sustainability of construction materials. As nano TiO$_2$ continues to be explored and optimized, its application promises to redefine the performance standards of infrastructure materials, paving the way for more resilient and long-lasting built environments.

References

[1]   Choi, Y. M., Hwangbo, S. A., Lee, T. G., & Ham, Y. B. (2021). Effect of particle size on the mechanical properties of TiO2–epoxy nanocomposites. *Materials, 14*(11), 2866. https://doi.org/10.3390/ma14112866

[2]   Florean, C. T., Vermesan, H., Thalmaier, G., Neamtu, B. V., Gabor, T., Campian, C., ... & Csapai, A. (2024). The influence of TiO2 nanoparticles on the physico–mechanical and structural characteristics of cementitious materials. *Coatings, 14*(2), 218. https://doi.org/10.3390/coatings14020218

[3]   Ghomi, E. R., Khorasani, S. N., Koochaki, M. S., Dinari, M., Ataei, S., Enayati, M. H., ... & Neisiany, R. E. (2023). Synthesis of TiO2 nanogel composite for highly efficient self-healing epoxy coating. *Journal of Advanced Research, 43*, 137–146. https://doi.org/10.1016/j.jare.2022.02.008

[4]   Jalal, M., & Tahmasebi, M. (2015). Assessment of nano-TiO2 and class F fly ash effects on flexural fracture and microstructure of binary blended concrete. *Science and Engineering of Composite Materials, 22*(3), 263–270. https://doi.org/10.1515/secm-2013-0211

[5]   Li, H., Zhang, M. H., & Ou, J. P. (2007). Flexural fatigue performance of concrete containing nano-particles for pavement. *International Journal of Fatigue, 29*(7), 1292–1301. https://doi.org/10.1016/j.ijfatigue.2006.10.004

[6]   Ravandi, R., Heris, S. Z., Hemmati, S., Aghazadeh, M., Davaran, S., & Abdyazdani, N. (2024). Effects of chitosan and TiO2 nanoparticles on the antibacterial property and ability to self-healing of cracks and retrieve mechanical characteristics of dental composites. *Heliyon.* https://doi.org/10.1016/j.heliyon.2024.e27734

[7]   Vijayan, P. P., & Al-Maadeed, M. A. S. (2016). TiO2 nanotubes and mesoporous silica as containers in self-healing epoxy coatings. *Scientific Reports, 6*(1), 38812. https://doi.org/10.1038/srep38812

[8]   Xu, M., Bao, Y., Wu, K., Shi, H., Guo, X., & Li, V. C. (2019). Multiscale investigation of tensile properties of a TiO2-doped engineered cementitious composite. *Construction and Building Materials, 209*, 485–491. https://doi.org/10.1016/j.conbuildmat.2019.03.112

CHAPTER 60

Effect of Thermomechanical Processing on Mechanical Properties of Ferritic Stainless Steel

S. Behera*

Department of Metallurgical and Materials Engineering, Indira Gandhi Institute of Technology, Sarang, India
*Corresponding Author: (Swarnalata Behera)
Email-id: swarnalata.behera@igitsarang.ac.in

Abstract

The present study fine-tunes the processing route of ferritic stainless steel mainly designed for achieving superior mechanical properties such as yield strength, Vickers hardness, Charpy impact toughness, and nanohardness by thermomechanical heat treatment schedules. Initially, the sample was heated at a temperature of 1200°C, held for 1 h, hot forged at a temperature of 1000°C to a thickness reduction of up to 50%, and then air-cooled at room temperature. The effect of thermomechanical processing on microstructure improved mechanical properties as well as physical properties have been studied. This study contains a Cr content of 19.0148% with a very small amount of nickel, and generally, it varies from 10.5% to 27% chromium and is characterized by its body-centered cubic crystal structure. Ferritic stainless steel has a higher chromium content but a lower carbon content compared to martensitic stainless steel. Ferritic stainless steels have a higher percentage of chromium content, making them more resistant to corrosion. However, they are less durable compared to austenitic grades. This steel has superior engineering characteristics, such as enhanced ductility and formability, compared to austenitic alloys. It undergoes a regulated heat treatment process, including specific sequences of heating and cooling, to modify its physical and mechanical properties, thus ensuring its suitability for specific engineering applications. Ferritic stainless steels possess distinctive characteristics that render them valuable in both the automotive sector and nuclear reactors.

Keywords: Stainless steel, Annealing, Microstructures, Mechanical properties, Nanohardness

1. Introduction

Nowadays, ferritic steel is used in a large scale of applications due to its higher corrosion resistance, better engineering properties like ductility and formability, and the ability to achieve the physical and mechanical properties required to meet desired engineering applications, making them ideal for use in harsh environments. Ferritic stainless steels are nonhardenable with good mechanical properties and moderate corrosion properties, consisting of iron–chromium alloys and following standard 400-series alloys as well as

DOI: 10.1201/9781003596776-60

modified versions of these alloys containing 10 to 30% Cr and 0.08 to 0.20% C. The addition of chromium increases the hardenability and strength of alloy steel along with corrosion resistance. Moreover, heat treatment, particularly heat alone, can alter the strength, structure, and physical properties of the material. For particular materials, alloys, and applications, it depends upon the heat treatment process and temperature. Ferritic steels have a body-centered cubic crystal structure, and ferrite is the main phase. After the addition of chromium, iron, and carbon, modernized stainless steel also contains some ferritic stabilizers, such as titanium, nickel, and molybdenum. The addition of chromium, molybdenum, niobium, and nickel enhances its strength and hardness, making it stronger and increasing corrosion resistance. Where high stress and friction are necessary, stainless steel is used because it is highly resistant to wear and tear, making it suitable for particular applications. Ferritic steel is capable of withstanding high temperatures without losing its strength or other properties and is used in high-temperature applications. Ferritic stainless steel is commonly used in kitchenware, chemical power plants, nuclear reactors, and the automobile sector. The work aims to study the mechanical behavior and microstructure of ferritic stainless steel before and after the thermomechanical process.

2. Literature Review

In 2009, Mohd Fahmi and Abdullah Sani studied the effect of the heat treatment process, annealing, and water quenching of stainless steel, particularly for watch manufacturing applications [1]. From some literature, it was found that theoretical and experimental work of the material with the heat treatment process of annealing temperature and time can statistically measure mechanical properties and conduct fracture analysis. The final microstructure and mechanical properties must predict the relationship with the heat treatment process. In June 2002, Kalyon et al. studied that stainless steel was formed by the combination of ship steel with stainless steel through welding [2]. Determination of the impact toughness of ship steel/stainless steel is essential; Charpy impact V-notch tests were conducted at different temperatures. Pierre Berthier (1821) was the first to characterize the resistance of corrosion of iron–chromium alloys and noted their resistance against attack by some acids, suggesting their use in cutlery. Such artifacts, which are corrosion-resistant, have survived from antiquity. A good example is the Iron Pillar in Delhi.

3. Methodology

3.1 Heat Treatment Schedule

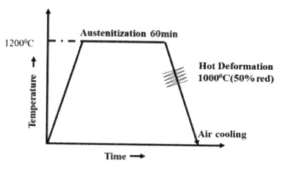

Figure 60.1: Hot deformation of the sample.

The as-received steel was initially cut into several small blocks, followed by annealing and soaking at 1200°C for 1 h, then immediately hot-forged at a temperature of 1000°C with a height reduction from 20 to 10 mm thickness. After hot forging, the sample was air-cooled. The chemical composition of the AISI 430 ferritic stainless steel, obtained by AES (atomic emission spectroscopy) analysis, is shown in Table 60.1.

Table 60.1: Chemical composition (in wt.%) of AISI 430 ferritic stainless steel.

C	Mn	Si	P	S	CCr	Ni	Mo	Al	Fe
0.05	1.45	0.25	0.03	0.002	19.01	8.28	0.33	0.01	69.50

For the enhancement of the mechanical properties of material, grain refinement is the most important factor; both strength and toughness improve simultaneously [3]. After hot deformation, the sample was carried out for microstructural characterization (Optical, SEM) and mechanical testing. Through an EDM machine, the sample was cut to a standard size for the tensile test, then testing was done through a UTM machine. A nanoindentation test was conducted to characterize the mechanical properties of steel with coarse grain (CG) and fine grain (FG) of the material.

3.2. Nanoindentation test

From the nanoindentation test, mechanical properties like Vickers hardness, nanohardness, and elasticity of the material can be measured. This test focuses particularly on the depth of penetration in nanometer resolution and the forces in mN. Hardness is measured by applying the formula applied load divided by the projected surface area of the material. When the load is applied continuously, it reaches the maximum value, then partial loading is performed. During partial unloading, the desired depth of penetration is attained [4].

4. Empirical Results and Discussion

4.1 Optical Microstructure

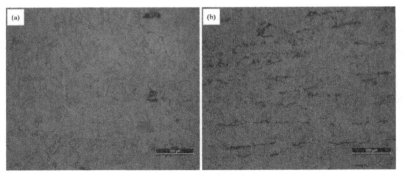

Figure 60.2: Optical microstructure of (a) Coarse grain and (b) Fine grain of ferritic stainless steel.

The optical micrograph, clearly shows that the ferrite grains are embedded with fine car-bide grains. The microstructure of the forged sample has a finer grain structure than the coarser grain one. Due to hot deformation and chemical uniformity, the microstructure of the forged sample has reduced porosity, which contributes to the enhancement of strength in the hot-deformed sample.

4.2. SEM

The features of the microstructure of ferritic stainless steel show dispersed particles after high-temperature hot deformation (hot forging) as investigated (Figure 60.3). During thermomechanical heat treatment, plastic deformation develops, and fragmented grains with low-angle misorientation boundaries are formed. Carbide particles of the MC type are shown in the SEM microstructure. The elemental composition of the matrix after heat treatment becomes nominal, which suggests that a significant proportion of carbide-forming elements remain in the solid solution but do not participate in the formation of dispersed particles.

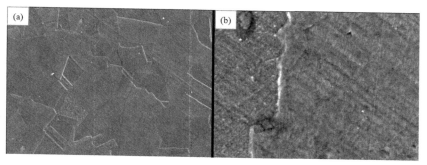

Figure 60.3: Scanning electron microscope of (a) Coarse grain and (b) Fine grain of AISI 430 ferritic stainless steel.

4.3. Tensile Test

Engineering stress–strain curves of ferritic stainless steel were investigated and plotted in Figure 60.4, from which the mechanical properties were obtained (Table 60.2). Due to the lower percentage of carbon, ferritic stainless steel is softer compared to austenitic and martensitic stainless steel and is ductile in nature. Elastic modulus, yield strength, ultimate tensile strength (UTS), reduction of area, and elongation % were obtained from the engineering stress–strain curve to characterize the strength and ductility of the material. The elongation % was higher in the case of CG material compared to FG material. Hot deformation eliminates internal cracks and voids, and there is chemical uniformity throughout the material. FG material has higher yield strength and ultimate tensile strength compared to CG material due to the fine precipitate of MC carbide particles. Assuming that yield strength is correlated with the grain size refinement obtained, according to the Hall–Petch relationship [5, 6].

$$\sigma_{ys} = \sigma_0 + kd^{-1/2} \tag{1}$$

Table 60.2: Mechanical properties of heat-treated ferritic stainless steel.

Tensile Properties	CG	FG
YS(MPa)	195	228
UTS(MPa)	406	412
Fracture Strain (%)	27.06	24.06

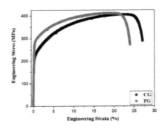

Figure 60.4: Engineering stress–strain curve
of ferritic stainless steel

4.4. Vickers Hardness and Nanoindentation Test

From Table 60.3, it is observed that the hardness of fine grain material is higher compared to coarse-grain material. Due to hot deformation, its yield strength and ultimate tensile strength increase due to the refinement of grain size, so hardness also increases in the case of fine grain material. From the nanoindentation test, the load vs. displacement is plotted as a 100 mN load is applied here, and the hardness is measured in terms of displacement. The nanohardness value of fine grain material is more compared to that of coarse grain material (Figure 60.5). The figure clearly shows that during the test, there is the first part of loading, then at the peak point, it holds maximum force, after which final unloading occurs.

Table 60.3: Vickers hardness and nanohardness of AISI 430 ferritic stainless steel.

Grains	Nanohardness	Vickers Hardness
CG	3087 MPa	174.6
FG	4819.6MPa	225.6

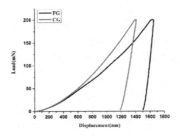

Figure 60.5: Nanoindentation test of AISI 430
ferritic stainless steel.

5. Conclusion

1. During thermomechanical processing, the simultaneous application of heat and deformation process results in the refinement of microstructure and increases the mechanical properties of steel.

2. Hot deformation eliminates internal voids or porosity, which leads to grain refinement and proper orientation of grain flow, producing a uniform grain structure that increases the hardness and strength of the material.
3. From the Vickers hardness and nanoindentation test, the hardness value of FG material is more compared to CG material.

References

[1]   Fahmi, M., & Sani, A. (2009). *Effect of heat treatment process on stainless steel for watch manufacturing application.*

[2]   Kalyon, A., Kaya, Y., & Kahraman, N. (2002). ANN prediction of impact toughness of ship steel/stainless steel plates produced by explosive welding. *Journal of Engineering Research, 8*(2), 266–284.

[3]   Vafaeian, S., Fattah-alhosseini, A., Mazaheri, Y., & Keshavarz, M. K. (2016). On the study of tensile and strain hardening behavior of a thermomechanically treated ferritic stainless steel. *Materials Science and Engineering A, 651*, 131–137.

[4]   Lucca, D. A., Herrmann, K., & Klopfstein, M. J. (2010). Nanoindentation: Measuring methods and applications. *CIRP Annals - Manufacturing Technology, 59*(2), 803–819.

[5]   Di Schino, A., Salvatori, I., & Kenny, J. M. (2002). Analysis of the strain hardening behavior of martensitic stainless steel. *Journal of Materials Science, 37*(19), 4561–4565.

[6]   Kashyap, B. P., & Tangri, K. (1995). Martensitic transformation in high-carbon steels. *Acta Metallurgica et Materialia, 43*(11), 3971–3981.

Printed in the United States
by Baker & Taylor Publisher Services